Developmental Dynamics

in Humans and Other Primates

Discovering Evolutionary Principles through Comparative Morphology

Adonis Science Books

Adonis Science Books are inspired by a scientific approach that goes beyond physical analysis and explores the qualitative and dynamic aspects of nature. It is neither a bottom-up (reductionist) nor a top-down (ideological) but rather a phenomena-centered approach. Its goal is to allow the phenomena to speak for themselves. Chemicals, trees, or animals are seen as expressions of dynamic qualities, of organizing principles that manifest in all of their parts, processes, and actions. Far from rejecting the empirical findings of modern science, this new approach incorporates these findings into a more comprehensive view informed by the essential qualities that give the phenomena their integrity and inherent value.

Adonis Science Books and other science books taking a similar approach to nature can be found on our website at **www.adonispress.org**.

Developmental Dynamics

in Humans and Other Primates

Discovering Evolutionary Principles through Comparative Morphology

Jos Verhulst

Translated by Catherine Creeger

ADONIS PRESS

This text has been revised and expanded from the original German edition *Der Erstgeborene: Mensch und höhere Tiere in der Evolution* published by Verlag Freies Geistesleben, Stuttgart, 1999.

Copyright © 2003 Adonis Press

Published by Adonis Press
320 Route 21C, Ghent, NY 12075

Library of Congress Cataloging-in-Publication Data

Verhulst, Jos,
Developmental dynamics in humans and other primates : discovering evolutionary principles through comparative morphology / Jos Verhulst ; translated by Catherine Creeger.

Rev. ed. of : Der Erstgeborene: Mensch und höhere Tiere in der Evolution. 1999.
Includes bibliographical references and index.
ISBN 0-932776-28-0 (hardcover)
ISBN 0-932776-29-9 (paperback)

All rights reserved.
No part of this book may be reproduced or used in any manner without the written permission of the publisher.

Printed in the United States of America

Contents

Foreword

Few debates between science and the public stir more controversy and emotion than those that center on evolution. Clearly, the majority of modern scientists accept that evolution has generated, in Charles Darwin's words, "endless forms most beautiful and most wonderful." But there is somewhat less agreement regarding the precise processes that have given rise to the rich diversity of life. For example, a number of Darwin's proposals, such as the slow and gradual tempo of evolution, have come under question by evolutionary biologists. In this book, however, Jos Verhulst goes further and takes issue with the very epistemological underpinnings of neo-Darwinism, that is, with its foundation in scientific materialism and reductionism. He does so not by standing on the sidelines and pointing a critical finger but by offering a synthesis of ideas of key historical players and by examining an almost exhaustive array of anatomical structures from a rather new perspective; consequently, an intriguing picture emerges of the traces left by the evolutionary process. As Verhulst leads us from example to example, two major themes weave throughout the text:

1. All organisms are a product of interrelated developmental events that express themselves in differential rates and timing.
2. The human form is the primary reference point for understanding comparative morphology among the primates and even among the vertebrates.

In what follows, I hope to elaborate on these themes and emphasize the significant contribution Verhulst has made to our understanding of organic evolution.

With the publication of Charles Darwin's *Origin of Species* in 1859, a biological revolution of immense proportions was ushered into Western society. Besides the content and social implications that it entailed, a shift in emphasis occurred in which *external* forces of natural selection operating on random variation were given precedence as explanatory agents for the diversity of life at the expense of *internal* processes of developmental integration (McKinney and McNamara, 1991). Furthermore, the notion of the archetype as an *informing principle* that generates biological form was supplanted by Darwin's concept of the *hypothetical ancestor* which, through common descent, gave rise to the biological diversity we witness today (Brady, 1987). In both cases, the focus swung from internal to external causation. To appreciate Verhulst's thesis, however, both points need re-examination.

In 1977, Stephen Jay Gould's *Ontogeny and Phylogeny* rekindled interest in the neglected study of the relationship between the development of an individual organism—its ontogeny—and the evolutionary history of that organism—its phylogeny. The phenomenon that linked the two was *heterochrony*: changes in the rate and timing of development of organs or organ systems of a descendent relative to its ancestor. Consequently, attention was refocused on development as playing a key role in evolution. McKinney and McNamara (1991) and, more recently, McNamara (1997) further underscored the importance of changes in development translating into morphological change within a lineage through time. Because every multicellular organism experiences a specific ontogeny (except those organisms that arise through budding or fission), and each organic structure undergoes its own specified developmental trajectory, it is evident that any given morphological state is in large part a result of heterochronic processes. In this regard, "It's all in the timing" is not an overstatement.

In his comparisons of humans with other primates, Verhulst goes to great lengths to see each selected anatomical feature in the light of heterochronic processes. The relative position of the foramen magnum, of the thumb, and of the aortic arch is an expression of differential developmental rhythms. Specifically, the foramen magnum, for example, in its position almost directly below the human cranium, is an expression of fetalization or paedomorphosis ("youthful form"). In humans its location changes little between the fetal and adult stages, while in anthropoid apes it shifts dramatically from beneath the skull to the back of the skull as these animals mature. Accordingly, regarding this feature, the young ape is in a more humanlike condition relative to an adult ape. As Verhulst shows, even variation in degree of curvature of our fingernails is indicative of differential rates and timing of development among our five fingers.

Many of these morphological phenomena arise from two somewhat paradoxical key elements in human ontogeny; a number of organs or parts of the body:

1. experience a pronounced *prolongation* of development and therefore
2. assume a paedomorphic, and often enlarged, morphology.

Conversely, animal development often exhibits *compression* of various growth stages, which results in specialized anatomical features. This can be seen clearly in the development of the larynx, discussed in Chapter ten. Compared to that of other primates, the human larynx is paedomorphic and thus unspecialized, and in addition it experiences a prolonged descent as it moves from the upper part of the throat in newborns to its ultimate deep location in adults. The condition in anthropoid apes comes closest to that of humans, but overall their larynx is morphologically more specialized and its descent is relatively foreshortened and minimal. Furthermore, human larynx development is correlated with various other characteristics, such as suppressed muzzle development and prolonged growth of the tongue, two features necessary for speech production.

Verhulst also examines the complexities of limb growth in humans. Evidently, the limbs exhibit shifting rhythms of growth with age in which there are alternating periods when the maturational sequence is mainly distal (fingertips) to proximal (shoulder) and then proximal to distal. Remarkably, these rhythms appear to converge on a seven-year cycle, which corresponds to major physiological and psychological benchmarks in the growth of a child. The integration and covariation of differential rhythms of development, both among organisms and among anatomical features within an organism, as Verhulst demonstrates in so many examples, point to the inadequacy of random variation and natural selection as explanations for evolution and the generation of biological diversity. As Verhulst's work implies, Darwin's mechanisms should not be abandoned but rather relegated to a level of secondary importance.

The second major theme of the book builds on the theme of heterochrony just discussed. Because *Homo sapiens* appeared last in the fossil record of hominids and other primates, it is customary to refer to modern humans as evolutionary modifications of presumed ancestral forms.

Verhulst, however, supports a radically different view, namely that human evolution was already prefigured in the appearances of early hominids and even other primates. In other words, modern humans and anthropoid apes, for example, did not descend from a common apelike ancestor—which Verhulst shows is anyway unsubstantiated—but rather the great apes descended from a hominidlike ancestor. In fact, the evolution of all primates points to the same relationship. The forward (i.e., humanlike) location of the foramen magnum in monkey fetuses, as well as many other features, makes biological sense only to the degree that it heralds a future evolutionary condition—that of modern humans. The notion of "preadaptation," as Verhulst maintains, explains nothing and in fact begs the question: For what is it preadapted? Thus, human evolution is paradoxical in that *Homo sapiens* appeared last in time but is also the guiding principle of primate evolution.

This scenario clearly stands in contrast to the conventional scientific approach that views causal relationships in a linear temporal dimension. Verhulst's thesis invokes a principle similar to Aristotle's final cause as a directive agent in evolution, a principle which is perhaps most evident in Goethe's world view. As described in Chapter one, Goethe sees evolution as the progressive emergence of a central gestalt, that being the generalized humanlike image. With its archetypal quality, this gestalt contains the potential for expression of one-sided, specialized animal forms, which take specific shape according to outer (i.e., environmental) circumstances. This "typological" explanation has unfortunately been misunderstood by many modern critics who utterly fail to grasp the dynamic quality of its expression and dwell instead on static representations. Thus, what they reject as typological thinking in regard to evolution is justified, but it is far from what is implicit in Goethe's thinking (for critiques and applications of Goethean science, see Amrine et al., 1987; Bortoft, 1996; Seamon and Zajonc, 1998; Kranich, 1999).

The scenario painted by Verhulst also revives, and gives new meaning to, a previously discredited evolutionary process, that of orthogenesis. In its original meaning, orthogenesis, as an evolutionary process, denotes a linear, unidirectional unfoldment toward greater perfection ending ultimately with the human being. The notion of perfection justifiably weakens the orthogenetic theoretical framework. If perfection implies degree of adaptedness, then it clearly becomes meaningless when comparing different organisms: Is a leopard frog any less perfect than a spider monkey? If perfection implies degree of biological complexity then it also is insupportable: paedomorphosis—resulting in less complex forms—has occurred repeatedly in innumerable lineages. Similarly, the idea of a linear, unidirectional unfoldment has its associated problems, again because paedomorphosis, in a sense, reverses evolutionary trends, and the phenomenon of adaptive radiation is more explosive than linear. What, then, is salvageable from the concept of orthogenesis?

First of all, a modified notion of unidirectionality can be retained if it is seen as the ultimate outcome of iterative, or rhythmical, processes. As a tide rises on a beach, water alternately advances and retreats but the eventual outcome will be a higher reach of the tide. Similarly, evolution "advances" with peramorphic forms—which develop beyond the ancestral condition—and "retreats" with paedomorphic forms—which develop less than the ancestral condition. However, I use the quotes to connote the admitted imprecision of the analogy because the two heterochronic processes are intimately linked. As McNamara (1997) explains, every organism is a mosaic of what he calls dissociated heterochrony: of peramorphic and paedomorphic features. Moreover, the former can arise only because the latter have occurred, resulting in a developmental tradeoff: one step back permits one step forward. According to McNamara (1997), what makes matters more complex is that every limb, for example, is both peramorphic and paedomorphic. Verhulst convincingly shows this to be the case in feature after feature and, as mentioned earlier in regard to human limbs, he demonstrates even greater degrees of complexity in that local growth rates of a given structure can rhythmically alternate their developmental trajectories over time. Thus, although evolution has ultimately reached the human form, it has not occurred by sequentially placing one foot in front of the other along a straight path.

Admittedly, staunch neo-Darwinists will take issue with the teleological assumption that the human being emerges as the result of specific interacting developmental processes toward which biological evolution has progressed, which is implied by orthogenesis. From a materialistic, reductionistic standpoint, resistance to this view is expected. However, after a careful reading of Verhulst, and open-minded consideration of the numerous examples he offers, it is clear that this assumption cannot be refuted and is in fact well supported. To his credit, therefore, Verhulst directs the reader to a new context for understanding human evolution—and evolution in general—and especially to the related complex heterochronic processes that constitute the growth patterns of humans and other primates. The implications are far reaching: the watchmaker, after all, may not be blind and there may be more "grandeur in this view of life" than even the genius of Darwin could grasp.

Mark Riegner
Environmental Studies
Prescott College

References

Amrine, F., Zucker, F., and Wheeler, H. (Eds.). 1987. *Goethe and the sciences: a reappraisal.* Boston: D. Reidel Publishing Co.

Bortoft, H. 1996. *The wholeness of nature: Goethe's way toward a science of conscious participation in nature.* Hudson, New York: Lindisfarne Books.

Brady, R. 1987. Form and cause in Goethe's morphology. Pp. 257-300. In: Amrine, Zucker, and Wheeler (Eds.) *Goethe and the sciences: a reappraisal.* Boston: D. Reidel Publishing Co.

Gould, S. J. 1977. *Ontogeny and phylogeny.* Cambridge: Harvard University Press.

Kranich, E. M. 1999. *Thinking beyond Darwin: the idea of the type as a key to vertebrate evolution.* Hudson, New York: Lindisfarne Books.

McKinney, M., and McNamara, K. 1991. *Heterochrony: the evolution of ontogeny.* New York: Plenum Press.

McNamara, K. 1997. *Shapes of time: the evolution of growth and development.* Baltimore: The Johns Hopkins University Press.

Seamon, D., and Zajonc, A. (Eds.). 1998. *Goethe's way of science: a phenomenology of nature.* Albany, New York: SUNY Press.

Preface

to the original German Edition

This book is the product of the many years that I, a chemist by education, spent teaching biology to Waldorf School students. As a high school student myself, some thirty years ago, I was profoundly affected by the fact that the biological sciences invariably view the human being as a meaningless product of animal evolution. Neo-Darwinists explain the course of evolution itself in terms of processes that are ultimately purely material and mechanical in character. As early as 1866, Ernst Haeckel, the greatest proponent of the materialistic theory of evolution on the European mainland, wrote:

> In nature, coincidence does not exist, nor do purpose and so-called free will. Rather, every effect is determined by prior independent causes, and each cause produces inevitable effects. In our view, absolute necessity takes the place of coincidence in nature, just as it also takes the place of purpose and free will.

Darwinism cannot see moral impulses as anything other than mere instincts or conditioned reflexes. As far as logically consistent materialists are concerned, moral rules are pure conventions. At the same time, however, human thinking is reflexive—that is, aware of its own contents—which means that as soon as we believe a moral impulse to be a mere instinct or reflex, it can no longer unfold as a truly moral impulse, and thinking of it as such becomes problematic. That is why the materialistic worldview has morally and emotionally destructive and debilitating effects.

But it does not necessarily follow that materialism is untrue. Reverence for the truth takes precedence over all else, even if that truth is unpleasant and difficult to accept and even if it destroys morality. Any shortcomings of materialism, however—if such shortcomings exist—should be clearly recognized and made conscious.

Is the materialistic image of the human being, or materialism as a whole, really true, and does Darwinism adequately explain human evolution? I am convinced that it does not; I outlined the rational basis for this conviction in my book *Der Glanz von Kopenhagen. Geistige Perspektiven der modernen Physik* ("The brilliance of Copenhagen: spiritual perspectives in modern physics," Verhulst, 1994a). If the materialistic image of the world and of the human being is false, the facts of biology should reveal its error. As a biology teacher, I thought it important to be able to track down this error, and I discovered some preliminary pointers in books by two anthroposophical authors, Friedrich A. Kipp (1980) and Hermann Poppelbaum (1973). This book can be seen as taking their ideas further.

From Kipp and Poppelbaum, the trail leads to Rudolf Steiner (1861-1925) on the one hand and to the Dutch physician and professor of comparative anatomy Louis Bolk (1866-1930), on the other. Especially in the German-speaking countries, Arnold Gehlen's book *Man: His Nature and Place in the World* (1986) made Bolk's ideas known to a certain extent. In the English-speaking world, Stephen Jay Gould's book *Ontogeny and Phylogeny* (1977a) reenlivened the discussion.

Although Bolk and Steiner were contemporaries, they were active in completely different professional fields and apparently remained unaware of each other's work. Yet the views they developed on the relationship between human beings and animals coincide remarkably. These views share the premise that the human being represents the original form, so to speak, from which animals not only developed but also diverged. The *idea* of the human being is central to animal evolution and manifests with increasing clarity as evolution progresses. In mammalian evolution the human being plays the role of something like an Aristotelian "final cause" or guiding factor. Basically, this book emerged from my attempt to test this hypothesis, which is present in seminal form already in the works of Goethe.

But how can we scientifically prove the actual presence of a factor that guides evolution toward the appearance of the human species? Bolk's "intrinsic evolutionary factor," like any final cause, does not contradict the laws of inanimate nature but works "between the lines." It manifests in any material phenomenon that does not contradict natural laws but cannot be derived from their workings alone. Purely scientific investigation of a corpse allows us to conclude that it is the dead remains of a living being.

Similarly, comparative research into human and animal morphology allows us to conclude that the human gestalt serves as the point of reference while animal forms constitute specialized modifications.

The natural sciences are in no position to investigate any immaterial reality that may exist. They can, however, research the material effects of immaterial factors. Scientists in this situation are like explorers who discover the tracks of an unknown animal in the snow. They can rationally conclude that since the snow itself did not produce the tracks, some animal must have done so. Based on details of the impressions, they can even draw certain conclusions about the unknown species. In this book, we will explore the possibility that a humanlike archetypal gestalt left traces on the "field of snow" of the facts and findings of comparative anatomy and paleontology.

Jos Verhulst
Antwerp, 1999

1

BASIC ISSUES IN HUMAN EVOLUTION

Introduction

Many cultures see the human and animal kingdoms as a continuum. As an example of this view, Heuvelmans and Porchnev reproduce a remarkable Ecuadorian fresco (see *Figure 1*) of a sequence of living things. From undifferentiated beginnings, wormlike or snakelike creatures emerge, which then grow legs and become more substantial, stand upright, lose their tails, and use tools.

Fig.1: Inca fresco from Ecuador (Heuvelmans and Porchnev, 1974; from Vollmer, 1828).

This image includes several striking details. The early stages of human evolution include animals without limbs (more about this in Chapter 3). The animal figures are sketched quite realistically, while the human form is reduced to an equilateral triangle, as if the artist wanted to emphasize that the structure of the human body is governed by a threefoldness. Also conspicuous is the fact that the human being is not considered the ultimate stage but is superseded by an angel-like figure flying toward the sun.

In Europe, ideas about the evolutionary relationship between humans and animals had been widespread ever since the eighteenth century. In subsequent sections, we will explore in greater detail several factors that contributed to such views.

By the early eighteenth century, the first descriptions of anthropoid apes reached Europe. Especially important was Edward Tyson's book *Anatomy of a Pygmie*, published in 1699, which discusses the anatomy of a young chimpanzee in great detail. Tyson was greatly impressed by the humanlike character of the "pygmy," which he invariably portrayed standing upright. In his classification system, he assigned it a position approximately midway between human beings and apes (see *Figure 2*).

Stephen Jay Gould (1985) mentions two reasons why Tyson was inclined to exaggerate the little ape's humanlike character. The first was the fact that Tyson studied a very young animal, and young anthropoid apes look much more like humans than adult specimens do. Gould comments:

> Tyson knew that his pygmy was a young animal, for extremities of the bones were still formed in cartilage and not fully ossified, but he regarded it as nearly full grown because he mistook the complete set of milk teeth for a permanent dentition (the baby teeth of apes do, in some respects, resemble the permanent teeth of humans). Thus, he did not realize how young an animal— a baby almost—he was dissecting . . . We have evolved by slowing down the general developmental rates of primates and other mammals. Thus, human adults resemble juvenile chimps and gorillas much more closely than adult great apes. Consequently, the skeleton of a baby chimpanzee will retain many humanlike characters that an adult would lose— including a relatively large head, . . . a more upright mounting of the head on the spine, . . . a more bulbous cranium, . . . weaker brow ridges, and smaller jaws. Tyson's plate of his pygmy's skeleton, a remarkably accurate figure, . . . shows all these humanlike features.

Fig. 2: Head of an infant chimpanzee and an adult (after Naef, 1926a). Naef's photos of chimpanzees are well known and have been reproduced or copied as drawings in countless books. Note the young ape's domed forehead (still without brow ridges), its still underdeveloped muzzle, the relatively low position of its shoulders, and its clearly visible neck.

Fig. 3: Chimpanzee fetus at approximately seven months (Bolk, 1918). The pattern of hair on the body is very similar to that of humans.

As Gould (1987) emphasizes elsewhere, scientists soon realized that young anthropoid apes were humanlike in appearance. As early as 1836 (in the minutes of the Paris Academy of Science), Etienne Geoffroy Saint-Hilaire discussed the human features of a young orangutan's skull and compared it to the shape of the skull of the adult ape, which he described as "fearsome and of a repulsive bestiality."

The conspicuously humanlike form of fetal anthropoid apes repeatedly made a deep impression on scientists (see *Figure 3*). The same exceptional resemblance to humans is often also very apparent in juvenile apes. We will return to this phenomenon again and again.

Tyson's second reason for interpreting the chimpanzee as especially humanlike was culturally determined. In his time, nature was seen as a continuum, a great chain of being in which each form represented both the metamorphosis of a lower creature and the starting point for transformation into a higher one. Tyson had been expecting the discovery of a highest animal that would fall somewhere between humans and apes. His thinking was dominated by the idea of unity within the animal world and between humans and animals.

This view was very prevalent in the eighteenth century. For example, the French natural historian George Buffon, who observed selected traits common to very different mammalian genera, had the impression that God intended to use only a single idea, which He then varied in all possible ways. Buffon spoke of an original, universal bodily structure that could be found throughout the animal kingdom (Wells, 1967). Buffon's contemporary, the physician and anatomist Petrus Camper, used to illustrate this common bodily structure by drawing a dog on the blackboard and then transforming it first into a horse and then into a human being.

Goethe, the Goethean Tradition, and Darwinism

Any history of evolutionary thinking must consider the achievements of Goethe. Although known primarily for his literary works, Goethe himself considered his scientific activities at least as important. He knew and quoted Buffon and Camper and carried on the work they had begun. From 1795 onward, he developed the concept of the archetypal animal, a spiritual form striving toward material realization and encompassing the potential for all animal genera. In Goethe's eyes, the archetypal animal is neither a material reality nor a sublimated physical force. It is a spiritual principle attempting to manifest in matter.

Such principles play an important role in human life. A series of increasingly successful attempts to realize a goal is understandable only on the basis of the intention behind them, which is an image in the consciousness of the person carrying them out. Like a circle or the number π, this intention is not material in any outward sense. No law describing material interrelationships offers any point of departure for understanding the phenomenon of consciousness. This was true in Goethe's time and is still true in ours. There are no laws of physics that contain any element that could possibly produce consciousness. Adding and subtracting even numbers, no matter how complicated the series may be, can never result in an odd number; similarly, the most complicated arrangements of particles, fields, and the like can never produce feelings, thoughts, or intentions. As the atomic physicist and Nobel prizewinner Eugène Wigner (1983) laconically remarked, it is an "obvious fact that thoughts, desires, and emotions are not made of matter."

Goethe quite rightly sees no reason to believe that human intentions can be reduced to purely physical processes. He sees just as little reason to restrict phenomena of intentionality to human consciousness. In his eyes, a "formative will" is at work in the animal kingdom, where it strives to allow specific archetypal forms to manifest in matter.

It is possible to scientifically ascertain whether something like Goethe's formative will actually is active in nature. The laws of inanimate nature do not totally determine the course of material processes but simply define the limits of possibility (Verhulst, 1994a). Whenever interrelationships among phenomena are in harmony with natural laws but cannot be exclusively derived from them, a Goethean immaterial, form-giving factor is at work. If the phenomena of evolution included only interrelationships that could be satisfactorily explained by material mechanisms such as Darwin's natural selection, there would be no reason to assume that any "formative impulse" is at work. Along similar lines, precise study of a corpse leads to the conclusion that although it is a dead object, it stems from a living being. While its attributes are in harmony with the laws of inanimate nature, its concrete structure cannot be reconciled with the view that it results from exclusively non-biological processes.

Goethe distinguishes two aspects of the formative will in animals:

- On the one hand, it encompasses the striving to bring a very specific central gestalt to expression in material form. We can call this particular effort *the will toward realization of the central gestalt*. Goethe leaves no doubt in our minds about the human-like character of the gestalt that is central to the animal kingdom. Inherent in this gestalt, however, is also the potential to give rise to an endless series of one-sided animal specializations. Goethe calls this central gestalt (including its possible specializations) the archetypal animal. Neither the human body, in which the possibility of animal one-sidednesses appears only weakly or sporadically, nor the inherently one-sided forms of individual animals are good examples of this archetype, according to Goethe.

- On the other hand, the formative will in animals encompasses the metamorphic tendency. Consequently, the central human-like image is subject to alterations and is never perfectly expressed in matter. Such one-sided alterations occur under the influence of outer, material circumstances. "Circumstances shape the animal to their own purposes, hence its inner perfection and outer adaptation" (Goethe, 1795). As examples of outer circumstances, Goethe mentions among others climate, altitude, and the animal's life element—aquatic and airborne animals assume different forms, for instance.

For two reasons, however, Goethe rejected the tendency—already prevalent in his time—to explain an animal's form exclusively as a consequence of its function:

- First, not all animal forms are possible. The only ones that can actually exist are the possible derivatives of the central gestalt. Goethe was convinced that any animal specialization, any exaggerated development of an organ, is possible only at the expense of loss or reduction in a different part of the body. Goethe considered it the chief task of the morphologist to discover the laws of metamorphosis, the limitations imposed by the central, fundamental principle. "Instead of claiming that a bull is provided with horns so that he can butt, we will study how it is possible for him to acquire horns in order to butt" (1795).

- Second, metamorphosis is not a purely passive process of adaptation to environmental influences (within limits imposed by the central principle). Each animal genus is inherently predisposed to respond to specific stimuli and succumb to predetermined specialization. The will to specialize, unique to each genus, modifies and hampers the expression of the central will and codetermines how that genus responds to environmental influences. We can compare this to how humans behave when confronted with sweets—what we do is determined not only by the presence of sweets but also by our desire for them.

Goethe accepted the idea of animal evolution but did not think of it as a purely material phenomenon. He always attempted to explain human and animal forms—and also the evolution of one animal form into others—in terms of spiritual activity striving toward these forms.

The scientific mainstream of the nineteenth and twentieth centuries, however, did not accept Goethe's premise. While Goethe saw the morphological relationship between humans and animals as a consequence of their common spiritual origin, nineteenth century Darwinism explained this relationship on the basis of purely material factors.

Charles Darwin's first major work, *On the Origin of Species*, first published in 1859, not only defends the idea that evolution really takes place but also provides an explanation for it, namely, the mechanism of natural selection, i.e., "the strongest or the most adaptable survives." Within any biological species, slight differences or variations appear repeatedly and spontaneously in different individuals or generations. (Darwin himself knew nothing about what caused such variations, since the theory of heredity did not succeed in clarifying their origin until the twentieth century.) Darwin postulated that in the struggle among individuals, the best adapted are most likely to pass their variations on to the next generation. This process of natural selection gradually alters the characteristics and range of the species, step by step and generation by generation. Spontaneous variation and natural selection, therefore, are the key concepts in Darwinism and the core of a purely materialistic explanation of evolution.

Although Darwinism initially encountered a great deal of resistance, Darwin's book proved to be the deciding factor in the general breakthrough and acceptance of the idea of evolution. By the time of Darwin's death in 1882, the idea of evolution had essentially triumphed. At the request of a group of members of Parliament, Darwin was buried in state in Westminster Abbey, as Isaac Newton had been before him.

On the Origin of Species did not yet address the problem of human descent. That remained for the quick-witted Darwinist Thomas Huxley to do several years later in 1863, when he published his pamphlet "Man's Place in Nature"—the first work in which the human being is explicitly considered a purely material product of animal evolution. Darwin expressed his own views on this subject in *The Descent of Man*, first published in 1871.

Huxley and Darwin saw human beings as descended from some type of anthropoid ape, and the idea of the "missing link," the transitional form between apes and humans, emerged. Darwin conjectured that humans were descended from an African ape. He placed great emphasis on the idea of an ascending line of primates ending with the human being. From this perspective, the large African anthropoid apes, the chimpanzees and gorillas (cf. *Figures 10* and *11*, p. 19), seemed to be the nearest animal relatives of the human being—a view that most modern scientists accept as a matter of course.

Darwinian theory was refined as time went on. In 1891, Arthur Keith first formulated the "brachiator theory," which is still accepted today and supplies a mechanical explanation of the development of upright posture. This theory postulates that human beings, gibbons, orangutans, chimpanzees, and gorillas are all descended from a common ancestor who developed the ability to travel by swinging from branch to branch. This ability required various adaptations, including the development of a shoulder joint with a much wider range of motion. (Lower primates, whose shoulder joints are much less mobile than those of humans or anthropoid apes, prefer to run along the branches on all fours.) Incidentally, it is worth noting that Keith himself abandoned the brachiator theory toward the end of his life.

Karl Snell (1806-1886)

Throughout the nineteenth century, a scientific undercurrent in opposition to Darwinism continued to elaborate on Goethe's views. As early as 1809, the German natural historian Lorenz Oken, in his *Lehrbuch des Systems der Naturphilosophie* (Textbook of the System of Natural Philosophy), conceived of the animal kingdom as a series of side shoots that had lost their connection to the central human gestalt. 1863 saw the publication of both Huxley's *Man's Place in Nature* and a book, *Die Schöpfung des Menschen* (The Creation of the Human Being), by Karl Snell (1806-1886), whose *Vorlesungen über die Abstammung des Menschen* (Lectures on the Descent of Man) were published posthumously in 1887. Both of Snell's works presented opinions diametrically opposed to the prevailing materialistic view and attracted little attention.

Snell placed himself squarely in the evolutionary camp, a scientific position that was not yet a matter of course at that time. His outline of evolution, however, differs in essential points from the materialistic view of Darwin or Huxley. Snell's arguments are based on two obvious facts:

- Human beings, unlike animals, are capable of thinking and moral activity. (Snell's collective term for these two attributes is "reason.")
- Thinking and morality are realities that cannot be explained in terms of the attributes of inanimate matter (see my comments on this subject in the Preface).

If all things originated in an undifferentiated beginning, it must have included the potential not only for inanimate matter but also for thinking and morality. The human line of evolution, therefore, was the one that developed this intrinsic kernel of rationality. Animals evolved in directions that branched off from this central line, forming specialized lines whose evolution was dominated by external environmental influences. Snell suggests that morality and the ability to think are actually universal potentials that the world as a whole strives to attain. Human beings, as guardians of these potentials throughout evolution, have therefore remained universal, undifferentiated, and untouched by the mechanisms of animal specialization. He observes:

> Each development that is enslaved by the outer circumstances affecting any form of existence in sensory reality is always accompanied by another development. In spite of becoming more concrete with each new instance, this second development always takes shape as the vessel and living organ for the infinite. In spite of penetrating deeper into the specific character of earthly life, it never breaks with the essence of reason, with the capacity for the universal and unconditional . . . The ancestors of human beings constitute the unbroken line of carriers of this potential for reason. If we trace the human family tree backward, we pass through a series of beings whose entire inner nature contrasts them with, and definitively separates them from, all other creatures because they alone preserve the potential for reason in all its fullness and unclouded clarity. If we then move forward in the succession of generations, we find both human beings and animals emerging from a primary genus that can be reckoned among the ancestors of human beings. From this perspective, no barrier exists between animals and human beings (1863).

Snell offers various observations in support of this view of evolution, four of which I will mention here:

- The human organism is conspicuous in its lack of specialization. "All that we can say about prototypic human beings with regard to their form and physique is that they never demeaned their organs by allowing them to become crude tools for any single work, that they neither allowed their nails and teeth to become terrible weapons nor distilled their body fluids into deadly poisons of anger, and that through an unbroken sequence of modifications, they developed the . . . five soft fingers into the universal human hand, which is not specifically adapted to any single type of work but is suitable for all work and capable of equipping itself with countless tools. Beyond that, we can only say that with all their aspirations, these beings never allowed themselves to settle down and become comfortable in any restricted surroundings, that they counteracted all incitements and temptations to enjoy the immediate present with a universal impulse that was ultimately not satisfied with any surroundings other than the infinite world of God and nature, and that in spite of all the urgent necessities of the present, they did not take their eyes off their great and distant goal" (1863).

- Certain organs that see little use in animals develop tremendous capabilities in humans. Snell gives the example of the human larynx, which is unspecialized in comparison to the larynx of other mammals. Only in humans, however, is the larynx used for speech. Another example is the hand. "Consider the hand of a gorilla. It has all the bones, muscles, and ligaments of the human hand and is also very similar to the latter in the location and arrangement of these parts. To judge by its structure, such a hand should be quite adequate for piano playing . . . The gorilla, having inherited it from an ancestor who put it to uses more appropriate to its structure, wasted this and many other aspects of a superior and productive inheritance by allowing them to go wild" (1887).

- Snell also points out the similarity (already discussed here) between young apes and the human being. "For example, Darwin concludes from the diagonal stripes we frequently notice on the feet of newborn horses that an ancestor of the horse had stripes like the zebra. This may be correct. But why then do we totally forget this rule when we notice that young orangutans look so much like human infants? In addition to this physical similarity, these young apes are also manageable and gentle and learn quickly. But as they grow, the animal element comes increasingly to the fore until finally they mature into ungainly, sullen, and hot-tempered beasts. According to the above-mentioned rule, we would have to conclude that a human bodily form and humanlike psychological states were dominant in the ancestors of the orangutan" (1887).

- Taking up Goethe's concept of metamorphosis, Snell introduces one final argument. ". . . Light and order are most easily brought into this doctrine by first placing the human being in the center, with the mammals round about, and then proceeding to prove how animal forms appear as modifications of the human form that result from many different elongations, foreshortening, displacements, and deformations . . . Just imagine taking the horse or the whale as our point of departure in trying to demonstrate unity of type. Would this even occur to anyone, and what would the result be? . . . The species that we will most naturally accept as the ancestral species is the one that seems to give rise to other forms most naturally and spontaneously when placed in their center" (1887).

In Snell's view, the unspecialized character of the human body and the fact that animals have undergone specialization in widely divergent directions suggest that the human being constitutes a natural point of reference from which mammalian forms fan out and are derived through metamorphosis. Qualitative tests of this hypothesis should be possible with the help of modern data processing techniques.

Wilhelm Heinrich Preuss (1843-1909)

In 1882, Wilhelm Heinrich Preuss published his book *Geist und Stoff* (Spirit and Matter), which contains ideas closely related to Snell's views:

> The human being is the firstborn of all creation, and zoogenesis is a secondary product of anthropogenesis . . . Admittedly, we human beings have risen from imperfect animal circumstances through evolution. Nonetheless, we were never animals, and it will not be possible to prove the presence of apes or quadrupeds in our ancestry.

Preuss considered not only animals but also the plant and mineral kingdoms to be specialized offshoots of the central human line of evolution:

> The human being is the goal of all telluric processes, and every other form appearing alongside the human form has borrowed its traits from the human being. The human being is the firstborn being of the entire cosmos. Not in our present form, admittedly, but in the form acquired during 273 days in the womb, beginning with ovum and sperm, the human race has experienced all the transformations that led from the very simplest stages of life to our present, perfected form . . . All inorganic matter, regardless of its tremendous mass, must be seen as a waste product of the breakdown of organic matter as a consequence of the life process . . . The great inorganic mass of the Earth is a metabolic by-product, both of organisms now inhabiting its surface and of their ancestors (1882).

Preuss goes on to state that the laws of nature change in the course of evolution:

> Just as reason itself was able to appear only when organic matter began to be transformed into inorganic substance, all the attributes that reason discovers in metamorphosed substance also appeared over the course of time. Thus we must relinquish the idea of the immutability of all so-called natural laws and accustom ourselves to seeing them as dependent on their times (1882).

The views of Snell and Preuss lead inevitably to the conclusion that the immutability of natural law cannot be presumed to persist for all time. If the spiritual and material aspects of reality formed a unity at the beginning of cosmological evolution, and if the spiritual aspect cannot be reduced to a function of material existence, then the initial state must have been something other than purely material. But since most modern research scientists subscribe to the uniformitarian view (i.e., that natural laws remain the same for all time), they necessarily reject perspectives such as those of Snell and Preuss, which ultimately lead to the view that a sacrifice of sorts is concealed behind the existence of animals, plants, and minerals. These kingdoms of nature separated themselves from the human line of evolution, entrusting the formerly universal capacity for thinking and morality to humans, and created circumstances in which human beings could continue to evolve. If this view of evolution is correct, it provides an objective foundation for ecological morality.

Conversely, if humankind is a purely coincidental by-product of natural evolution, there is no reason to see any objective value in maintaining further evolutionary possibilities for humans or for nature as a whole, and any reasons we present can be based only on sentimentality or on our collective human egotism. If, however, scientific research leads to the conclusion that material evolution developed out of a central, spiritual evolutionary process, a more objective view of morality may develop. The reality of human moral responsibility for nature cannot be logically deduced from the laws of inanimate nature. Moral responsibility is always spiritual in character and cannot possibly assume material form. It can never achieve objective existence in a purely material world. It does exist, however, within the spiritual reality that Snell and Preuss saw as guiding evolution. The ecological consciousness of many individuals may in fact be founded on an unconscious sense that human beings do indeed have an objective, moral responsibility for the natural world. Incidentally, it is not surprising to discover Preuss' strong empathy for nature in general and animals in particular. On the subject of vivisection, for example, he writes:

> I cannot accept that the path to human health and well-being must be paved with thousands of miserably martyred animal corpses . . . I am convinced that the gain for medicine as a whole would be greater if a hospital for sick animals were established and serious efforts were made to observe and cure their illnesses and injuries (1882).

Related ideas, although for the most part less clearly expressed, are also found in the works of other authors. At the Prussian Academy of Science in 1900, H. Klaatsch spoke about the human being as the central gestalt for mammals and primates (Klaatsch, 1920; Poppelbaum, 1973). Cleland concluded a public lecture in Glasgow with the following words:

> To me the animal kingdom appears not an indefinite growth like a tree, but a temple with many minarets, none of them capable of being prolonged—while the central dome is completed by the structure of man. The development of the animal kingdom is the development of intelligence chained to matter; the animals in which the nervous system has reached the greatest perfection are the vertebrates; and in man that part of the nervous system which is the organ of intelligence reaches . . . the highest development possible

> to a vertebrate animal, while intelligence itself has grown to reflection and volition . . . I believe, not that man is the highest possible intelligence, but that the human body is the highest form of animal life possible, subject to the conditions of matter on the surface of the globe, and that its structure completes the design of the animal kingdom (1887).

In the early twentieth century, this tradition in the understanding of evolution found an important proponent in Louis Bolk, whose views we will discuss in Chapter 2.

The Primitive Aspect of the Primates

In this section, we will consider the mammalian order of the primates. A brief overview of primate taxonomy (see Fleagle, 1988) will help us understand this order better. (See also *Table 1*, p. 20 and *Figure 12*, p. 21.)

Human beings are assigned to the mammalian order of primates, which (according to a traditional classification scheme) includes two suborders, the prosimians (Prosimii), and the monkeys, apes, and humans (Anthropoidea). The prosimian suborder in turn includes two so-called infraorders, the Lemuriformes and the Tarsiiformes. *Figure 4* shows a typical lemur (*Lemur catta*) from Madagascar; *Figure 5* shows the Eastern tarsier, a little insectivore from Southeast Asia. The suborder Anthropoidea also includes two infraorders, the platyrrhines or New World monkeys and the catarrhines or Old World monkeys. Platyrrhines always have twelve (i.e., 4 x 3) premolars (bicuspids), while the catarrhines, like humans, have only eight (4 x 2).

Figure 6 shows one of the largest New World monkeys, the spider monkey (genus *Ateles*). Like many other platyrrhines, it has a prehensile tail. In contrast, catarrhines never have true prehensile tails.

The catarrhines are further subdivided into the lower catarrhines (Cercopithecoidea) and higher catarrhines (superfamily Hominoidea, anthropoid apes and humans). The Hominoidea never have tails, but lower catarrhines (such as the familiar baboon shown in *Figure 7*) almost always do.

Fig. 5: The Eastern tarsier (Tarsius spectrum), *a small nocturnal animal (note the large eyes) from Southeast Asia; one of three species. (Weight ca. 130 g, length of head and torso 8.5-16 cm, length of tail ca. 13-27 cm).*

Fig. 4: The ring-tailed lemur (Lemur catta), *a native of Madagascar. Unlike other prosimians, which are primarily arboreal, this species is primarily a ground-dweller. (Weight ca. 2.6 kg; length of head and torso 30-60 cm.)*

(Drawings in Figures 4-11 are by W. Genard.)

Fig. 6: The spider monkey (Ateles *spp.*)*, a platyrrhine with a strong prehensile tail.* Ateles *has no thumb, hence its scientific name. It is one of the largest platyrrhines (weight ca. 7-9 kg).*

Fig. 7: The baboon (Papio *spp.*)*, the largest of the lower catarrhines, inhabits the savannas of Africa. The male (weight ca. 20-30 kg) is nearly twice as heavy as the female.*

Fig. 8: The gibbons (Hylobates, *six species) are the smallest anthropoid apes (weight: 5-6 kg). They have exceptionally long arms and are excellent climbers. All of the gibbons, including the larger but closely related siamang* (Hylobates syndactylus), *weight ca. 10-11 kg), are from Southeast Asia.*

Fig. 9: The orangutan I (Pongo pygmaeus), *the only large anthropoid ape from Asia, is native to the rainforests of Borneo and Sumatra. The male weighs 80-120 kg, the female 30-60 kg. The orangutan has very long arms and very short thumbs and big toes. These sluggish creatures are arboreal and only reluctantly descend to ground level.*

Fig. 10: The gorilla (Gorilla gorilla) *is the largest anthropoid ape and the largest of all the primates (male: ca. 200 kg, female: 90 kg.). This rare animal, found only in a few parts of sub-Saharan Africa, is a ground-dwelling species and almost exclusively herbivorous.*

*Fig. 11: The chimpanzee (*Pan troglodytes) *is native to Africa (male: ca. 50-60 kg, female: 33-47 kg). The chimpanzee is usually considered the nearest relative of humans (see Figure 12). A slightly smaller species, the pygmy chimpanzee or bonobo* (Pan paniscus), *is native to the Congo.*

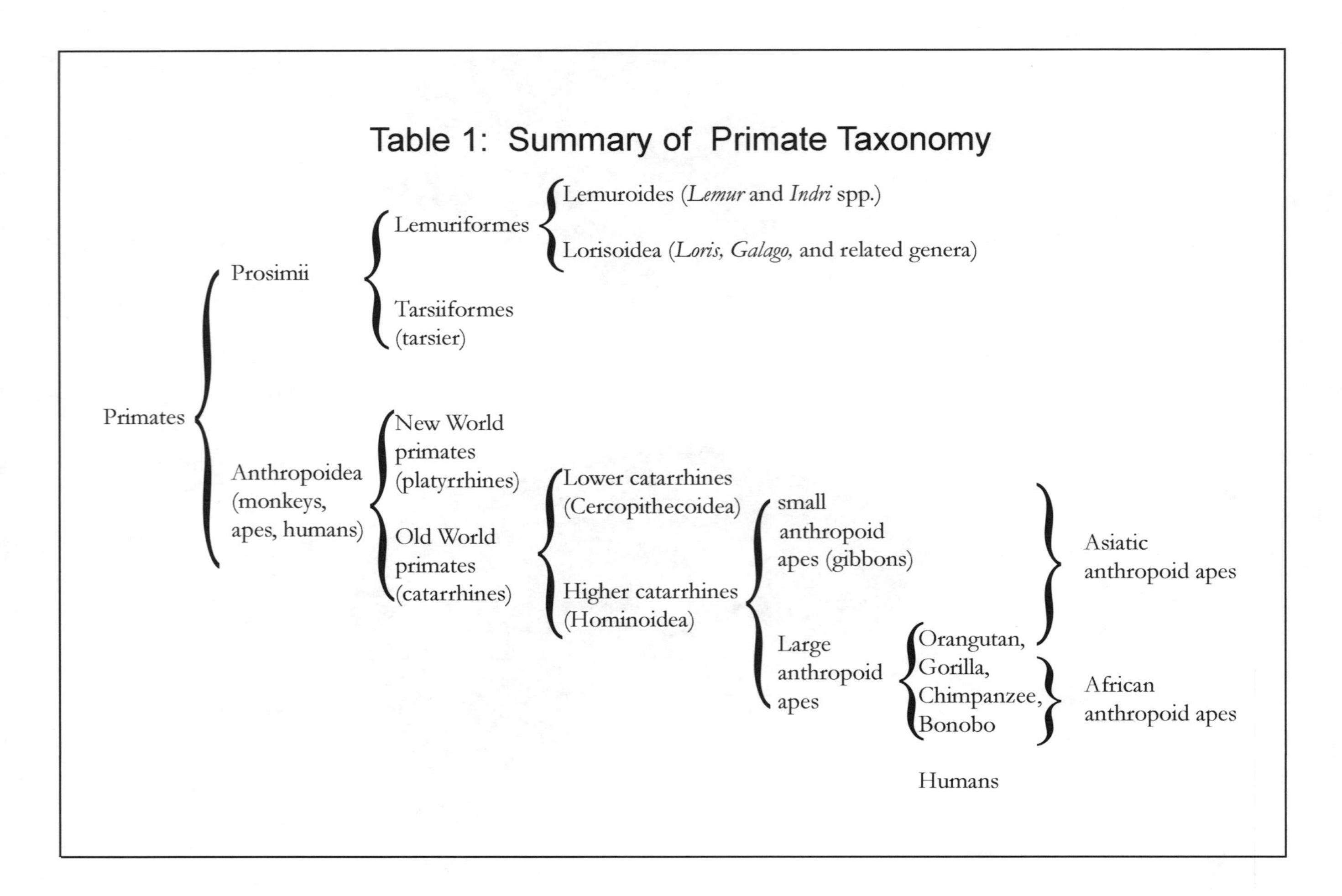
Table 1: Summary of Primate Taxonomy
Primates
Prosimii
Lemuriformes
Lemuroides (Lemur and Indri spp.)
Lorisoidea (Loris, Galago, and related genera)
Tarsiiformes (tarsier)
Anthropoidea (monkeys, apes, humans)
New World primates (platyrrhines)
Old World primates (catarrhines)
Lower catarrhines (Cercopithecoidea)
Higher catarrhines (Hominoidea)
small anthropoid apes (gibbons)
Large anthropoid apes
Orangutan,
Gorilla,
Chimpanzee,
Bonobo
Humans
Asiatic anthropoid apes
African anthropoid apes

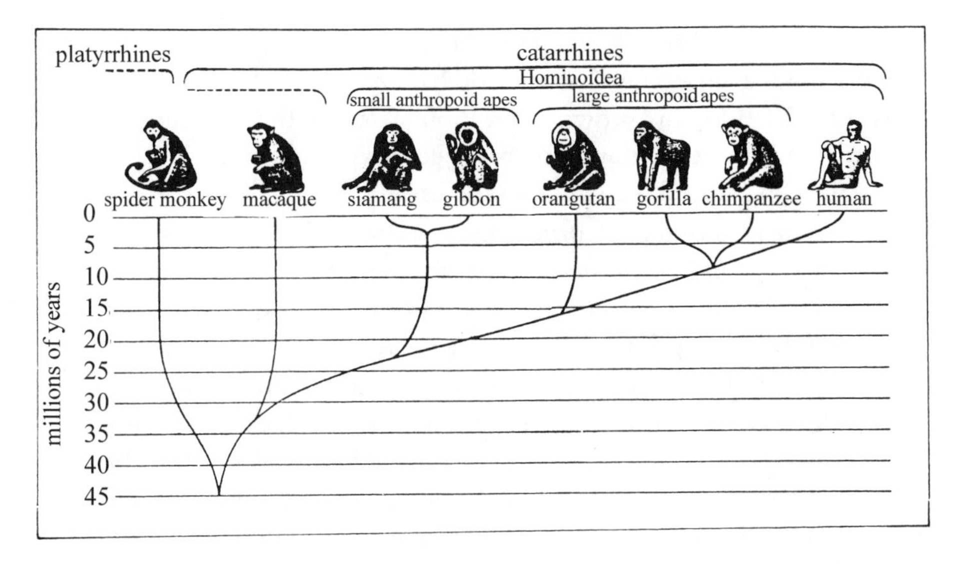

Fig. 12: Presumed phylogenetic tree of the higher primates, according to a modern version (Walker and Teaford, 1986).

A further distinction is made between the small and the large anthropoid apes. The former—which include gibbons (*Figure 8*) and their larger relative, the siamang—are all from Southeast Asia. The large anthropoid apes include three genera. The orangutan (*Figure 9*) lives in Sumatra and Borneo, and the gorilla (*Figure 10*) and the two chimpanzee species (*Figure 11*) live in sub-Saharan Africa. The small anthropoid apes and the orangutan are known collectively as the Asiatic anthropoid apes, the gorilla and the chimpanzee species as African anthropoid apes. The latter are generally considered the animals most closely related to humans.

The modern description and classification of primates emerged step by step rather than all at once. The first edition of Linnaeus' *Systema* was published in 1735. In keeping with the spirit of the times, this first attempt at a taxonomy of the entire plant and animal kingdoms included humans in the animal kingdom. Linnaeus described a mammalian order that he called the Anthropomorpha, which included not only human beings but also apes and sloths. In the tenth edition (1758) of his work

Linnaeus changed both the name and the composition of this order, calling the animal grouping that included humans the "primates" (in the sense of a firstborn, or "crown of creation"). The sloth was removed from the order, but the lemurs and one bat species were added. In 1779, Johann Blumenbach removed the bat from the list.

Obviously the animal order that was to include humans was not clearly defined and delineated. What characteristics of animals allowed them to be classified as primates? As it turns out, this question is difficult to answer because primates are distinguished from other mammals by their very lack of specialization and by their undifferentiated structure. We find no hoofs, wings, trunks, horns, etc. among primates. Zoologists have repeatedly encountered examples of this unique universality. Bernard Grzimek (1970) points to the relative simplicity of the primate stomach and to the rudimentary character of their kidneys (in comparison to the segmented kidneys of many ungulates and beasts of prey. F. Wood Jones noted in 1929 that there is no single specialized characteristic distinguishing the primates as such. W. E. LeGros Clark (1965) describes this point as follows:

> It is peculiarly difficult to give a satisfying definition of the Primates, since there is no single distinguishing feature which characterizes all the members of the group. While many other mammalian orders can be defined by conspicuous specializations of a positive kind which readily mark them off from one another, the Primates as a whole have preserved rather a generalized anatomy and, if anything, are to be mainly distinguished from other orders by a negative feature—their 'lack' of specialization. Thus they have mostly maintained a generalized structure of their limbs, preserving the primitive pentadactyly (i.e. five fingers and five toes), as well, also, as keeping intact certain elements of the limb skeleton (such as the clavicle or collar-bone) which tend to shrink or disappear in some other groups of mammals. Again, the grinding teeth (molars) preserve, on the whole, a simple and primitive structure, particularly when compared with the complicated and highly specialized teeth which have been developed in some orders (such as the elephants, ungulates and carnivores).

Clark lists a number of evolutionary tendencies typical of primates. With regard to the primate skull, for example, he notes:

- A gradual increase in the size of the skull cavity, which also becomes increasingly round.
- The eye socket must be bony, that is, it is surrounded at least partially by a bony vault above it (as in lemurs but also in horses; see *Figure 156*, p. 306) or by a bony wall that completely shuts it off from behind (as in apes and humans).
- Gradual development of stereoscopic vision (completely developed in higher primates), i.e., both eyes always look in the same direction.
- A tendency toward reduction in the size of the zygomatic bone.
- Decreased muzzle development; i.e., the facial skull increasingly tends to be located below, rather than in front of, the braincase.
- The foramen magnum tends to move from the rear to the underside of the skull. (The foramen magnum is the opening in the skull through which the spinal cord connects to the brain.)

It is clear from this list that the human being represents the extreme stage of most of these tendencies. We will discuss several examples in detail later. But does the human being really represent the zenith of primatehood, so to speak, or was Clark's and others' classification of animal species subjectively oriented toward the human being? Fossil finds tend to confirm that the evolutionary tendencies observed by Clark do indeed correspond to reality. Human encephalization (the increasing size of the brain, which is of course directly related to Clark's first tendency, above) is the endpoint of an evolution typical of primates from the very beginning. The earliest primates, which lived 60 million years ago, already had disproportionally larger brains than other mammals, and this tendency persisted throughout primate evolution (Jerison, 1979).

J. H. Schwartz (1987) points out that Clark's approach is less than compelling, because "evolutionary tendencies" cannot help us identify an individual primate fossil as such. Through a process of elimination, we are forced to conclude that the fossil animal is a primate only because there is no evidence of any of the specialization typical of any other mammalian order. As an alternative, Schwartz proposes a definition based on a combination of two "hard" criteria, namely, a primate is a mammal with both of the following characteristics:

- The nail of the first digit of the foot is either flat or absent.
- The eye socket is bony. (Note that this characteristic is also found in the horse, for example.)

The fact that Schwartz, after lengthy searching, chooses to emphasize marginal, relatively unspecialized, and somewhat contrived characteristics is indicative of the low level of specialization of the primate order as a whole.

The Most Primitive Primate

The primates are the least specialized mammals, and the human being is the least specialized primate. For this reason, Ernst Mayr and Konrad Lorenz state that human beings specialize in nonspecialization (see Washburn, 1963 and Heberer, 1959). To A. Abbie, this persistent refusal to succumb to the "temptations" of specialization is the primary characteristic of the human line of evolution, which is why his article is entitled "No! No! A thousand times no!" (1948). M. Voit cites several examples of the very primitive body structure of human beings (for example, the structure of the fetal membranes and the adrenal glands) and relates the unspecialized character of the human being to the unspecialized character of primates in general:

> As mammals became adapted to specific conditions of life, two complexes of characteristics, in particular, underwent extensive variation, namely, the configuration of the extremities, which depends on mode of locomotion, and the structure of the teeth and gastrointestinal tract, which is influenced by eating habits. The development of these two areas of the body was the primary factor in the very rich diversification of the placental mammals into orders of insectivores, rodents, bats, carnivores, ungulates, etc. And with

> regard to the extremities and the digestive system in particular, primates and especially humans have retained very primitive proportions (1921).

In the thinking of the founders of the materialistic theory of evolution, the unspecialized character of primates in comparison to the other mammals, and of the human being in comparison to the other primates, plays virtually no role and is granted no real explanation. The fact that the only mammal capable of propounding a theory of evolution is also the least specialized mammal is dismissed as trivial.

Dollo's Law

The irreversibility of evolution is an established fact. The characteristics that the ancestors of an animal genus once possessed and then lost never return in the same form. For example, terrestrial vertebrates are assumed to be descended from fish. Many land animals had descendants, such as sea turtles and whales, that returned to the sea again but never regained the gills that allowed their earlier ancestors to absorb oxygen from the water. The dolphin, for example, is a better swimmer than most fish, but it remains a mammal—a warm-blooded, viviparous air-breather.

This insight was the basis for the French-Belgian paleontologist Louis Dollo's formulation of the eponymous principle of the irreversibility of evolution:

> An organism never returns to all details of an earlier body structure, not even if its conditions of life are exactly the same as those of its ancestors. Its past can never be completely erased, . . . and traces of the intermediary stages of its evolution will always remain visible in the organism (1905).

Dollo discusses many examples, such as that of the kangaroo, whose foot structure indicates that it is descended from arboreal species, or the saurian quadrupeds *Triceratops* and *Stegosaurus*, whose pelvic structure points to bipedal ancestors. Dollo quite rightly states that the principle of irreversibility applies only to complex characteristics, for which the probability of reversion to the ancestral characteristic approaches zero (Gould, 1970). Human beings are most closely related to anthropoid apes and can therefore share with them specialized characteristics that are absent in the lower catarrhines. However, if we assume the Darwinian line of descent, then, according to Dollo's law, it should be impossible for humans to share specialized characteristics that are present in lower catarrhines but absent in anthropoid apes. In the course of evolution human beings should never revert to lower catarrhine characteristics previously abandoned in the evolution of anthropoid apes. Thus it should also be impossible for humans to share detailed characteristics with platyrrhines or tarsiers when every trace of these traits is absent in catarrhines.

Does a comparison between humans and the other primates confirm Dollo's law? Yes and no, as we will see below. Surprisingly, most of the traits humans share with the higher apes are also present in lower apes. Indeed, we find that humans share different single characteristics with different primate species, both higher and lower. W. L. Straus (1949) comments:

> In a considerable number of important characters the human being can only be regarded as essentially generalized or unspecialized . . . In these characters humans find their counterparts not in the anthropoid apes but in animals that are clearly to be regarded on both paleontological and comparative-anatomical grounds as more primitive, namely, such primates as the monkeys and prosimians, and even mammals of other orders. In other words, in many characters, and particularly in those that define an anthropoid ape, the anthropoids (and the great apes especially) can only be considered as far more specialized than the human being.

This remarkable fact, which we will elucidate later, was discovered very early on, especially within the evolutionary school of thought that opposed Darwin's and Huxley's conceptions and developed a more spiritual view of the phenomenon of evolution. In his book *Man and Apes*, published in 1874, St. George Mivart emphasizes that specifically human morphological traits are not shared exclusively with any one group of primates. (For details on the history of this debate, see Straus, 1949, or Tuttle, 1974.) For example, human beings share a whole series of traits only with the platyrrhines and especially with the more highly developed members of this group, such as spider monkeys (*Ateles* spp.) or capuchins (*Cebus* spp.).

A number of pronounced specializations—such as very rudimentary thumbs and tails disproportionally longer than those of any other primates—suggest that spider monkeys (*Figure 6*) are far removed from humans. However, a number of remarkable correspondences between humans and the most highly evolved platyrrhines have long been observed. The brain structure of spider monkeys has several humanlike traits that do not reappear in the lower catarrhines, which are more closely related to humans (Hatschek, 1908). Researchers have also found that the kidneys of humans and spider monkeys share idiosyncrasies found in no other primates (Goodman, 1982; Straus and Arcadi, 1958). Van den Broek (1908) mentions no fewer than forty anatomical characteristics that reveal the humanlike body structure of the genus *Ateles*. For example, *Ateles* species have flat nails, in contrast to the more clawlike nails of other platyrrhines, and their thigh muscles have two heads, a trait otherwise found only in humans and anthropoid apes. (*Ateles* species also walk upright fairly well.) This muscle is more humanlike in spider monkeys than it is in orangutans. The cervical section of the spine, which is longer in humans than in any other primates, is also disproportionally longer in anthropoid apes than in lower primates. Again, the exception is *Ateles*, in whom the cervical spine is proportionally nearly as long as in humans (Schultz, 1961).

Another platyrrhine genus, the capuchins (*Cebus* spp.), share a different complex of traits with humans:

- After correction for body weight,[1] capuchins have the highest ratio of brain weight to body weight of all primates except humans (Martin, 1990).
- Like humans, capuchins are exceptionally long-lived (Fleagle, 1988).
- Capuchins are notably inventive in their search for food.

---

1. All else being equal, the brains of smaller animals are heavier in relationship to body weight than those of larger animals, that is, there is a negative allometry between brain weight and body weight. This effect must be taken into account when comparing the brain weights of animals with different body weights.

- The thumbs of capuchins, in contrast to those of other platyrrhines, are quite clearly opposable (see Chapter 5).
- With the exception of African anthropoid apes, capuchins seem to be the only nonhuman primates that use tools—stones for cracking nuts, for example (Antinucci and Visalberghi, 1986).
- The correspondences between the digestive tracts of humans and capuchins are also striking. With the exception of humans and orangutans, capuchins are the only primates with thick layers of dental enamel. (African anthropoid apes, in contrast, have thin enamel.) The digestive tract (esophagus, stomach, intestines) of humans and capuchins is disproportionally smaller than in the other primates (Martin, 1990).

A comparative study of skin structure in primates reveals that certain platyrrhines are unexpectedly closely "related" to anthropoid apes and even to humans. In a study of twenty-one skin characteristics, *Ateles* emerges as most closely related to humans, while another platyrrhine genus, the woolly monkey (*Lagothrix*) proves to be very closely related to the chimpanzee. (*Ateles* and *Lagothrix* are considered the most highly evolved platyrrhines.) The authors of this study write, "The grouping with hominoids may be fortuitous, but we find it intriguing that the highest grade of New World primates should group with the highest grade of Old World primates" (Grant and Hoff, 1975). The similarity, especially with respect to the epidermis and sweat glands, is quite astonishing.

Genet-Varcin (1974) comments that human beings share a greater number of primitive characteristics with the platyrrhines than with all of the species of catarrhines. A few examples are:

- The catarrhines have hairless "sitting pads" (ischial callosities) on their buttocks, a phenomenon that is also evident, although to a lesser extent, in the anthropoid apes. In humans, however, this structure is totally absent, as it also is in platyrrhines and lemurs (Mivart, 1874).
- The length of the thumb in proportion to the other fingers is similar in humans and most platyrrhines (with the exception of

Ateles and *Brachyteles*, whose thumbs are absent or rudimentary). In contrast, the thumbs of anthropoid apes are disproportionally shorter.

- In platyrrhines as in humans, the skull is relatively unspecialized. Its shape is fairly round, the muzzle is shortened, and it has neither a sagittal crest nor brow ridges (Genet-Varcin, 1974; see also *Figure 13*). For this reason, a fossil platyrrhine ape with a conspicuously generalized (and therefore humanlike) skull was given the name *Homunculus* (Jerison, 1979). Incidentally, Straus also points out that skull sutures, which ossify late in both platyrrhines and humans, generally close much faster in anthropoid apes.
- Oxnard (1969) has ascertained that shoulder structure is more humanlike in certain platyrrhines than it is in catarrhines. (Within the latter group, shoulder structure is more humanlike in species other than the African anthropoid apes generally considered the nearest animal relatives of humans.)

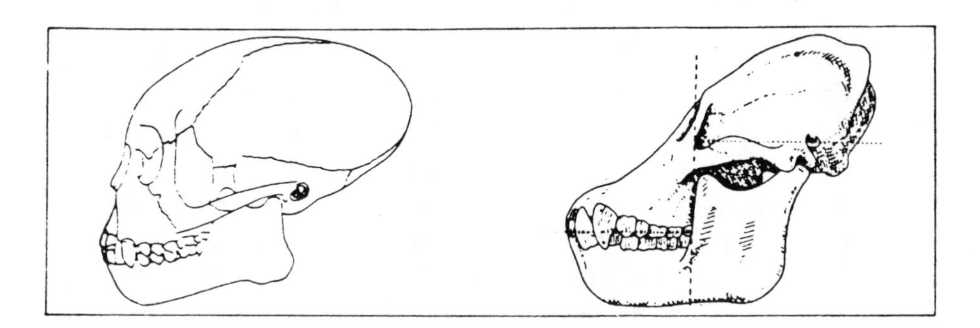

Fig. 13: A comparison of the skull of a platyrrhine monkey (capuchin, left) and an anthropoid ape (orangutan, right). (Frechkop, 1949 and 1954). The platyrrhine skull is much more generalized and "humanlike" than that of the anthropoid ape, although the latter is considered more closely related to humans. In the orangutan, all of the teeth lie in front of the braincase (dotted line) as a result of disproportionate muzzle development. The orangutan also has a conspicuous sagittal crest.

The only possible conclusion that can be drawn from these comparisons of individual characteristics is that the similarities between humans and platyrrhines (especially the most highly evolved species in this group) cannot be due to specializations already present in their common ancestors, because these characteristics are not present in the lower catarrhines, which are considered more closely related to humans. Thus these similarities must indicate a common primitive endowment that was present already before the evolutionary split between humans and apes occurred. Both humans and platyrrhines retained this endowment with regard to specific characteristics, while the more specialized catarrhines are further removed from the original form.

We also encounter similarities between humans and lower catarrhines with regard to characteristics not shared by the anthropoid apes. One curious example is the typically human phenomenon of progressive baldness, which generally does not occur in apes but does appear in one species of macaque. Closer study reveals that the human process of going bald is exactly paralleled in stumptailed macaques, with the one exception that baldness in the monkeys is not limited to one sex (Montagna et al., 1966; Uno et al., 1985).

Furthermore, Schultz points out numerous instances of characteristics common to gibbons and humans but not shared by the large anthropoid apes, which are more closely related to humans. Two examples are:

- The typical parabolic shape of the human ribcage is also found in gibbons (*Figure 14*). In contrast, the ribcage is much narrower in the lower apes and more conical in the anthropoid apes. Humans and gibbons have five lumbar vertebrae between the thoracic vertebrae and the sacrum, while chimpanzees have only three. In the large anthropoid apes, the ilia extend all the way to the ribs, while in humans and gibbons the distance between the pelvic bones and the ribs is much greater, making the back much more flexible. It is often argued that the low position of the human sacrum is a specifically human adaptation to upright posture, but in comparison to the length of the torso, the sacrum in macaques is as low as it is in humans. Hence, the sacrum's higher position in chimpanzees, rather than its lower position in humans, must be considered the specialized location.

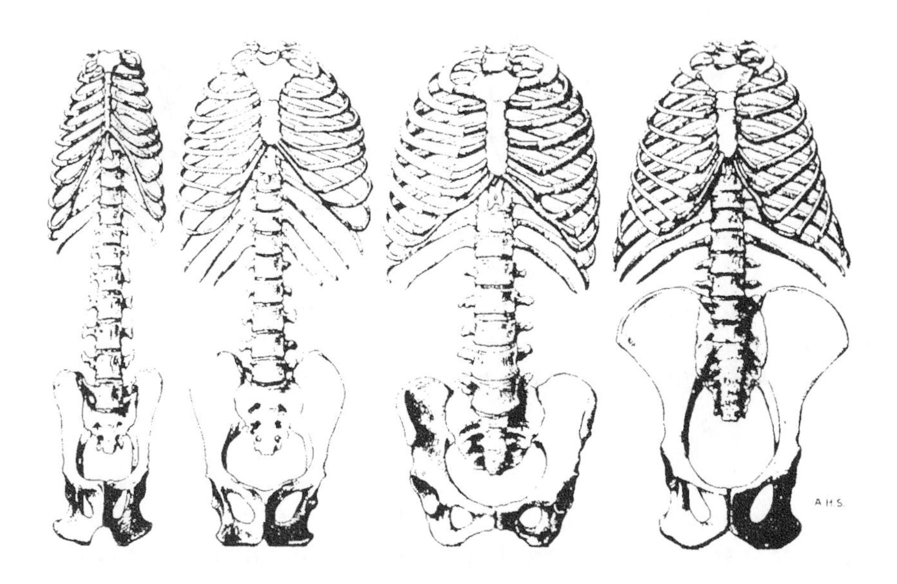

Fig. 14: A comparison of the ribcage, spine, and pelvis of the macaque, gibbon, human, and chimpanzee (left to right) (Schultz, 1961). The chimpanzee species is considered most closely related to humans, while the gibbon is a distant relative, yet humans and gibbons share a number of conspicuous characteristics that are absent in chimpanzees. The similar form of the ribcage, the longer lumbar spine, and the lower pelvis are immediately apparent. Although the low, broad human pelvis is often considered a specialized and species-specific trait, Schulz (1968) points out that most of its unique features are already clearly present during fetal development. The torso in humans and anthropoid apes is much broader than in the lower primates. Note, too, that human vertebrae are much more massive than those of the apes. The sacrum is extremely broad in humans, narrow in monkeys, and midway between the two in anthropoid apes (see Chapter 6).

- The chin is a typically human characteristic; in all adult apes, or at least large anthropoid apes, the lower maxilla recedes in front. Only the siamang also has a chin—a phenomenon pointed out by both Bolk (1924) and Frechkop (1948), among others (see *Figure 15*). Frechkop (1954) comments in this connection, "The catarrhines and platyrrhines seem to have divided certain human characteristics or "genes" among them. Thus we sometimes see humanlike chins in gibbons and siamangs, while certain spider monkeys (*Ateles*) retain humanlike foreheads."

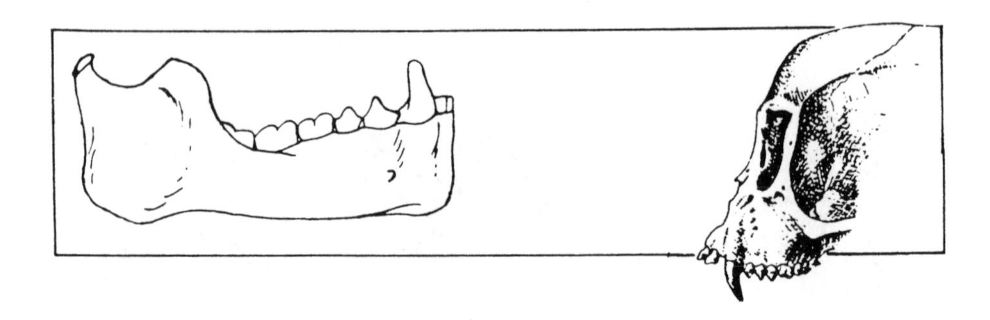

Fig. 15: Left: the humanlike mandible of a siamang (Bolk, 1924). Compare it to the receding lower jaw of other primates (see Figure 53, *p. 113). Right: The humanlike face and forehead of the spider monkey* (Ateles *spp.) (Schultz, 1926b).*

This same argument can be taken still further. Some characteristics that humans share with mammals in general are absent in anthropoid apes and certain hominid fossils. Bromage (1987) writes:

> The skulls of most mammals are organized in such a way that a line can be drawn connecting three points: the base of the brain, a protuberance on the upper jawbone called the maxillary tuberosity, and a point between the upper central incisors. Furthermore, if lines are drawn from the ear opening to the maxillary tuberosity and to the center of the eye socket, they make an angle of about 45°. These two patterns result from the way the skull develops at important growth sites and boundaries, and they are shared by most mammals, including modern humans. However, the apes and the earliest hominids are exceptions to this rule. They do not conform to the normal mammalian pattern.

Ontogeny and Phylogeny

How can human beings possibly share certain characteristics only with relatively distant relatives such as the platyrrhines? This problem is best understood by comparing the forms of different skulls. A comparison of *Figures 2* (p. 3) and *13* (p. 29) reveals that the morphological characteristics (high forehead, flat facial skull) shared by humans and capuchin monkeys (*Cebus* spp.) are also present in young chimpanzees; that is, these characteristics are also shared by anthropoid apes, but only during their juvenile period. What does this mean?

During the nineteenth century, the possible connection between the embryonic and infantile development of an individual animal ("ontogeny") and the evolutionary history of its species ("phylogeny") was the subject of lively debate. Ernst Haeckel (1834-1919) was the chief proponent of the radical view summarized in the form of the "fundamental biogenetic law" ("Ontogeny recapitulates phylogeny in a shortened form."). Although sometimes called Haeckel's law, this law in its fundamental form was formulated long before Haeckel (Gould, 1980; Thomson, 1988). According to this law, the entire evolutionary series (of adult parental forms) of an animal species is repeated in shortened form in the morphological changes of the developing animal. Haeckel claimed, for example, that a "fish stage" with typical gills can be found in the embryonic development of mammals or reptiles and that this embryonic developmental stage reflects a specific phase in the evolution of terrestrial animals, which evolved from fish at some point in time.

Haeckel's fundamental biogenetic law came under vigorous attack in the twentieth century and endured frequent criticism during the nineteenth century as well. A possible alternative to the fundamental biogenetic law, inspired by Aristotle and discovered by Karl Ernst von Baer (1792-1876), can be formulated as follows: The traits that appear earliest in the course of embryonic development are the most universal and are characteristic of the broadest category of animals; these early traits are followed by specialized traits belonging to ever more restricted groups of animal species, and traits unique to the individual species appear last.

According to von Baer's law, the embryo of a higher animal resembles the embryo rather than the adult form of a lower animal. Although neither the fundamental biogenetic law nor von Baer's law holds good in all instances, von Baer's law is generally considered more correct. Although clear examples of Haeckelian recapitulation exist in the animal kingdom, a short form of von Baer's law sums up the actual course of development fairly well in most cases of ontogeny: "The embryonic development of a given species recapitulates the embryonic phases rather than the adult phases of its predecessors."

We are now in a better position to return to our discussion of the primate skull. According to Haeckel's fundamental biogenetic law, the chimpanzee would have to be descended from humans because its skull passes through a humanlike phase in youth. If von Baer's law is valid, however, we must conclude that the skull shape that persists in humans and in *Cebus* species has a more general character and that the chimpanzee has abandoned this general form. In both cases it is safe to say that the human form is more original and that the shape of the adult anthropoid skull is derivative and specialized. Characteristics that humans share exclusively with distant relatives must always be more primary characteristics. Closer relatives such as anthropoid apes have lost these characteristics by continuing to evolve away from the original pattern.

It is astonishing to note, as we saw earlier, that human morphological traits are distributed not only among different primates but also among various other animals. This phenomenon suggests that the human form represents the original primate endowment to a very great extent. Different ape species have each retained a unique selection of these original traits while suppressing others. The original pattern is best preserved in *Homo sapiens*, the most primitive primate.

2

FETALIZATION AND RETARDATION

Fetalization

At this point, the research of Louis Bolk (1866-1930) will prove helpful in our further discussion of human evolution. Bolk was a professor of anatomy at the University of Amsterdam. His controversial appointment at the young age of thirty-two was fully justified in retrospect by his subsequent thirty-two years of scientific activity.

Bolk was a reticent man who devoted himself completely to his work and never married. In 1918 his cancerous right leg was amputated, and two other operations followed, but he bore his suffering with inner fortitude and never complained.

Bolk made significant contributions in several areas of anatomy and was very influential in professional circles. His visionary mind was usually occupied with several different subjects at once, but his *theory of retardation* was the one that made him famous.

Sometime around 1900, Bolk studied an adult male with various physical anomalies that were all remnants of normally transitory stages in fetal development. At the end of his life, Bolk said that this case firmly anchored the idea of the persistence of fetal characteristics in his subconscious.

In 1915 Bolk published a paper on the foramen magnum in humans and apes. In ape and human fetuses, the opening of the foramen magnum is on the underside of the skull. Humans retain this fetal characteristic, while in apes the foramen magnum gradually shifts toward the rear of the skull (see *Figure 16*). Bolk therefore considered the position of the foramen magnum in humans a fetal trait that is retained into adulthood.

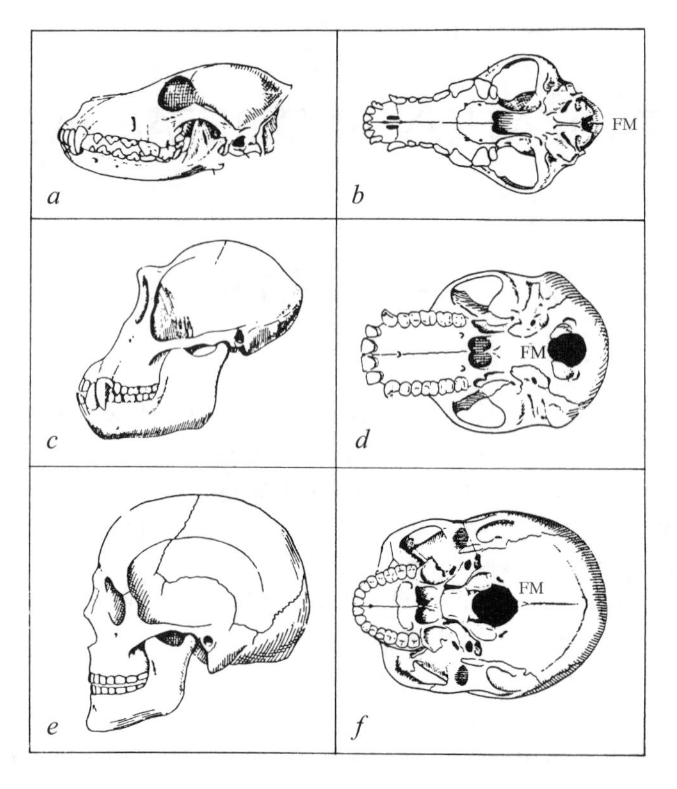

Fig. 16: Side view (left) and view from below (right) of the skulls of a dog (a, b), a chimpanzee (c, d) and a human (e, f). As the brain cavity increases in size, the muzzle retreats and the facial skull becomes flatter. The foramen magnum (FM) is located at the extreme rear of the dog skull, in the center of the human skull, and in an intermediate position in the chimpanzee (Clark, 1965). Bolk points out that during the fetal phase, the foramen magnum occupies a central position in both chimpanzees and humans. In humans, this position is retained, but in apes the foramen magnum shifts to the back as growth and development continue. (For details, see Chapter 6.)

A short time later, Bolk saw an approximately seven-month-old chimpanzee fetus and was impressed by its humanlike appearance (see *Figure 3*, p. 3, and *Figure 17*, p. 38).

> With the exception of the head, the entire body of the ape fetus is still hairless, and the distribution of hair on its head is exactly the same as in a human newborn. The hairs are close together, quite long, slightly wavy, and chestnut brown . . . Note also that the forehead is well developed and rises steeply . . . and that the nose is unmistakably present and projects freely (1918).

As we know, Bolk's observations were not new. The humanlike appearance of fetal and juvenile apes had already been noted in the nineteenth century. Keith, for example, reports that a seven-month-old chimpanzee fetus exhibited in a London museum was interpreted as a human infant by most inattentive observers (1949). Bolk, however, posed the question of the significance of this phenomenon more emphatically than anyone else.

Beginning around 1918, Bolk, inspired by his earlier observations, began to develop his so-called theory of fetalization or retardation. According to Bolk, the human adult can be considered a sexually matured ape fetus, and the generalized and fetal appearance of the human body is its essential anatomical characteristic. As it happens, the basic characteristics distinguishing humans from apes are often fetal characteristics that are lost as apes mature but retained into adulthood in humans.

One example is the pattern of hair growth (see *Figure 17*). As apes mature, hair gradually covers more and more of their bodies. In contrast, humans remain in a fetal-like stage, with hair more or less restricted to the head. Bolk calls human development "restrained" or "conservative" and ape development "progressive" or "propulsive." In his view, human beings represent a basic type that apes resemble less and less in successive developmental phases.

Bolk called such persistent fetal characteristics in humans *primary qualities*; we will call them Type 1 characteristics. During the earliest developmental stages (in the fetal phase or early childhood), these traits are common to humans and animals; they later disappear in animals but persist in humans. A high forehead and a flat facial skull (in contrast to the protruding muzzle of adult anthropoid apes) are examples of Type 1 characteristics (see *Figures 18* and *19*), as is the human pattern of hair growth.

In 1926, Bolk published a detailed study comparing patterns of hair growth in a 24-cm-long chimpanzee fetus and a 23-cm gorilla fetus (see *Figure 17*). In the chimpanzee fetus, the pattern is the same as in humans, but in the gorilla fetus the forehead is already covered with hair. Furthermore, both fetuses have hair on the parts of the face where beards or mustaches grow in human males of some races. In his conclusion, Bolk mentions that the pattern of hair growth in fetuses of African anthropoid apes briefly corresponds to that of adult humans.

According to Bolk, retention of fetal characteristics is not an exclusively human phenomenon; the same tendency is present to a clear but lesser extent in anthropoid apes. Here, too, we can take hair pattern as an example. Schultz (1940) researched the development of the chimpanzee fetus and discovered that the hair on the head is longer than that on the rest of the body throughout most of the fetal phase. At birth, hair growth in chimpanzees is still minimal on the chest, the abdomen, and the insides of the limbs. Schultz (1937) observed that hair growth in macaques also begins at the head. One hundred days after conception, the head of the macaque fetus is hairy while its torso and limbs are still hairless. In both humans and macaques, eyebrow hairs are the first to break through the skin. In short, hair pattern in these lower apes passes through the same initial stages observed in both humans and chimpanzees, but in macaques the hair covering spreads much more quickly; at 145 days, fur covers most of the body of a macaque fetus. Not only in humans but also (to a lesser extent) in chimpanzees, the spread of hair is clearly delayed in comparison to macaques.

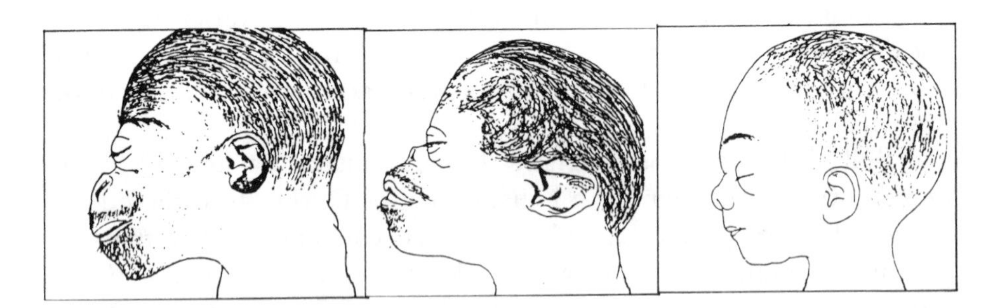

Fig. 17: Side view of the head of fetuses of approximately the same size. Left to right: gorilla, chimpanzee, human (Bolk, 1926b).

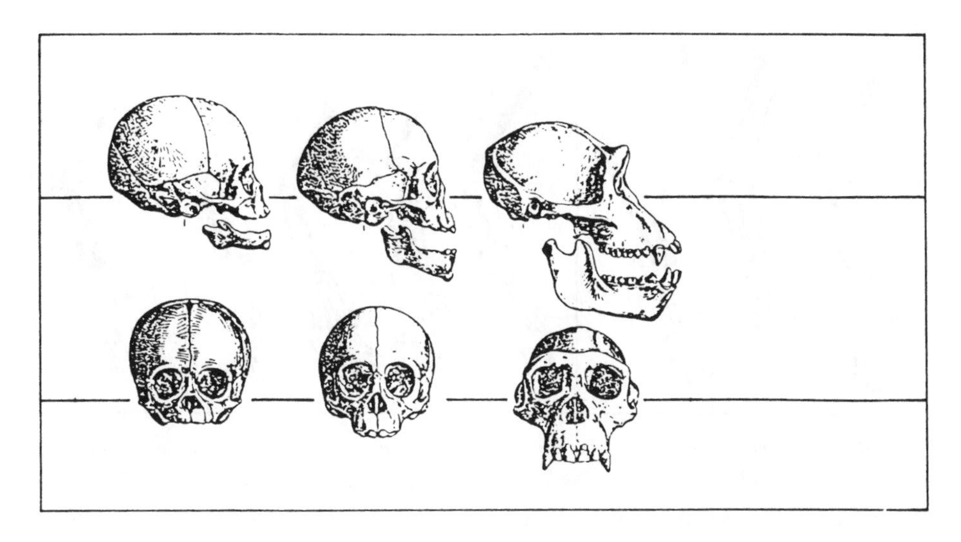

Fig. 18: Female chimpanzee skulls. Left: fetus (214 mm); center: juvenile (74 days); right: adult specimen (Schultz, 1940). The proportions of the skull are much more humanlike in younger animals than in the adult chimpanzee. In the latter the forehead has disappeared, brow ridges and sagittal crests have emerged, and the muzzle has developed.

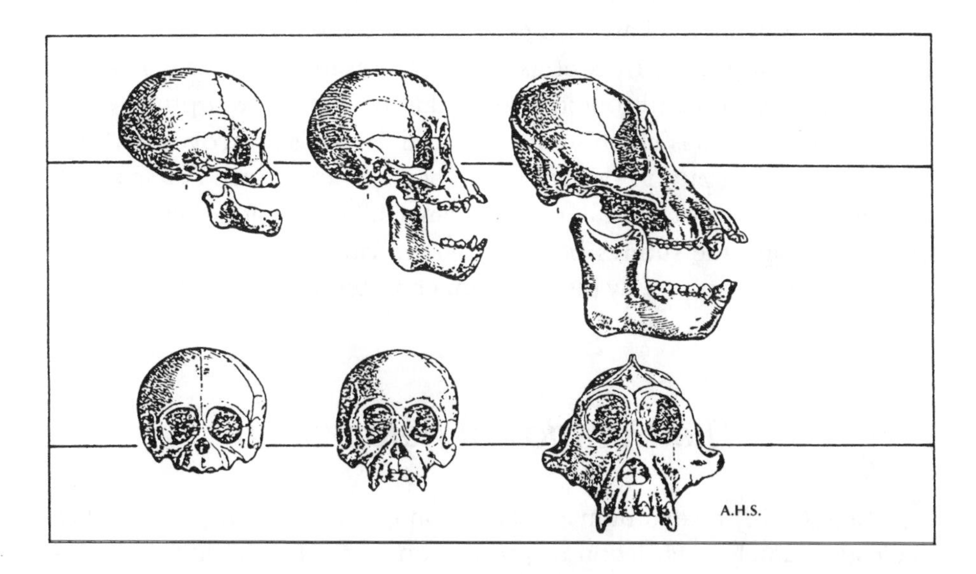

Fig. 19: Male orangutan skulls. Left: newborn; center: juvenile; right: adult specimen (Schultz, 1941).

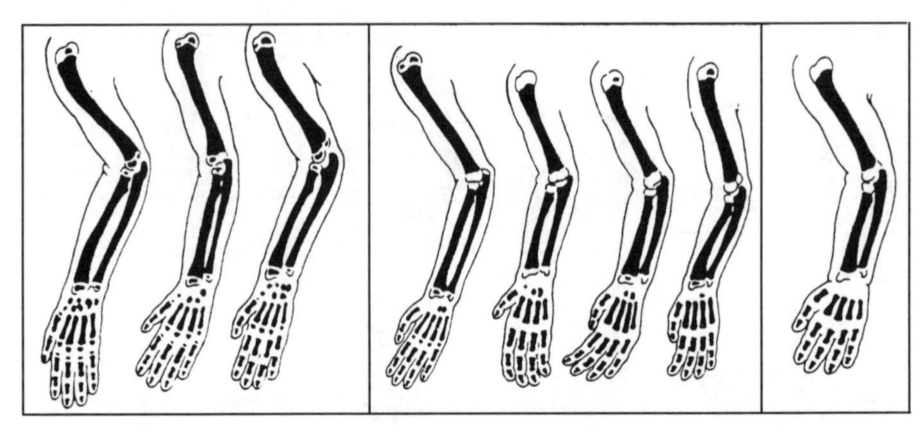

Fig. 20: Drawings from x-rays of upper limbs of primate newborns. Left to right: macaque, gelada, baboon; gibbon, orangutan, chimpanzee, gorilla; human. In humans, ossification centers are still absent from the base of the hand and the cartilaginous ends of the longer bones. Anthropoid apes (center box) have a few ossification centers in these locations, while the more primitive catarrhines (left box) have more. In general, the fetal cartilaginous skeleton persists longer in humans than in apes (Schultz, 1969a).

Figure 20 shows a second example, also from Schultz's research (1969). The skeleton of a human newborn generally still contains a great deal of cartilage; in the base of the hand, for example, ossification centers have not yet appeared. In contrast, some ossification centers are already present in newborn anthropoid apes, and in newborn lower primates the ossification process is much more advanced. Bolk also discovered a similar pattern with regard to ossification of skull sutures. Even at a relatively advanced age, the flat bones in the human skull have not grown together completely, and the sutures between the individual bones remain clearly visible. These joints usually disappear much faster in anthropoid apes.

The Paradox of Fetalization

Bolk (1918) related human fetalization to another conspicuous characteristic, namely, retardation or protraction of the human life span. The phases of human life succeed one another exceptionally slowly:

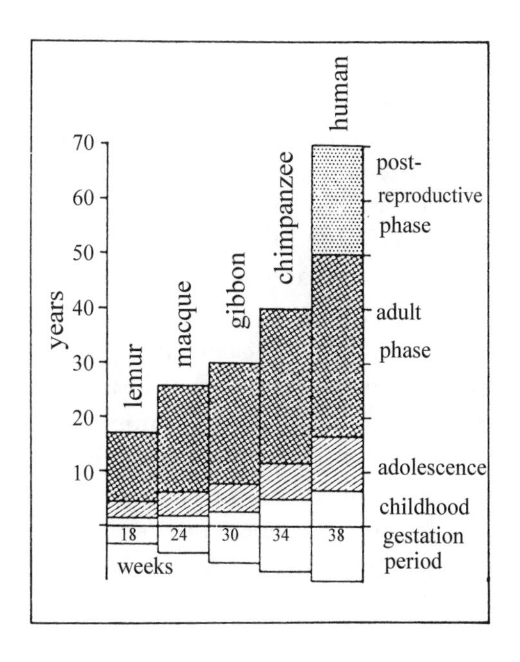

Fig. 21: The length of different phases of life in humans and some other primates. Only humans have a significant post-reproductive phase. (Schultz, 1969a).

There is no other mammal that grows and matures as slowly. What other mammal can enjoy the prime of life as long as a human being? Furthermore, the aging process that follows this slow blossoming and long period of maturity is more protracted than in any other mammal. What other animal enjoys such a long post-reproductive life?

In comparison with mammals of similar size, humans actually do have an exceptionally long life span (see *Figure 21*), and human development seems to be delayed or postponed. Bolk called this the phenomenon of *retardation.* A. Portmann (1967) provides the following examples of the much more rapid growth of large mammals. Deer are mature at three years of age, lions at six or seven years. Elephants reach adulthood at fourteen or fifteen years, male gorillas at ten. In comparison to these animals, human growth is conspicuously delayed. A.F. Richard (1985) has shown that while life span increases with body size, among animal species with similar body weights, larger-brained species live longer. Beasts of prey, for example, have larger brains than even-toed ungulates and also live longer. The life span of a lion, with a body weight of about 120 kg, is about 24 years, that of a puma (70 kg) 20 years, while the life span of a gnu (165 kg) is approximately 17 years and that of a giraffe (750 kg) 28 years. As a group, primates have larger brains than other animals and also live longer. Anthropoid apes can easily live to be 30 or 40 years old; humans, whose body weights are comparable, live even longer. In fact, the post-reproductive phase of life is a typically human phenomenon. There is no true menopause in animals (Pavelka and Fedigan, 1991).

One of the most remarkable mammalian characteristics is that the ratio of the average duration of the respiratory cycle to that of the cardiac cycle is constant among mammal species. W. R. Stahl's calculations yield a ratio of 1:3.9; that is, one breath takes roughly as long as four heartbeats (1962, 1965). We find this same ratio in human beings. Furthermore, Stahl discovered that the total number of respirations in a mammal lifetime is also constant. Between birth and death, a mammal breathes an average of approximately 190 million times. This number is independent of body weight. Small mammals breathe faster, but their lives are correspondingly shorter.

There is one great exception to this rule. Stahl discovered that primates live two or three times as long as expected on the basis of body weight. A primate life span, therefore, encompasses more breaths and heartbeats. At nearly 18 breaths per minute and an average life span of 70 years, human beings breathe nearly 660 million times between birth and death and thus surpass all animals. Anthropoid apes, however, also live longer than "normal." Incidentally, Stahl also established a parallel between duration of life and relative brain size.

Bolk believed that the phenomena of retardation and fetalization are closely related. Infinitely retarding (postponing) a characteristic is tantamount to eliminating it. Again, the spread of hair in primates can serve as an example. The spread of hair is retarded in chimpanzees in comparison to macaques. Macaques are born with a complete coat of fur, but newborn chimpanzees are still partially hairless. In anthropoid apes, a complete coat of fur develops only in the course of childhood. In humans, however, this retardation persists still longer, and while hair growth does spread over the body to some extent during the growth phase, a full coat of fur never develops. Hence, a phenomenon that is merely delayed in anthropoid apes is eliminated in humans. Ever-increasing retardation eventually eliminates more and more adult characteristics and therefore leads to fetalization. According to this point of view, if retardation persists indefinitely, it should ultimately eliminate *all* characteristics and thus also the creature in question. Bolk clearly recognized this paradox and appended these somewhat apocalyptic remarks:

> Nor will the human race of today persist for all time, for it too is subject to this omnipotent natural law, according to which each individual, species and group is granted only a limited period of existence. Life alone is eternal and immutable, but the forms it creates are transitory. The thought always fascinates me that in the future the human race will succumb to the same cause to which it owed its emergence in the past. Consider, for example, the possibility that progressive inhibition of our life forces, when it exceeds a certain limit, will reduce our vitality, our resistance to harmful external influences— in short, our ability to maintain ourselves. The further the human race progresses along the path of humanization, the more it approaches the fatal point where further progress means annihilation (1926a).

Progressive Emergence of the Human Gestalt

Bolk, however, did not stop short at this rather depressing conclusion. He gradually came to realize that fetalization does not explain all human traits. Upright human posture, for example, involves a number of morphological differences between humans and apes, some of which cannot be explained as consequences of fetalization.

One such difference is the position of the spine in relationship to the ribcage (see *Figure 22*). A. Schultz researched this phenomenon in great detail, making castings of the interior of the ribcages of humans and apes. In both ape and human newborns, the spine barely protrudes into the chest cavity, while in adult humans its location is much closer to the center of the ribcage. In functional terms, this central position can be directly attributed to human upright posture because it allows for better distribution of weight around the vertically oriented, weight-bearing vertebral column. In this respect, the spinal position that persists in apes is more similar to the initial position common to apes and humans, and we certainly cannot call the more central adult human position a Type I or primary characteristic. It must also be noted, however, that the same shift in the position of the spine is also evident in adult anthropoid apes. Schultz comments:

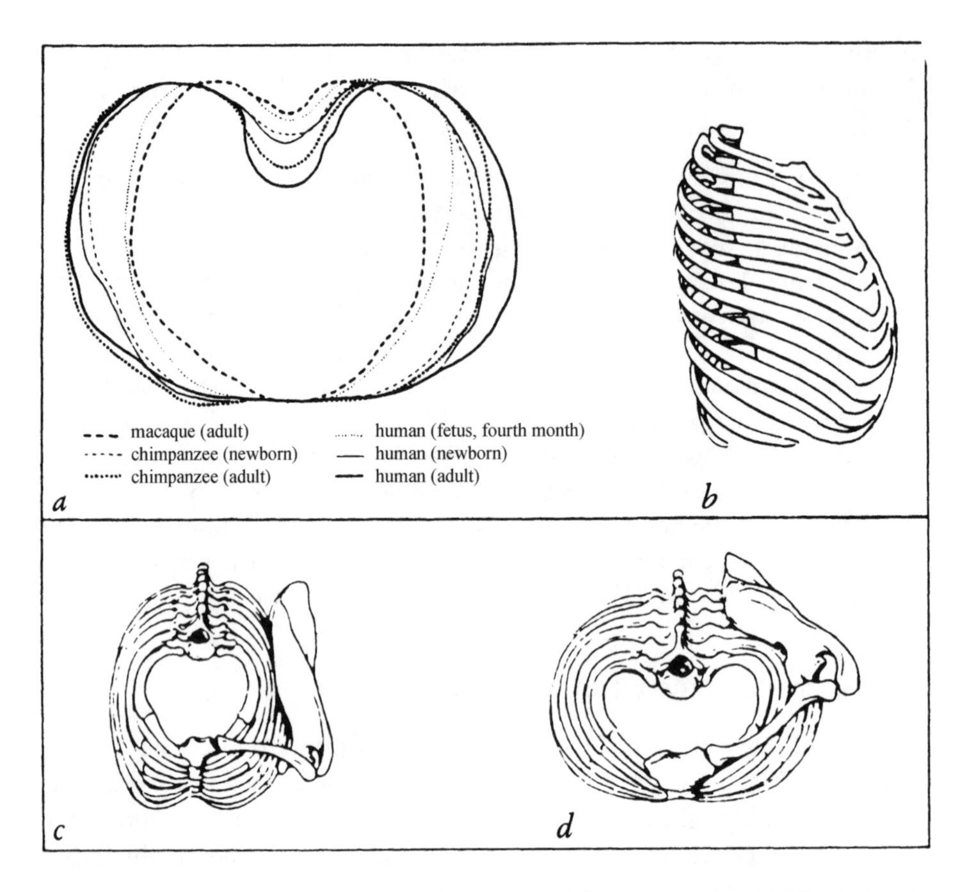

Fig. 22: (a): Outline of the interior of the chest cavity in different species (corrected for size) at the level of the sixth pair of ribs. (The vertebral column is above, the sternum below.) As we see, the ribcage of the rhesus monkey (macaque) is relatively round, and its vertebral column barely protrudes into the chest cavity. In both humans and chimpanzees, the spine shifts toward the center as growth progresses, while the relative width of the ribcage increases. This phenomenon is related to the upright orientation of the human body; weight is distributed evenly around the vertebral column. In adult chimpanzees (and also in the other anthropoid apes) this trait is already astonishingly well-defined. The central position of the spine in adult humans is a Type II characteristic. *(See definition of Type II characteristics on p. 48. See also* Figures 43, *p. 92 and* 66, *p. 132.)*

(b): Side view of bones of adult human ribcage during exhalation. Note the position of the spine.

(c – d): View from above of the ribcage, collar bone, and shoulder blade of the macaque (c) and human (d). In humans, the ribcage is flatter, the vertebral column protrudes deeper into the chest cavity, and the shoulder blade has shifted toward the back (Schultz, 1950a).

> In adults of the great apes, as in the chimpanzee…the relative position of the spine very closely approaches that of the human adult, showing in this respect a surprising degree of 'preparedness' for erect posture. In the early stages of development of all primates this [shallow] position of the spine, relative to the thoracic cavity, is very much as it remains into adult life in monkeys. In this respect, even as late as at birth, man is still more monkey-like than typically human. The adult human being represents merely the extreme of an ontogenetic innovation which as such is clearly recognizable in all higher primates (1949b).

The central position of the vertebral column, which offers no mechanical advantage in quadrupeds, becomes significant only in combination with the upright posture specific to humans. Even when an anthropoid ape walks, its chest cavity remains suspended below the spine (see *Figure 74*, p. 145), and yet the skeletal structure of this higher animal clearly prefigures human posture.

But how does the upright human stance come about, if not through fetalization? Bolk's answer is as profound as it is succinct. "Anthropogenesis did not begin because the body was becoming upright. The body began to stand upright because its form was becoming human" (1926a). This statement is characteristic of Bolk's way of thinking. He is no Darwinist. He views organic evolution not as the unintended workings of purely material influences but as an active principle:

> To me, evolution is not a result but a principle. Evolution is to the unified whole of organized nature as growth is to the individual. Evolution, like growth, is subject to influence by outer factors that modify forms but never create them. I believe that the essence of evolution as such still evades analysis, because evolution is a function of life as a whole rather than of individual life. The totality of life constitutes an organism with its own laws of growth and differentiation. What we call evolution is differentiation as it manifests in this macrocosmic organism (1926a).

Like Goethe, Bolk believed that the evolution of organisms is evidence of an intrinsic drive that is influenced but not explained by the phenomenon of natural selection, just as a detour influences a driver's route without fully accounting for it. Bolk vigorously defends this general view as it relates to the emergence of the human species. With regard to the fetal character of the pattern of hair growth in humans, for example, he writes:

> How can a bodily characteristic acquired by humans in the course of their evolution possibly influence the individual development of chimpanzees? . . . The fetal characteristic we are observing here cannot result from adaptation to outer circumstances, because no chimpanzees or their ancestors have ever had naked bodies with hair limited to the head. This phenomenon, therefore, must be a manifestation of a more deep-seated evolutionary principle. A weighty conclusion forces itself upon us: the factors causing loss of body hair but retention of head hair in humans were already active during the fetal development of anthropoid apes. Hence these causes cannot be external in character, nor can they have exerted their influence for the first time in anthropogenesis. An intrinsic evolutionary factor must exist, a factor that is already active in principle in anthropoid apes but manifests fully only in humans . . . Taken to its ultimate conclusion, this viewpoint leads to the conviction that the inevitability of human emergence was inherent in the earliest, lowest living thing (the primal organism, if you will), which gradually evolved into the human species as inevitably as a fertilized ovum ultimately develops into an adult animal . . . I am fully aware that this conclusion has led me onto very slippery and dangerous terrain. I know that I expose myself to the objection that because I introduce an inherent, endogenous evolutionary factor presumed to codetermine the emergence and morphological transformations of animal species over time, I am forced to view evolution as a predetermined process and as a given arising out of itself . . . The objection is justified, but I will not shrink from this logical conclusion (1918).

To Bolk, the human form is not a coincidental by-product but a given, a guiding principle inherent in organic evolution from the very beginning. An independent tendency toward humanization is discernible in animal evolution; we see the human form emerging ever more clearly from intractable organic raw materials. The human pattern of hair growth is soon abandoned by macaques but persists longer in chimpanzees, and in humans the pattern that apes abandon after the very earliest stages of life finally emerges completely and permanently. Thus it can be said that the human gestalt becomes ever more explicit as evolution progresses. In animals, the initial attempt at becoming human is overwhelmed by specialization, while humans remain true to the basic pattern.

Goethe viewed the archetypal animal as a spiritual reality struggling to achieve ever-purer expression in matter, and Bolk also does not conceive of his "intrinsic evolutionary factor" in materialistic terms. How could the human gestalt possibly be present in material form during the earliest beginnings of organic evolution? For Bolk, the emergence of the human being is the Aristotelian "final cause" of evolution (Verhulst, 1994a). Retardation as such is a negative process. But by neglecting the process of animal specialization, creatures whose development is delayed make space for a complementary, constructive process, namely, ever-increasing expression of the "intrinsic factor" that drives evolution. The full power of this factor, its full expression in matter, unfolds only in human beings.

Lack of specialization must have been the common thread throughout the human line of evolution. Specialization is a two-edged sword. An animal that adapts to a given situation in life takes advantage of specialization, while an unadapted animal enjoys fewer advantages in the same situation. Once initiated, however, specialization is irreversible, as Dollo's law states. Specific abilities are acquired at the expense of others. Birds, for example, sacrifice their "hands" for the sake of the ability to fly. Animal specialization entails both gains and losses—the loss of flexibility, of the ability to master new and different situations. Human evolution is characterized by a constant refusal to succumb to the pressure of natural selection or to seek refuge in safe adaptation to a specific environment. Undoubtedly, such thoughts underlie Bolk's conviction that natural selection cannot explain the primary dynamic of organic evolution.

If the human gestalt has been present as a guiding principle since the very beginning of evolution and has been expressed with ever-increasing clarity in living things, new human characteristics must have been added and become explicit in each successive stage. Each human characteristic can appear for the first time only after a certain point in evolution. For example, since only mammals have hair, the human pattern of hair growth could not possibly come to expression in earlier stages when mammals did not yet exist.

Type I (fetal) characteristics are present in primates (or lower forms of life) only in the earliest ontogenetic stages, but they persist in humans. Such primary characteristics as a well-rounded cranium or flat facial skeleton achieved expression in very early stages of the human line of descent, but in animals they often disappear again because they are overwhelmed by specialization. Hence, differences between humans and animals are due to the fact that specialization in animals *suppresses* the original Type I human characteristic.

Other, more developed characteristics, however, are not fully expressed in animals and appeared only in later stages of human evolution, when even the highest animals had separated from the evolutionary line leading to the human being. In contrast to the primary, fetal characteristics of Type I, we will call these more developed traits *Type II characteristics*. ("Consecutive characteristics" is the roughly equivalent term Bolk uses.) They differ in two respects from Type I characteristics (as well as from Darwinian specializations):

- Type II characteristics appear in the highest animals but generally not in the earlier developmental stages of either animals or humans. Type II characteristics are permanent in adult humans and (in less pronounced forms) in adult higher animals. Hence, differences between humans and animals arise as the result of more complete expression of these characteristics in humans.

- Although Type II characteristics appear in less pronounced forms in certain higher animals, their purpose and their functional significance become clear only in connection with specifically human traits such as speech, the use of tools, or upright posture.

The position of the vertebral column in relationship to the ribcage in humans is an example of a Type II characteristic. In humans (and anthropoid apes) this characteristic appears only late in the course of life. It is most developed in humans and makes sense only in the context of human upright posture.

The appearance of Type II characteristics resolves Bolk's fetalization paradox. Increasing humanization is not identical to increasing fetalization. As such, fetalization is a negative process; it means that the animal tendency toward specialization is disregarded. This restraint, however, allows new human characteristics to manifest more explicitly. As evolution progressed along the human line of descent, ever-new human characteristics achieved expression. Characteristics that manifested relatively early in the course of human evolution are Type I characteristics. They are also found in animal species that remained part of the human line of descent beyond this point. Such a characteristic, however, may also be overwhelmed by specialization later, after the animal diverges from the human line of descent. In this case, the characteristic appears only in the fetal or juvenile stage of the species in question.

An animal that diverged from the human line of evolution relatively early may stubbornly retain a specific Type I humanlike characteristic, while another species that diverged later may eventually abandon that same characteristic. For example, the adult capuchin monkey, a New World monkey that belongs to a lower group of primates, retains the human form of the skull better than the adult chimpanzee, which is more closely related to humans but abandons the common juvenile skull form.

In contrast, Type II characteristics became apparent only late in human evolution, at a stage when only the highest animals and prehistoric hominids had not yet fully separated from the human line of descent. Traces of Type II characteristics are found only in the most highly evolved animal species, develop only gradually in human individuals, and fully reveal their functional significance only in humans.

Animals that diverged from the human line of evolution at an early stage reveal fewer humanlike characteristics. They also have a longer period of animal specialization behind them. These two points can be summed up in this principle: *The less animal specialization occurs, the more human traits become explicit.*

The Paradox of Retardation

Several very astute modern critics have noted another paradox associated with Bolk's theory (Dullemejer, 1975; Shea, 1989). Even if all phases of human life, from conception to death, are equally retarded, that does not mean that fetalization must occur. Postponement of an event or process does not necessarily result in its complete elimination; the total duration of a human life could simply increase, with all characteristics simply appearing later. Hence, fetalization as such—a process that ultimately eliminates specific developmental steps—is not an inevitable consequence of retarded or delayed development. To understand why Bolk nonetheless saw a connection between retardation and fetalization, we must explicitly state his theory in a way that incorporates his anthropocentric perspective, which is very clearly presented in his book *Hersenen en Cultuur* (*Brain and Culture*, 1918).

According to Bolk, the human being is a generalized and largely unspecialized mammal. Bolk views the emergence of the human species as evolution's intrinsic goal and ultimate cause. While he sees Darwin's natural selection as irrelevant to the emergence of humans simply because the human being is an unspecialized creature, Bolk does acknowledge the role of natural selection in the emergence of animal lines of descent, which he understands as side branches that diverge from the central human trunk of the phylogenetic tree:

> I would by no means deny the significance of selection and adaptation in the emergence of countless forms of life, but it seems inadequate to explain the progressive thrust of evolution from the simplest organism to the human being. Selection and adaptation may well have contributed to the proliferation and detailed diversification of animal forms and may even have played a major role in the latter, but I view evolution from below to above as an *a priori* given, a fundamental property of life itself . . . While higher life forms constantly underwent detailed variation as a result of outer circumstances, evolution (as a characteristic of life itself) continued on an uninterrupted course and remained true to its function, eventually reaching the stage of anthropogenesis, during which the anthropoid apes emerged as morphological variations (1918).

In another passage, Bolk writes:

> If the difficult struggle for survival, to which many of its proponents attach such great significance for the evolution of humankind, had played any role, . . . the human race would have remained an animal species. In the form in which it is actually fought in nature, the battle for survival could never have led humanity to moral insights or ethical consciousness. The nature of the struggle for survival, as it appears in the natural world, is a struggle between power and greater power in which the greater power is always right and weakness is a crime . . . As grounds for humanization, as a driving force behind the higher evolution of humanity, this struggle, in my opinion, is not important in the least (1918).

Clearly, we would do better to restate Bolk's theory in explicitly anthropocentric terms, summarizing his theory of retardation as follows:

- In relationship to human development, animal development is shortened or *compressed*. Conversely, human development can also be seen as protracted or *retarded* in comparison to animal development.

- In animals, the developmental stages corresponding to the *later* stages in human development are either abbreviated or wholly or partially eliminated; their place in ontogeny is preempted by specializations characteristic of the species in question.

And in fact, the abbreviation of animal development is generally more pronounced during later stages. Gould writes, "A mouse day equals about 4 human days early in embryogenesis; this eventually increases to about 14 human days as human development rates slow down" (1977a). The human post-reproductive stage, which can last several decades, is unparalleled among animals. Changes in later developmental phases are less destructive of the pattern of development as a whole and are therefore more easily incorporated. By contrast, the consequences of earlier interventions, for example during embryonic development, are more drastic and life threatening.

In this anthropocentric perspective, animal ontogeny is abbreviated because animals must eliminate many steps from a humanlike developmental pattern in order to replace them with a smaller number of specialized animal steps. Let's compare this to a pile of wooden blocks of various shapes which, when fitted together, exactly fill their box. If we replace only one of these blocks with a mismatched piece with the same volume but a different shape, we can no longer fit the whole set into the box. Furthermore, the box will be incompletely filled, and the coherence and consistency of the original, complete set will be lacking. Including new, mismatched blocks means that the box can be filled faster (abbreviation or compression occurs because fewer pieces have to be placed), but its contents will be less evenly and consistently packed. Later developmental phases correspond to the upper layers of blocks in the box, where replacing an original block with a mismatched one has less impact on the total packing pattern, since the consistency of the lower layers is not affected. Replacing a block in the lowest layer, however, threatens to reduce the entire packing pattern to chaos. Substituting a mismatched or "specialized" block in the upper layers has less impact on the consistency of the whole. Thus it is not overly disruptive to the development of the chimpanzee when its muzzle begins to dominate over the other parts of its head; but this specialized development precludes a large number of developmental steps that are realized in the human being: further development of the brain and of the organs of speech which in turn allow for the acquisition of a broad spectrum of cultural capacities.

Formulated like this, the connection between retardation and fetalization becomes clearer. Human development is both delayed and (partially) fetalized for one and the same reason, namely, the absence of "mismatched blocks" or specialized animal developmental steps in the human ontogenetic pattern. Thus the theory points directly to the existence of the two types of human characteristics that we outlined earlier:

- Type I characteristics are a sign of the absence of animal traits in human development. Bolk calls these characteristics examples of fetalization, and they are often called "neotenic" characteristics in the literature. In our metaphor of the blocks, these characteristics correspond to the absence of mismatched pieces in the box.
- Type II characteristics denote the presence of primal human traits that have disappeared in most animal species due to compressed development. They correspond to original blocks in the upper layers, where they are most likely to be replaced when mismatched blocks are inserted.

Bolk placed great emphasis on Type I characteristics but did not adequately recognize the true nature of Type II characteristics, which are also often interpreted as examples of human specialization. As we have already demonstrated (p. 48), however, Type II characteristics differ from Darwinian specializations in that they are prefigured in the most retarded animal species (that is, those whose development is the least compressed) but reveal their function only in the fully manifest human gestalt.

Bolk's intimation of the relationship between fetalization and extension of the life span (retardation) in human beings made a decisive contribution to our understanding of human evolution. *Figure 23* (p. 55ff.) summarizes these relationships once more, elucidating and confirming Bolk's theses.

An adult organism reveals a great many anatomical and physiological traits whose individual paths of development can be traced, beginning with the zygote (fertilized egg). Each ontogeny, therefore, can be depicted as a bundle of many different developmental paths (shown in *Figure 23* as lines running from the zygote stage to the adult stage). Each developmental path, in turn, can be depicted as a series of elementary processes of development or growth. In phenomenological terms, these basic processes cannot be further subdivided. Abstractly speaking, we can also call them "quanta" of growth or development.

Figure 23a shows a general, idealized pattern of development in which all developmental paths are fully actualized. So that it can be compared to the animal pattern of development, we have divided this general pattern into the six equal segments A, B, C, D, E, and F, with each segment containing the same number of basic "developmental quanta."

In contrast, *Figure 24b* shows a compressed animal ontogeny. What has happened here in comparison to the pattern in *Figure 23a*?

First, the workings of natural selection have broken off the growth pathways at random. A few developmental paths have been able to reach completion, but others were suppressed during the early development of the animal organism. (This situation is comparable to atomic decay in a radioactive substance, in which the same percentage of atoms is breaking down at any given moment.) Because of this process, earlier developmental processes in a compressed ontogeny have a greater chance of survival, while later developmental stages retain decreasing numbers of developmental processes from the original pattern.

Fig. 23 a-c (opposite): Schematic representation of patterns of development: (a) general or universal, (b) animal, (c) humanlike. Each ontogeny is represented as a bundle of fundamental developmental paths that unfold between conception and adulthood. Each path in turn is subdivided into a series of successive elementary developmental processes or "quanta."

All developmental paths and quanta are present in the general pattern (a), which is equivalent to Goethe's archetypal animal.

In an animal ontogeny (b), portions of the developmental paths are broken off. We can imagine part of this process as follows: From the perspective of the universal pattern, a scatter of breaks (apparently statistically random) occurs as a result of natural selection. This means that the fundamental developmental paths can be assigned approximate half-life periods, as in atomic physics. The developmental pathways break down, and the successive phases A, B, C, D, E, *and* F *of the animal ontogeny contain exponentially decreasing numbers of the developmental processes present in the universal pattern. In addition, the animal ontogeny also includes specialized developmental processes (circles), which are fewer in number than the original developmental quanta that have been eliminated.*

Human ontogeny (c) eliminates fewer original developmental processes and includes almost no specialized developmental quanta. On the whole, human ontogeny encompasses more developmental processes than animal ontogeny, because the number of original developmental processes eliminated must always be greater than the number of specialized processes incorporated.

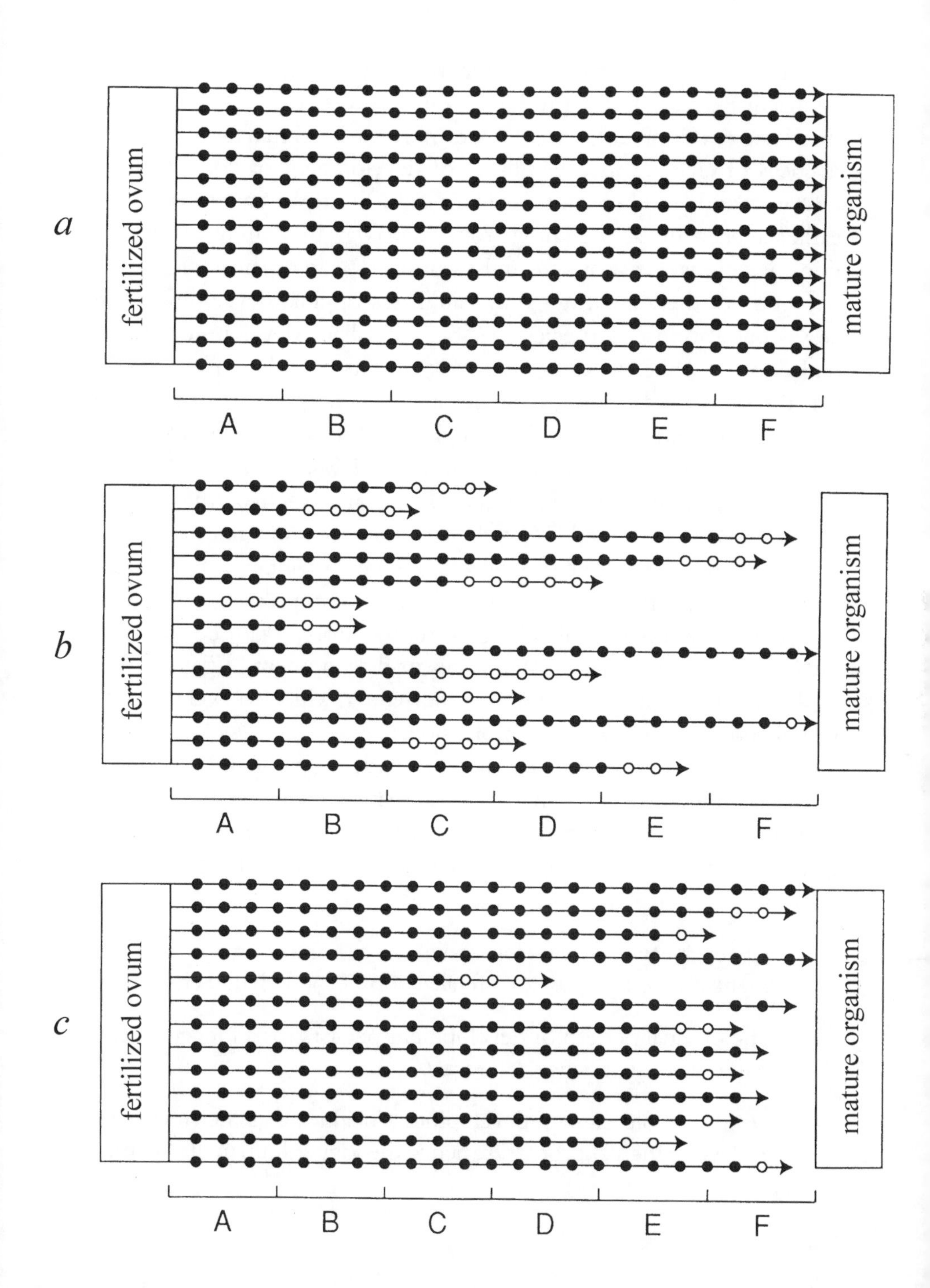
a
fertilized ovum
mature organism
A
B
C
D
E
F
b
fertilized ovum
mature organism
A
B
C
D
E
F
c
fertilized ovum
mature organism
A
B
C
D
E
F

The decrease over time is roughly exponential, as is also the case in radioactive decay. From the biological perspective, it also makes sense that later developmental phases are more easily eliminated. The consequences of changes in later phases of development or construction are less profound and therefore less likely to be life-threatening than those of earlier alterations, as we know from teratology (the study of malformations or serious deviations from the normal type in organisms).

Second, the animal not only eliminates original developmental processes but also inserts species-specific, specialized developmental quanta (indicated by white circles in *Figure 23b*) into its ontogeny. As we said earlier, the substitutions are less numerous than the processes they replace, just as fewer replacements than originals fit into our carton of blocks. Thus in comparison to the original situation (*Figure 23a*), animal ontogeny is shortened or compressed in such a way that each of the successive phases A, B, C, D, E, and F (which all contain the same number of developmental processes in the generalized ontogeny) contains fewer developmental processes than the previous phase. The compression that occurs in the development of an animal organism, therefore, is not equally distributed but is concentrated in later stages of development. *Figure 24* (p. 57) explains this unequal pattern of foreshortening, which we will call the primal phenomenon of animal compression.

Figure 23c summarizes the situation with regard to human development. In comparison to animal development, humans retain more of the original developmental processes and insert very few species-specific quanta. Thus human ontogeny differs from animal ontogeny in three respects:

- Fewer specialized developmental processes occur; the consequence is *paedomorphosis*, or fetalization (Type I characteristics).

- The original pattern of development persists for a longer time, resulting in *hypermorphosis* (see p. 67ff.).

- On the whole, human development includes a greater number of developmental processes (hence the longer human life span).

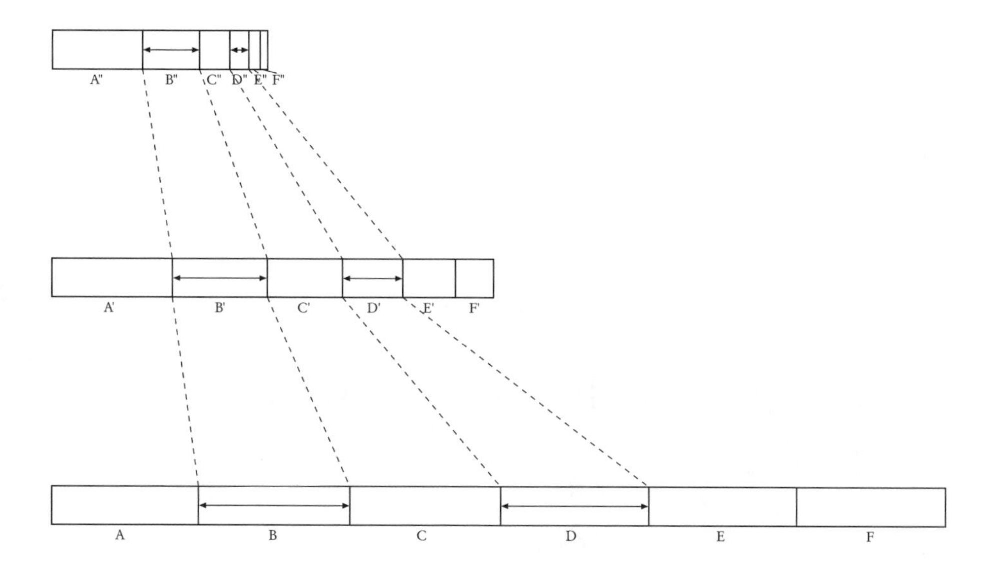

Fig. 24: The effect of compression on ontogeny.

Below: a generalized life span represented as a timeline divided into six equal segments A, B, C, D, E, and F (see Figure 23a*).*

Center: a moderately compressed ontogeny. The segments B and D, which are the same length in the generalized ontogeny, have been transformed into segments B' and D'. The later developmental segment D' is more highly compressed than the earlier segment B' because more elementary developmental processes have been eliminated from the later time period (primary phenomenon of animal compression).

Above: A highly compressed ontogeny. Here, development in time period D" is still more compressed in comparison to development in the segment B".

As a general rule, B/D < B'/D' < B"/D".

3

GROWTH AND RETARDATION

Introduction

In several respects, Bolk's formulation of his theory remained fairly unspecific, which hindered the general acceptance of his insights. Bolk often attributes the uniqueness of the human species exclusively to the phenomenon of fetalization, overlooking Type II characteristics. In *Hersenen en Cultuur* (Brain and Culture), for example, he writes:

> Throughout its growth, the human body deviates less from its fetal form than the bodies of anthropoid apes do; it retains more fetal morphological characteristics . . . After a certain point, further transformation stops in human beings; . . . only growth and the resulting increase in size continues. In anthropoid apes, however, transformation continues for a while longer, and different proportions develop; only at a later point in development does the existing form become fixed (1918).

What Bolk had in mind, however, was more than simply the retention of morphological characteristics. In his introduction to the third edition of *Hersenen en Cultuur*, which appeared shortly after Bolk's death, the famous neurologist Cornelis Ariëns-Kappers wrote, "Some [critics] have equated this continuation of fetal tendencies and adaptive possibilities with cessation [of development] in a specific period, but when Bolk spoke of retardation, he meant the retention of fetal tendencies rather than coming to a developmental standstill at a particular point" (1931). Similarly, a more recent commentator notes, "Bolk's critics always clung to the idea of the retention of fetal characteristics. Bolk, however, understood this to mean the retention of fetal capabilities, which is something different, because retention of a fetal dynamic may still lead to alteration in a morphological characteristic" (Dambricourt-Malassé, 1988).

And in fact, Bolk did not consider only morphological characteristics in substantiating his theory. For example, when he attempted to use the theory of retardation to explain certain aspects of human growth, he encountered unexpected difficulties because retardation seems to affect growth in two apparently contradictory ways:

- In a retarded organism, growth often proceeds more slowly, which means that the organism tends to remain *smaller*. Bolk expressly pointed to this effect. "At the end of gestation, a horse (which is then 12 months old) weighs 45 kg, and a calf (which is 11 months old) weighs 40 kg. At the end of the first nine months of life, however, a human being weighs only 3.5 kg. And while the horse doubles its birth weight in 60 days and the calf in 45 days, human beings take no less than 160 to 170 days to double their birth weight" (1927).

- Another consequence of retardation, however, is that growth as such continues for a longer period of time, which means that the organism tends to become *larger*. Bolk also observed this effect. "When the completion of the final phase of growth is delayed as a consequence of retardation— that is, when the duration of growth is prolonged— the inevitable consequence is a gradual increase in body size. In this instance, the principle of retardation sheds light on a fact that virtually all experts accept, namely, that the prehistoric ancestors of the human species were smaller" (1926c).

In most cases, larger bodies in animals are the result of specialization. Ariëns-Kappers (1942) gives the example of equine species, which have become both larger and increasingly morphologically specialized in the course of evolution. The earliest ancestor of the horse was approximately the size of a dog, and its limbs and teeth were still relatively unspecialized. As the ancestors and relatives of the horse became ever larger in the course of evolution, their teeth became more specialized and they lost four of their original five digits. In such cases, larger adult bodies are achieved by accelerating rather than prolonging growth. D'Arcy Wentworth Thompson notes the striking contrast between lion cubs, whose weight doubles in the first month of life and again during the second month, and human infants, who take five months to double their weight the first time and two years the second time. Thompson comments, "It is on the whole true…that the rabbit is bigger than the guinea-pig because he grows faster, but man is bigger than the rabbit because he goes on growing for a longer time" (1942).

Human growth is characterized by prolongation, not acceleration. The end result is an increase in size. In comparison to the lower primates, humans and anthropoid apes are notable not only for their protracted growth periods but also for their size. Among the primates (with the exception of the extinct *Gigantopithecus*), only gorillas— which grow much faster— are larger than humans, and average large apes are ten times heavier than lower primates (Martin, 1990). Within the relatively unspecialized primate order, therefore, increased retardation does indeed seem to be related to an increase in body size.

Retardation: The Primal Phenomenon

The growth of the body is an extremely complicated phenomenon:

- Some organs grow more quickly, others more slowly.
- Different parts of the body begin and end their growth at different points in time, and growth rates change constantly.
- Growth can occur in different ways. In some organs, the entire perimeter grows at the same time, while other organs grow only in certain parts, such as the ends of the longer bones in mammals (the epiphyseal plates).
- Tissues may grow either through cell replication or through increases in the size of individual cells.

- The growth patterns of different parts of the body are interdependent.
- External circumstances also play a role; human beings, for instance, grow faster in summer than in winter.

What happens when this complicated nexus of processes is retarded? Is there a common law governing both the postponement of successive developmental stages and the phenomena listed above?

Growth can be seen as consisting of a large number of individual elementary growth processes. Each of these processes begins at a specific point in time after conception and ceases at a later point. These two points can be viewed as two separate events, either or both of which may be postponed (retarded). Like morphological or physiological stages of development such as the first dentition or the onset of puberty, the beginning and end of elementary growth processes can be postponed through retardation. Retardation does not alter the unique character of such events but simply postpones them. With regard to elementary growth processes, this means that the growth rate itself is not changed. Only the beginning and the end of growth are postponed.

In the previous chapter, we saw that animal compression involves primarily the later stages of a humanlike pattern of development. This principle also applies to the phenomena of growth. The beginning and end of later-occurring growth processes are more retarded in humans (or more compressed in animals). We can express this primary phenomenon in a qualitative mathematical statement: *Retardation takes place in such a way that time intervals, calculated from the time of conception, are stretched in roughly exponential fashion*.

The essential point here, however, is not whether this process is exactly exponential but rather that the phenomenon of increasingly pronounced prolongation or retardation of the later phases of ontogeny exists. The exponential function is intended only as an indication, but it is justified by the fact that in many instances this function actually does provide a fairly precise approximation of events. *Figure 25* illustrates this principle as follows. Consider two events A and B, which are separated from C, the moment of conception, by two and four units of time, respectively. When retardation postpones both events, the time segments CA and CB are both multiplied by a "retardation factor." (For *Figure 25*, a factor of 1.3 was selected.) Thus the retarded events A' and B' occur only after 2.46 and 6.06 units of time, respectively. As we see, retardation causes longer postponement of later events than of earlier ones. We can consider events A and B in *Figure 25* as the beginning and end points of an elementary growth process. Clearly, although retardation postpones this process, it then lasts longer.

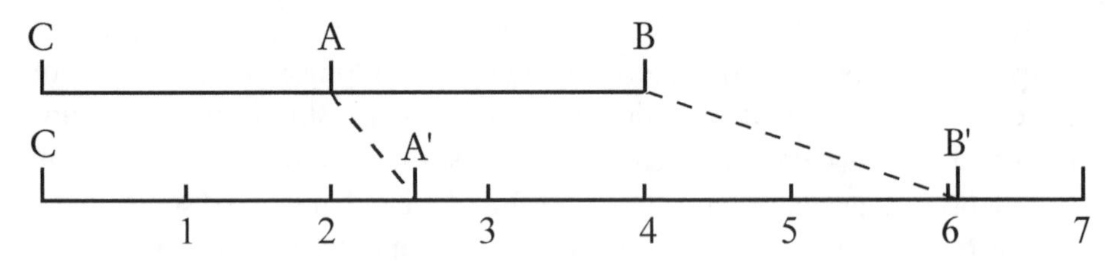

Fig. 25: Two events (A) and (B) (upper line), occurring two and four units of time after conception (C), respectively, are retarded by a factor of 1.3 (lower line), so that the retarded events (A') and (B') occur after 2.46 and 6.06 units of time, respectively. Through retardation, the time interval between (A) and (B) has increased from 2 to 3.6 units of time.

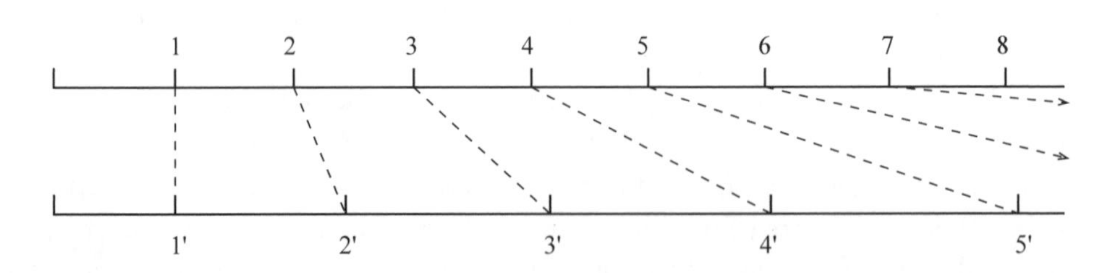

Fig. 26: Expansion of a timeline through retardation by a factor of 1.3. The first unit of time is equally long on both scales, while the fifth retarded time segment is already as long as two segments on the nonretarded scale. As a result of retardation, the growth of later-developing organs lasts longer than the growth of organs that develop earlier.

Figure 26 shows that later-appearing elementary growth processes last longer as a result of nearly exponential expansion or retardation and therefore produce relatively more growth than earlier processes. Conversely, this means that compression has more impact on later growth processes than on those occurring in early stages of development. Hence, *in an organism with retarded growth, organs or parts of the body that develop relatively late become disproportionally larger than parts that develop earlier.* We will return to this important principle later.

The situation with regard to total body size can be summarized as follows. Through protraction or retardation, growth lasts longer and leads to greater end results. Bolk was right, therefore, in believing that the longer duration of growth and the greater final size of human beings are consequences of retardation. In contrast, growth in certain animals is accelerated as a result of specialization. Their large size is due to accelerated rather than prolonged growth. In these animals, we do not find that late-emerging organs ultimately become disproportionally larger than organs that begin their development earlier.

An Example: Growth in Men and Women

Comparing growth in men and women shows how retardation of a complex growth pattern ultimately results in greater size.

Within the human growth period as a whole, three distinct phases can be delineated. During the first months and years of life, human beings grow very quickly. During childhood, a rhythmical growth pattern emerges. During puberty, growth accelerates markedly— a phenomenon that occurs only in higher primates and is especially pronounced in humans (see *Figure 27*). Women pass through these successive phases faster than men (Butler et al., 1990; Jolicœur et al., 1988).

Greater retardation of the growth pattern in men ultimately leads to greater size. We must not jump to the conclusion, however, that female development is foreshortened in relationship to male ontogeny in every respect. For example, women live longer, and in comparison to men they remain paedomorphic or fetalized in several respects such as the development of body hair.

Growth phases: average age (in years) at which growth peaks occur

Girls	*Boys*	
4.6	4.8	(preschool period)
6.7	7.8	(elementary school period)
8.6	9.0	(mid-childhood)
10.0	10.8	(prepuberty)
12.0	14.0	(puberty)
19.2	21.0	(cessation of growth)

(These numbers are averages. Acceleration of growth almost invariably occurs during the elementary school period and in puberty, but the smaller growth spurts may go unnoticed in individual cases.)

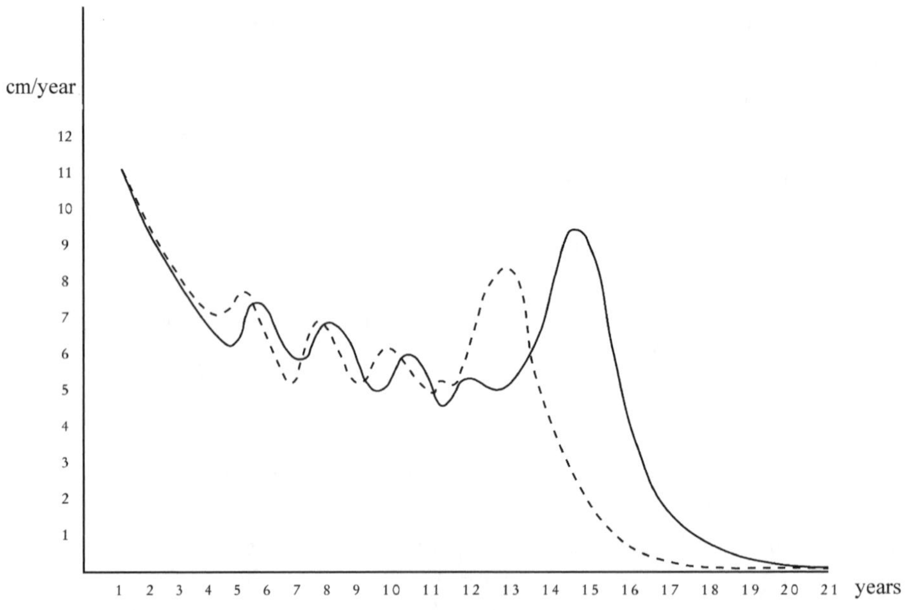

Fig. 27: Fluctuations in growth rate in boys (solid line) and girls (broken line). Amplitude and timing of the growth peaks were determined by averaging the values for these peaks in individuals, not by plotting a combined curve for a large number of individuals, which would have allowed the smaller peaks that occur in different ages in different individuals to cancel each other out (Butler et al., 1990).

Cephalocaudal Growth and Hypermorphosis

In its earliest stage of development, the embryo consists only of the so-called embryonic disk, which corresponds roughly to the torso and especially the chest of the adult body. This rudimentary body then begins to develop at opposite ends, where the future head and buttocks will be. Once the head forms, growth tends to focus on the upper end. The head takes shape rapidly, while the hind end initially lags behind. In the embryonic and early fetal stages (*Figure 28*), the head is very large in comparison to the torso. The torso itself is dominated by the swelling where the heart develops, while the sacral region remains relatively small.

Both the head and the torso, however, already have a large part of their development behind them when the limbs first appear as short, shapeless stumps. During the prenatal and early childhood phases, the limbs remain relatively short in comparison to the head and torso, catching up only gradually (see *Figures 28* and *29*). But the upper and lower limbs do not develop at the same rate. In the embryo, rudimentary arms appear somewhat earlier than legs. The fingers separate before the toes do (see *Figure 28*). Nails appear first on the fingers and then on the toes, reaching the fingertips in week 32 of gestation, the tips of the toes in week 36 (England, 1983, 1990). Hair appears first on the head and then spreads downward (Pinkus, 1963). Throughout prenatal development, the arms remain longer than the legs (see *Figure 29*). After birth, the length of the legs in proportion to the arms increases steadily.

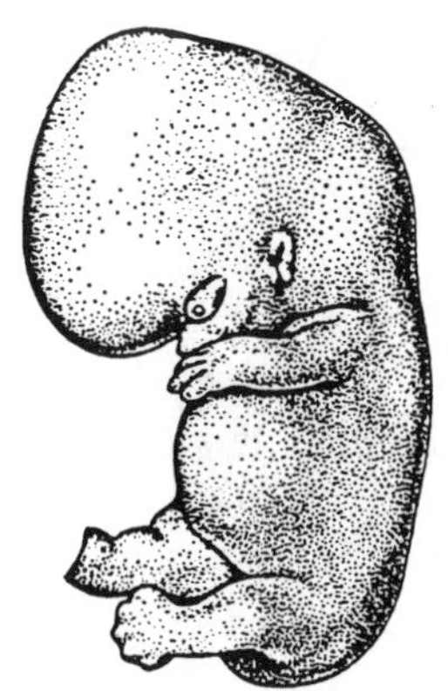

Fig. 28: Human fetus at seven weeks (length 18 mm from the top of the head to the buttocks). The limbs are still very short in proportion to the body. In comparison to the legs, the arms are off to a developmental head start— the fingers are already separated, but the toes are not (Langman, 1966).

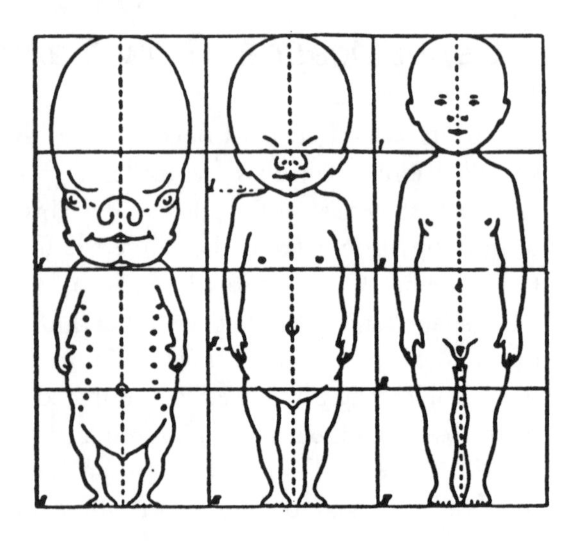

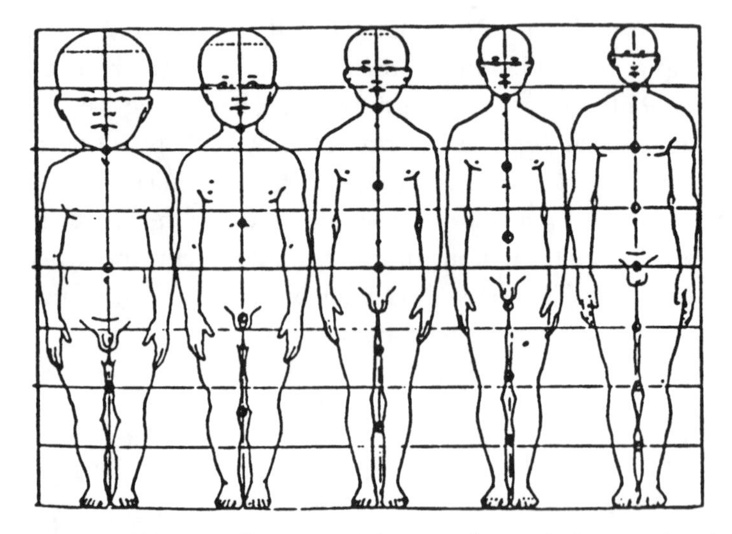

Fig. 29: The classic illustration of changes in human body proportions in the course of growth (Stratz, 1909).
Above: Prenatal development. Left to right: fetus at 2 months and 5 months; newborn.
Below: Postnatal development. Left to right: newborn, at 2, 6, 12, and 25 years of age.
Especially obvious is the catching up that occurs not only in the torso and limbs relative to the head but also in the limbs (especially the legs) relative to the torso. In the head, the facial skull catches up relative to the cranium.

In humans, as in all mammals, catching up occurs in a well-defined series of changes over the course of growth. The growth of the torso catches up with that of the head, and the lower limbs catch up with the upper limbs. The overall pattern of these catching-up gestures can be summed up as the *law of cephalocaudal development* (Kingsbury, 1924). According to this law, parts of the body closer to the top of the head have a developmental "head start" on body parts farther away from the head. The forehead, for example, is initially very large in proportion to the facial skull. In the torso, the ribcage is initially more developed than the hip area, which is much narrower. Around week 9 of gestation, the width of the fetus at the hips is only 58 percent of its shoulder width (Schultz, 1924). During childhood, the law of cephalocaudal development predominates in the development of body proportions in humans, and we can safely state that the growth tendency that manifests throughout the body is *cephalocaudal* (i.e., moving from the head downward). During the first years of life, the focus of successive waves of growth shifts consistently downward.

Hypermorphosis

As a consequence of the cephalocaudal growth tendency, some parts of the body begin their growth later but then catch up with earlier-developing parts. We have already seen that, in a retarded organism, growth can persist longer in organs or body parts that begin growing later than others (see page 63). This principle will help us understand certain differences in body proportions between humans and animals.

As our first example, let's look at the relative lengths of the limbs and torso. Primates are retarded in comparison to other mammals, and their limb growth is delayed in comparison to torso growth. Thus, as expected, we find that primates tend to have longer limbs. On the whole, the limbs of primates (in comparison to their torsos) are 1.1 to 1.6 times as long as those of other mammals (Alexander et al., 1979). Because humans and anthropoid apes are generally retarded in comparison to lower primates, we would expect to find that their limbs are disproportionally longer, as is indeed the case. Schultz (1933) lists the following average relative arm and leg lengths (the length of the arm/forelimb or leg/hind limb expressed as a percentage of torso length).

	Relative arm length	*Relative leg length*
lower catarrhines	118.0%	106.5%
gibbon	237.8%	146.5%
siamang	233.4%	146.5%
orangutan	181.9%	119.2%
chimpanzee	175.0%	128.0%
gorilla	171.1%	123.8%
human	150.1%	171.0%

As we see, the retarded anthropoid apes have relatively longer lower limbs than the lower catarrhines. In humans, where the phenomenon of retardation is still more pronounced, the legs are longer than in anthropoid apes. The extremely long legs of humans are a consequence of very pronounced retardation. In a highly retarded growth pattern, the lower limbs have the opportunity to develop fully.

The increasing length of the limbs relative to the torso and of the hind limbs relative to the forelimbs is evident in the phylogenetic tree as early as the fish. In these animals, the pairs of fins corresponding to the limbs of terrestrial vertebrates are typically quite modest in size in comparison to the torso, and the pectoral fins are clearly more developed than the pelvic fins.

Of course, strong development of the lower limbs is completely in harmony with the upright posture of human beings, which is also supported by a whole series of other retardation phenomena, as we will see later.

The growth of the arms also catches up with that of the torso. Retardation magnifies the effect of this phenomenon, so humans have relatively longer arms than monkeys. But why do anthropoid apes, especially gibbons and orangutans, have much longer arms than humans? (See *Figure 30*.) As is evident from the considerations below, specialization related to climbing rather than a retarded growth pattern accounts for the extremely long upper limbs of these animals.

- Increased arm length is already very pronounced in newborn anthropoid apes and thus does not result from catching up that occurs over the entire growth period. In contrast, human legs are quite short at birth and *gradually* catch up with the rest of the body (see *Figures 29, 30* and *31*; cf. also *Figures 25* and *26*). Schultz (1926a) has shown that the hind limbs of lower catarrhines grow faster after birth than the forelimbs. That is, these animals undergo the same general catching-up process that has been observed in humans. In anthropoid apes, this tendency is either less pronounced or absent, and in orangutans and gibbons (the most specialized climbers) the arms actually grow somewhat faster than the legs immediately after birth.

- In all primates, the lower limbs undergo a catching-up process relative to the upper limbs, in harmony with the law of cephalocaudal development. Where growth is retarded, the lower limbs should ultimately add more length than the upper limbs, and in humans as compared to lower apes, this phenomenon actually does occur. In anthropoid apes, however, the forelimbs actually grow longer in relationship to the hind limbs. Human proportions, unlike those of anthropoid apes, can be qualitatively explained as the result of simple retardation of the universal pattern of growth (see *Figure 30*).

Stephen Jay Gould (1977b), borrowing from Abbie (1948), uses the term *hypermorphosis* for differences in body proportions caused by retardation. Gould (1977a) defines hypermorphosis as "extension or extrapolation of ancestral allometries. With a delay of maturation, differentiation can proceed beyond its ancestral level." Thus compression of development also leads to selective suppression of growth processes that appear in later stages of development. But why is this suppression selective? Gould comments, "Hypermorphosis assumes the dissociation of maturation and somatic differentiation: ancestral rates of differentiation are maintained and delayed; maturation permits their extrapolation beyond ancestral conditions . . . If both [maturation and differentiation] are delayed at the same rate, we obtain a proportioned giant" (1977a). From the conventional perspective, however, there is no reason at all to expect the dissociation that

Gould mentions. Why should later developmental phases be selectively suppressed as ontogeny is compressed? In other words, why should the retardation phenomenon (as described on p. 60 ff.) occur? We can understand this phenomenon only if we accept human development as the primal state of affairs from which animal developmental patterns were later derived through compression that selectively impacts the later phases of the humanlike developmental plan.

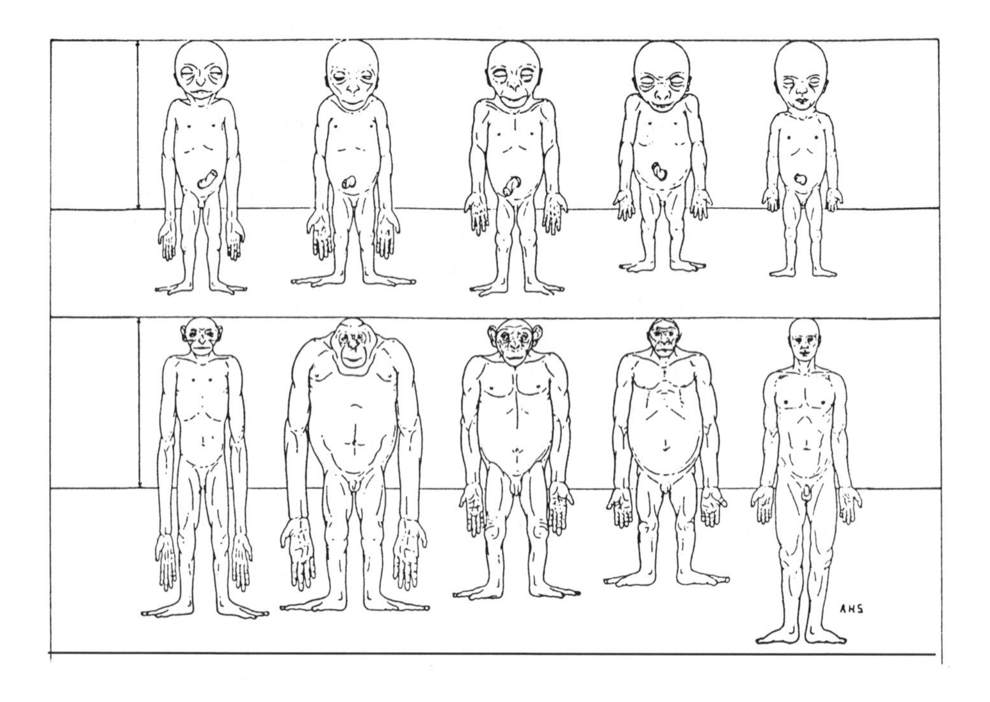

Fig. 30: The higher primates (Schultz, 1926a). Fetus (at comparable stages of development) and adult figures of (from left to right): gibbon, orangutan, chimpanzee, gorilla, and human. For purposes of comparison, all are shown without hair and with the buttocks at the same level. Note that:

- The hands and feet of the human fetus are much smaller than those of the anthropoid ape fetuses.

- In the human adult, the legs are longer than the arms, while the opposite is true of the anthropoid apes.

- The orangutan and the gibbon, typical arboreal species, have especially long arms.

- The human adult has a longer neck than the anthropoid apes.

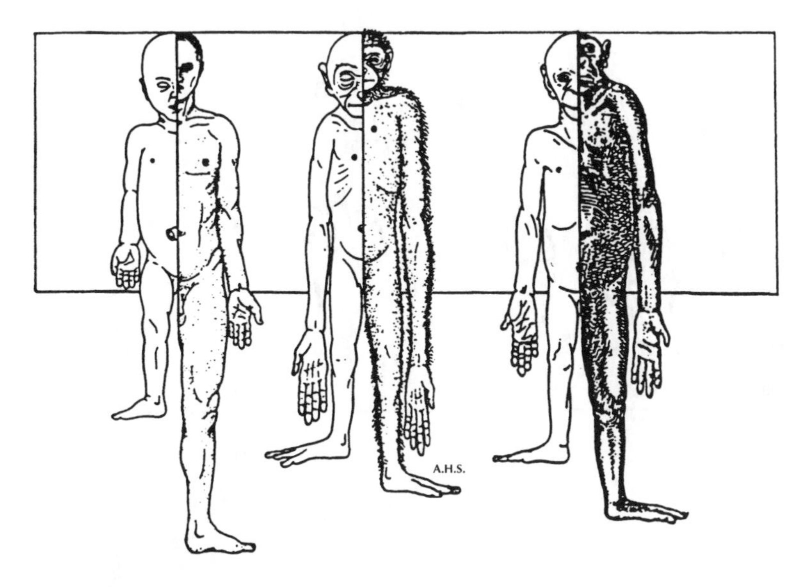

Fig. 31: Newborn and adult human, gibbon, and chimpanzee (left to right). (Schultz, 1925 and 1931).

Distoproximal Growth

In the early fetal stage, the growth pattern that appears *within* human and mammalian limbs is the opposite of the cephalocaudal growth tendency; that is, the limbs develop from their outer end toward the torso. This direction of growth, in which the growth of body parts farthest from the head accelerates first, is called the *distoproximal* growth pattern. In fetal development, the cephalocaudal growth pattern determines the overall proportions of the head, torso, and limbs, while the limbs' internal proportions are characterized primarily by the opposite, distoproximal pattern.

Figure 32 depicts this phenomenon with regard to the arm. At the point in time when four indentations— the first sign of finger separation — appear in the rudimentary arm, the disk of the hand is huge in proportion to the rest of the arm. Even once the fingers have actually separated, the hand is still longer than the forearm and the upper arm is still totally submerged in the torso.

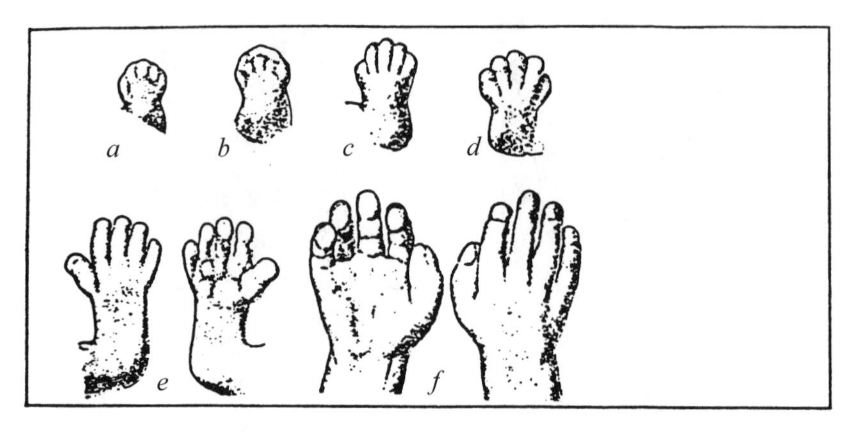

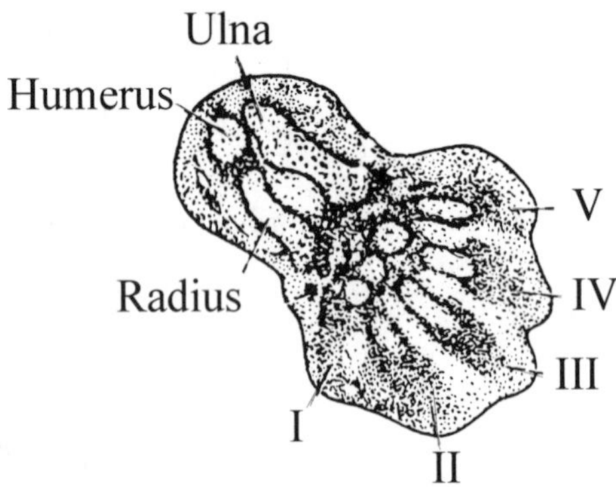

Fig. 32: Above: Stages in the embryonic development of the human hand and arm. (a): arm bud, with indentations in the finger zone already visible; torso length of the embryo = 12 mm. (b): arm bud/15 mm. (c): arm bud/17 mm; the hand is huge in proportion to the arm, and the upper arm is barely developed. (d): arm bud/20 mm; the palm is small in comparison to the fingers. (e): front and rear views of the arm; the fingers are separate, the thumb does not yet rotate, and the arm (especially the upper arm) is still very short. (f): front and rear views of the hand/52 mm; the thumb rotates (Retzius, in Patten, 1968).

Below: Rudimentary arm skeleton of a 15.5 mm embryo. The humerus (only the distal end is shown) is short and still contained within the torso. The radius and ulna are still short in comparison to the disk of the hand, which dominates the forearm (Jarvik, 1980).

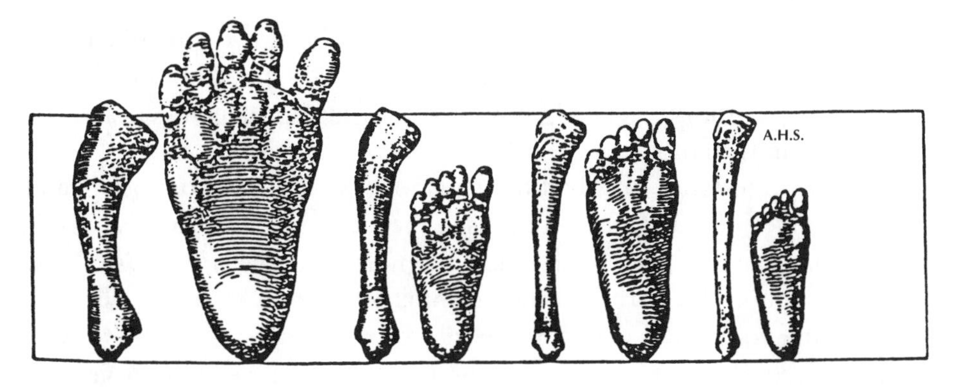

Fig. 33: Relative sizes of the foot and tibia in the course of human development. Left to right: fetus (week 9), fetus (week 15), newborn, adult. In the earlier stages, the distal ends of the lower limbs are overdeveloped — the foot is very long in comparison to the tibia, and within the foot itself the development of the toes is more advanced than that of the heel. The relative sizes of the foot and tibia gradually change in favor of the tibia, but a rhythmical alternation is apparent in this development — at birth, for example, the foot is relatively larger than in week 15 (Schultz, 1926a).

Figure 33 shows the effect of the same law on the lower limbs. Although leg growth is delayed in comparison to the rest of the body, the feet develop quite quickly (at age six, for example, the feet have already achieved 70 to 74 percent of their ultimate size, while the body as a whole is only 66 to 67 percent of its adult size). In its early stages of development, the foot is much larger than the tibia. Within the foot itself, the heel is initially less developed than the toes. As growth progresses, rhythmical shifts occur in the relative sizes of the foot and the tibia, ultimately favoring the tibia. This rhythm is probably related to the opposition between cephalocaudal growth processes, which determine overall proportions, and the distoproximal growth pattern, which is locally predominant in the limbs.

Thus we see that two opposing growth tendencies are active in the developing human body (and in mammalian development in general). The cephalocaudal growth tendency works throughout the entire body, beginning with the head; as a result, body parts farther away from the head develop later. This tendency is also active in the limbs. The opposing distoproximal tendency, however, works in the opposite direction, beginning at the extremities. In the limbs, the directions of these two growth tendencies are polar opposites, which sometimes results in rhythmical alternations in growth. In general, however, the distoproximal tendency is dominant in the limbs.

The appearance of rhythm in a zone where two opposing and approximately equal forces meet is a frequent phenomenon. A rising cloud of cigarette smoke provides a simple example (*Figure 34*). Two forces work on the warm, ascending smoke— an upward force, which is due to the colder and therefore heavier surrounding air, and the downward force of gravity. As long as the upward force clearly prevails, the smoke rises vertically. Higher up, the smoke begins to cool and falters in its ascent, tracing a rhythmically winding path as the difference in strength between the two forces at work on it decreases. Higher still, where both forces are equally strong, the force of levity no longer prevails. When its effect is no longer visible in pure form, the movement becomes chaotic.

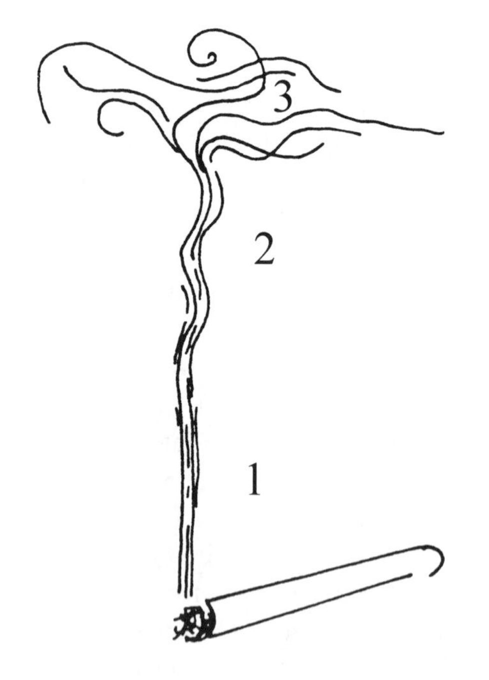

Fig. 34:

Three zones in an ascending column of smoke:

(1): zone of vertical ascent (the force of buoyancy is stronger than gravity).

(2): rhythmical zone (buoyancy prevails only weakly).

(3): chaos (buoyancy and gravity cancel each other out).

Hypermorphosis and the Distoproximal Growth Pattern

Because distoproximal growth predominates in limb development, the upper arm must catch up with the growth of the forearm; the same is true of the entire arm with regard to the hand. An analogous process occurs in the lower limbs. We would expect retardation of this type of growth to lead to hypermorphosis. That is, where development is highly retarded, as it is in humans, the parts of the limbs closer to the torso should become larger. And in fact they do. Here are two basic examples:

- In comparison to other primates, humans have very small hands and feet relative to the length of the limbs (see *Figure 31*, for example). In humans, foot length is smaller in proportion to leg length than in any other primate. The weight of the skeletal hand or foot expressed as a percentage of the weight of the skeletal arm or leg illustrates how exceptional these human proportions are (calculated from results obtained by Schultz, 1962).

	Skeletal foot weight as a percentage of skeletal leg weight	*Skeletal hand weight as a percentage of skeletal arm weight*
human	13.2	14.5
gorilla	16.1	16.0
chimpanzee	15.4	17.5
orangutan	22.6	18.6
gibbon	18.2	18.1
macaque	19.7	15.0
baboon	17.0	13.2
proboscis monkey	18.2	19.3
spider monkey	19.5	17.7
capuchin monkey	24.3	17.9
tree shrew	25.5	15.3

The ratio of skeletal foot weight to skeletal leg weight clearly increases in less retarded primates, a trend that is less apparent with regard to the upper limbs. Anthropoid apes have fairly heavy bones in their hands and baboons have heavy arm bones—deviations from the general trend that are probably related to these animals' means of locomotion. Anthropoid apes are brachiators, adapted to travelling by swinging from branch to branch, and their hands are correspondingly solid in build, while baboons are true quadrupeds whose arms (but not hands) help support their weight. In any case, in comparison to the weight of the total skeleton, human beings have lighter hand bones than all apes (Schultz, 1962).

- The brachial index (the length of the radius expressed as a percentage of the length of the humerus) and crural index (the length of the tibia as a percentage of the length of the femur) are lowest in human beings. The list below (Schultz, 1956) gives some representative values.

	Brachial index	*Crural index*
human	76	83
gorilla	80	85
chimpanzee	91	83
orangutan	99	88
gibbon	108	85
macaque	97	91
patas monkey	106	98
colobus monkey	100	91
northern owl monkey	94	97
capuchin monkey	96	94
spider monkey	101	95

It has been noted that the ratio between the logarithms of the lengths of the femur and tibia of different primates approaches a constant and is greater than one (Aiello, 1981; Aiello and Dean, 1990). In this case, the primary phenomenon of retardation actually does assume an exponential form.

The Seven-Year Rhythm in Human Growth

The idea that human ontogeny is characterized by a seven-year rhythm is very old. The concept of a seven-year rhythm in human biography was generally accepted in ancient Greece; Aristotle said that this view was basically correct and important for education (for details, see Roscher, 1906). The seven-year rhythm may play a role in certain social circumstances (Verhulst and Onghena, 1996, 1998a, 1998b). It corresponds with the closure of human skull sutures; see *Figure 35* (Verhulst and Onghena, 1997). In human female sexuality, the seven-year rhythm evidently plays a role even in later years (see Chapter 9).

Growth in humans occurs chiefly during the first three seven-year periods of life. Each of these three seven-year periods has its own distinct characteristics, as shown in *Figure 36*.

The first seven-year period is characterized by rapid growth of the nervous system. Around age seven, the nervous system is 90 percent complete, and the brain (according to some scientists) has already achieved its final size (Jolicoeur et al., 1988). The sense organs (for example, the eye—— with the exception of the lens— and the middle ear) generally follow the same growth pattern as the nervous system.

The second seven-year period is characterized by the greatest development of the lymphatic system. Its organs achieve extraordinary size around age ten, when they are much larger than they are in adults. The thymus, for example, is both very active and very large during the second seven-year period, and the great size of the tonsils in schoolchildren is well known.

The third seven-year period is characterized by the explosive development of the reproductive organs. It remains a mystery why the growth of the larynx also accelerates during puberty, especially in boys, whose vocal cords can double in length within a year, causing their voices to "break." The growth of the bridge of the nose is also pronounced during this period. Stoddart (1990) writes:

> Since the earliest of times a relationship—or supposed relationship—between the nose and sex has been widely accepted by peoples of many cultures. Celsus, who followed in the doctrinal

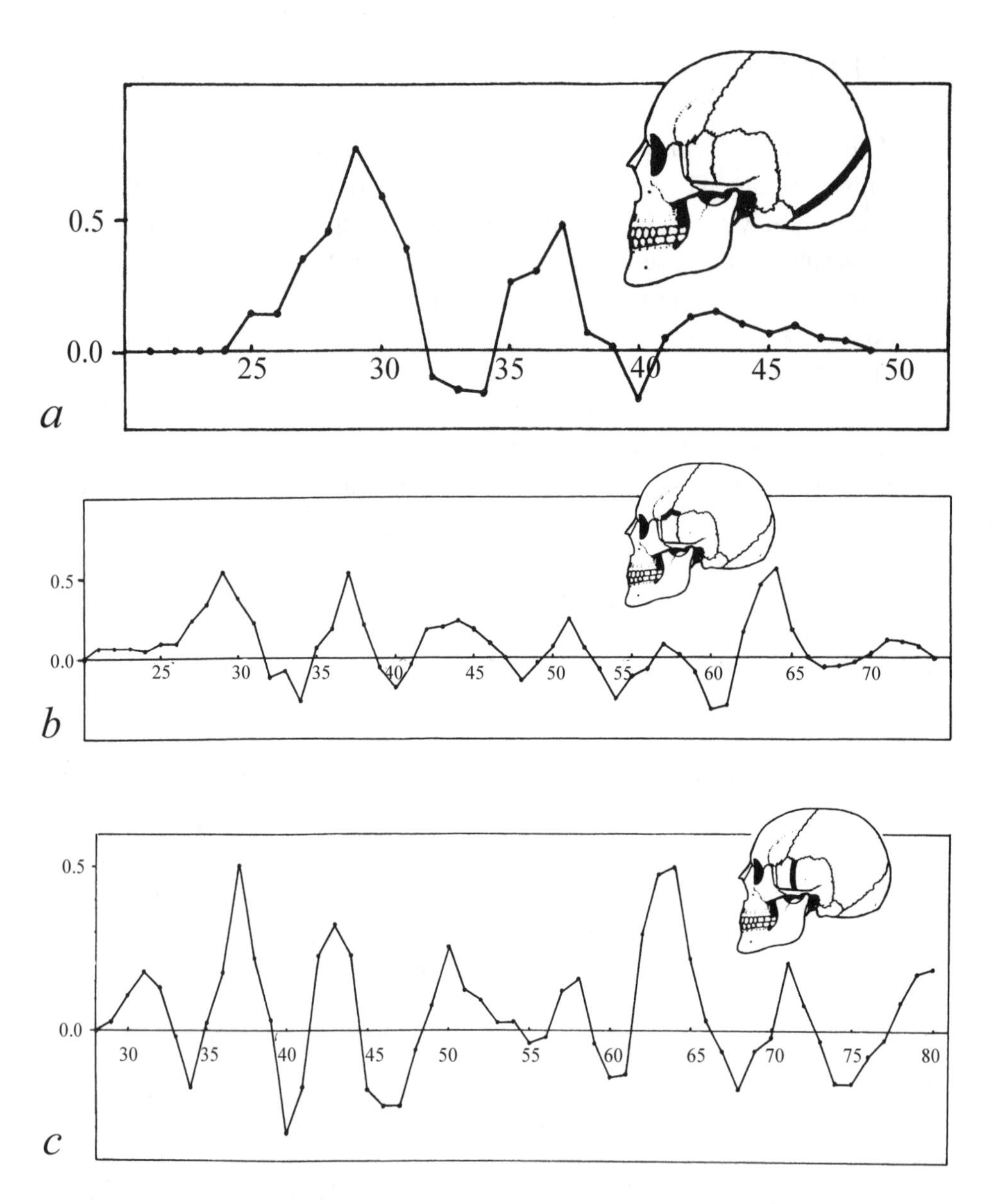

Fig. 35: Graph of closure rates of three human skull sutures.

(a): lambdoid suture; (b): sphenofrontal and sphenoparietal sutures; (c): sphenoquamosal suture. The horizontal axis represents age in years, the vertical axis the closure rate (a value of 1 was assigned to the closure rate of a year in which 25 percent of the suture closes).

> footsteps of Hippocrates 2000 years ago, admonished men to 'abstain from warmth and women' at the first signs of a cold or catarrh, since venery was thought to irritate the nose. Early physiognomists drew parallels between the size of the nose and the male member . . . The supposed virility associated with a large nose doubtless led to the practice, so well described by Virgil in the *Aeneid*, of punishing adulterers by nasal amputation.

According to Stoddart, more than 220 publications on this naso-genital connection appeared between 1900 and 1912. Wilhelm Fliess, a visionary much admired by the young Freud and, among other things, the inventor of the doctrine of biorhythm, treated venereal diseases by inserting cocaine into specific spots in the nose; he also treated Freud's nose (Gardner, 1981). In spite of such excesses, a mysterious connection does seem to exist between the upper respiratory organs and sexuality, a connection that is expressed in the accelerated growth of the organs of both systems during puberty.

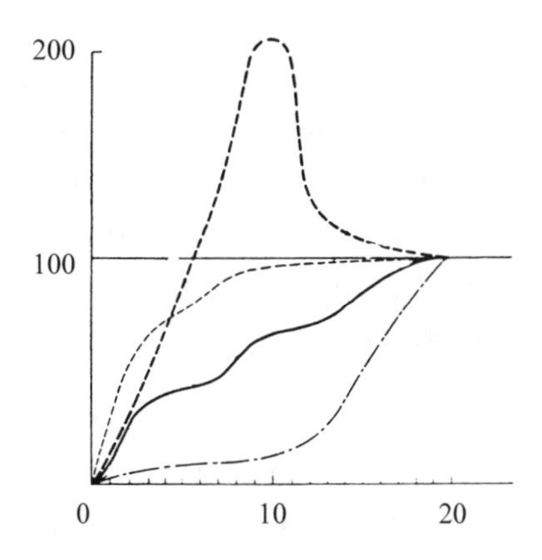

Fig. 36: Graph of the course of growth of different organ systems in humans. Horizontal axis: age in years; vertical axis: cumulative growth as a percentage of the end result in the young adult. Light dashes: nervous system; bold dashes: lymphatic system; dots and dashes: reproductive system and secondary sex characteristics; solid line: muscles and bones. The nervous system grows primarily during the first seven-year period, the reproductive system and gender characteristics during the third. During the second seven-year period the size of the lymphatic system increases dramatically, temporarily exceeding the end result in the adult. The speed of growth of muscles and bones increases at each transition from one seven-year period to the next (Tanner, 1962).

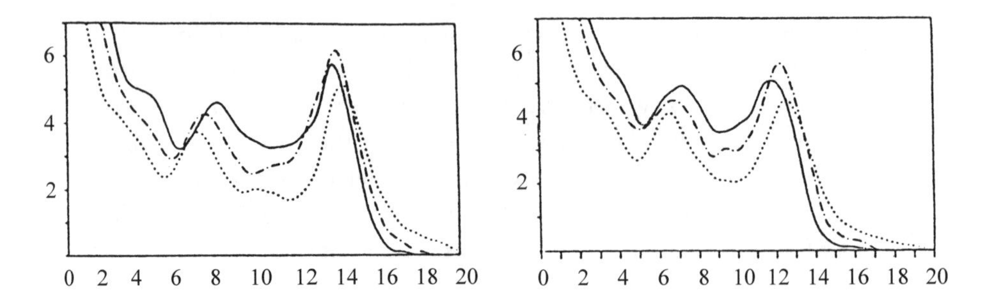

Fig. 37: The seven-year rhythm in human growth. In boys (left) and: girls (right).

Relative growth rate (yearly growth as a percentage of the final result in the adult) of: leg length: solid line; arm length: dots and dashes; torso length: dots. The curve reveals two growth spurts, the elementary growth spurt around age seven and the pubertal growth spurt at the beginning of puberty (around age 14 in boys, age 12 in girls).

The elementary growth spurt becomes evident first in the torso, then in the arms, and last in the legs, while the pattern of the pubertal growth spurt is reversed, occurring first in the legs, then in the arms, and finally in the torso, where growth continues until approximately age 20 (Gasser, Kneip, Binding et al., 1991).

Conversely, a clear contrast exists between the nervous system and the reproductive system. While nearly 90 percent of the nervous system's growth occurs during the first seven-year period, 90 percent of the reproductive system's growth occurs during the third seven-year period.

Gasser, Kneip, Binding et al. (1991) thoroughly researched the growth of the torso and limbs (see *Figure 37*). Their results provide interesting insights into the seven-year rhythm in human growth. Growth ceases around age twenty, and two growth spurts divide this period into three times seven years. The rhythm is very regular in boys and somewhat accelerated in girls.

Growth spurts do not begin in all parts of the body at the same time. The elementary growth spurt (around age seven) is cephalocaudal in its direction, while the pubertal growth spurt (which occurs around age twelve in girls and fourteen in boys) proceeds in the opposite (i.e., caudocephalic or distoproximal) direction. The following list gives the age (in years) when the speed of growth peaks in each of these growth spurts.

	Growth spurt:	*Elementary*	*Pubertal*
Girls:	torso	6.5	12.5
	arms	6.7	12.2
	legs	7.5	11.7
Boys:	torso	6.7	14.2
	arms	7.1	13.8
	legs	7.8	13.8

Gasser, Kneip, Ziegler et al. (1991) discovered a comparable pattern with regard to the breadth of the shoulders and pelvis. Here, too, they observed cephalocaudal progression during the elementary growth spurt and caudocephalic progression during the pubertal spurt. The list below shows the age (in years) when the speed of growth peaks.

	Growth spurt:	*Elementary*	*Pubertal*
Girls:	shoulder breadth	6.9	12.5
	breadth of pelvis	7.2	12.4
Boys:	shoulder breadth	6.8	14.2
	breadth of pelvis	7.2	14.1

Overall body proportions apparently develop in cephalocaudal succession until the end of the elementary growth spurt, that is, waves of growth begin in the head and move downward. Beginning in the third seven-year period, however, this tendency is reversed; waves of growth begin in the limbs and move toward the head. The second seven-year period, which lies between the two major growth spurts, seems to be a time of rhythmical equilibrium in which the two growth tendencies struggle to gain the upper hand. Beginning at puberty, distoproximal or caudocephalic growth, formerly restricted to the limbs, begins to govern the proportions of the entire body.

Strangely enough, there are a number of indications suggesting that a proximodistal or cephalocaudal growth pattern begins to prevail in the limbs themselves, beginning with puberty (Davenport, 1935). For example, Nath and Chacko (1988) researched the maturational sequence in the limbs of girls aged eight to eighteen in Udaipur (Rajasthan, India). They report, "The forearm exhibits an advanced maturation over upper-arm length between 8.0 and 13.0 years, indicating the existence of a caudocephalic direction of maturation. This gradient is reversed to cephalocaudal beyond the age of 13 years." In the legs, the maturational sequence is somewhat complicated, but there too the thigh achieves its ultimate length earlier than the lower leg. Various other studies could be mentioned in this context. Harris et al.(1992), for example, researched changes in length of finger and metacarpal bones in adults between the ages of 21 and 55. During this observation period, the two distal (i.e., peripheral) finger bones became somewhat longer, while the metacarpal bones actually shrank slightly.

Thus the cephalocaudal (proximodistal) and caudocephalic (disto-proximal) growth patterns seem to switch roles in the first and third seven-year periods (see *Figure 39*), while the second seven-year period is a period of oscillation and transition. In this connection, for example, Nath and Chacko (1988) write, "At the ages of 8.0, 11.0, and 15.0 the two extremities exhibit a cephalocaudal maturational trend while at the other ages the maturational trend is caudocephalic, indicating an advanced maturation of the leg segment over the arm segment as a whole." Beginning with the third seven-year period, the growth of the body as a whole follows a caudocephalic pattern, as *Figure 37* clearly suggests. Growth ceases earlier in the legs than in the arms and later still in the torso, as many researchers confirm (Hulanicka and Kotlarz, 1983). Clearly, the pattern of human growth is characterized not only by a noteworthy seven-year rhythm but also by an obvious reverse symmetry.

Incidentally, the seven-year pattern is found not only in human growth but also, to give one interesting example, in the calcium phosphate content of human blood plasma (see *Figure 38*). Calcium phosphate plays an important part in building up bone tissue. During the first seven-year period, the blood plasma is supersaturated with calcium phosphate. Concentrations sink to around the saturation point during the elementary growth spurt and oscillate around this point throughout the second seven-year period, sinking to their final levels during the third seven-year period.

(Like the pubertal growth spurt, the decline in calcium phosphate content occurs earlier in girls.) It is apparent that in humans, age-related changes in calcium phosphate levels are symmetrical to the characteristic saturation point. Presumably, delayed bone mineralization in humans and the absence of certain bony parts such as the penis bone and os cordis—both of which are still present in chimpanzees and can be considered Type I characteristics—are related to the typically low calcium phosphate levels of human beings.

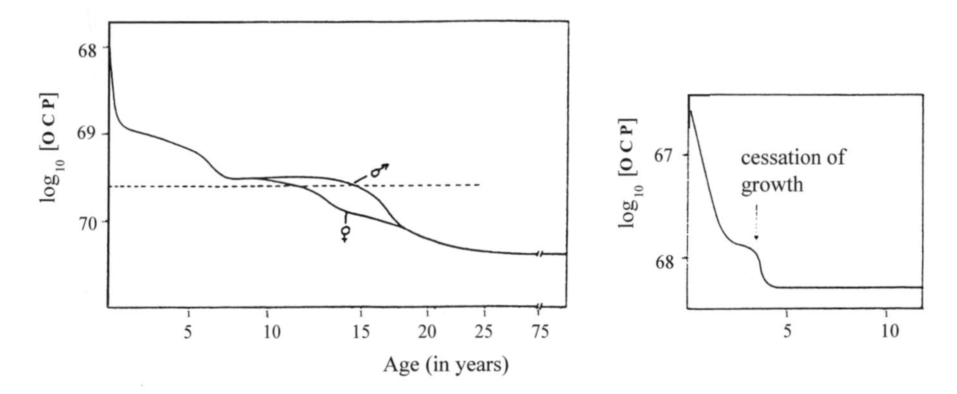

Fig. 38: Left: The logarithm of the ion-activity product for calcium phosphate in human blood plasma. The gradual decrease of this product reflects a seven-year periodicity. During the second seven-year period, the ion activity product approximates the saturation point (broken line). The duration of the decline, like the duration of growth, is compressed in girls.

Right: The analogous curve in bovines, whose development is extremely compressed in comparison to humans. Later phases of development, during which calcium phosphate levels sink, are temporally compressed and largely eliminated. In bovines (and other mammals), final, life-long calcium phosphate concentrations remain above the saturation point (Driessens et al., 1989).

Human body proportions are essentially determined in early childhood, when growth is fastest. During these years, the development of the proportions of the body as a whole follow a cephalocaudal pattern, while proportions within the limbs follow the opposite, caudocephalic pattern (see *Figure 39*).

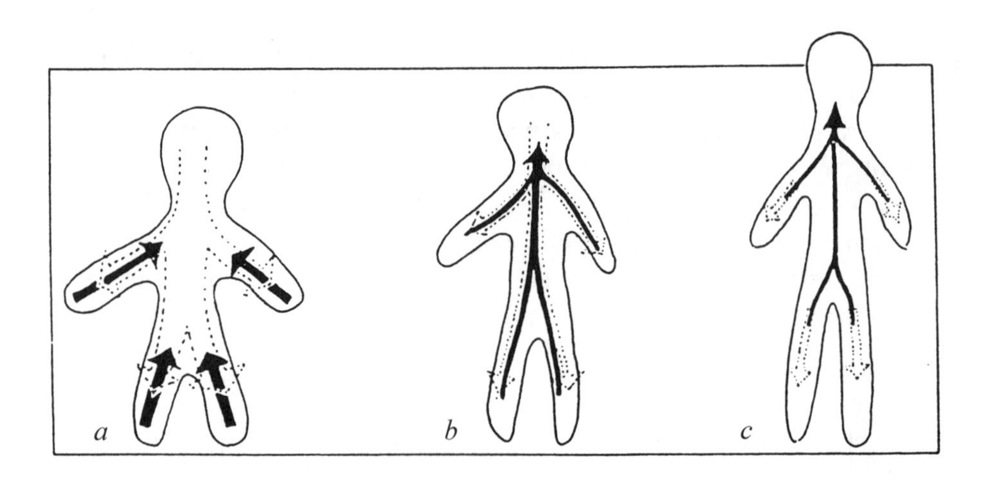

Fig. 39: Schematic representation of directions of growth in humans. (a): During the first seven-year period, all body proportions grow in cephalocaudal succession; within the limbs, however, the prevailing trend is reversed, i.e., caudocephalic development is observed. (b): During the second seven-year period, the two growth tendencies struggle for dominance throughout the body. (c): In the third seven-year period, growth in the body as a whole follows a caudocephalic pattern, but cephalocaudal growth becomes increasingly important in the limbs.

Because growth is strongest during the earlier years of life, we must take "situation a" as our starting point in clarifying the consequences of retardation.

Thus, the caudocephalic pattern appears later than the cephalocaudal pattern, manifesting primarily in the limbs, whose development begins after that of the head and torso and appears in the context of general cephalocaudal growth. At the stage of development when the limbs emerge, the caudocephalic growth tendency is added to the overall cephalocaudal growth pattern. It is essential to note that ontogenetic compression of the humanlike growth pattern in animals impacts primarily the later, caudocephalic waves of growth, which are compressed and therefore have relatively fewer consequences. The relatively short limbs (especially hind limbs), larger feet, and shorter humerus and femur in animals are basically expressions of the same retardation phenomenon that selectively impacts (i.e., compresses) later waves of growth.

It is difficult, however, to explain the existence of the conspicuously regular seven-year rhythm in human development on the basis of Darwinian selection mechanisms. It would be a great coincidence if such selection mechanisms (which are supposedly responses to randomly changing environmental circumstances) were to result in a situation in which the only living thing capable of asking questions about growth curves is forced to ascertain that its own growth curve reveals an exceptional degree of regularity. Considering the absence of any striking seven-year rhythm in the natural world, it is also difficult to conceive of the human growth pattern as an adaptation in the Darwinian sense. The seven-year rhythm in human development is best seen as a fundamental biological given, as a concrete expression of Bolk's "intrinsic evolutionary factor." Animal patterns of development result from compression or foreshortening of this seven-year pattern, which primarily affects later-developing organs, namely, the hind limbs and the proximal portions (i.e., those closer to the body's center) of the limbs. In general, therefore, mammals have shorter limbs than humans and the distal portions of their limbs (those more distant from the body's center) are disproportionally longer.

Retardation and the Idiosyncrasies of Human Body Structure

Several idiosyncrasies of the human body become more understandable if we apply the concepts we have developed here. In this section, we will discuss three well-known human characteristics. These initial discussions are based on inference and speculation, but I am convinced that they constitute an interesting area of research worthy of further study. Our consistent point of departure will be the view that *the human body most closely approximates the primal ancestral gestalt common to humans and animals.* On the basis of this assumption, we can test whether differences between humans and animals can be understood as the result of compressed development in animals.

Fatty Tissue

Human bodies contain exceptional amounts of fatty tissue. In adult males, the proportion of body fat is approximately 15 percent, in women 27 percent. In older men and women, these figures increase to 28 and 40 percent, respectively. In comparison to mammals in general and primates in particular, these values are very high. Many mammals living in the wild have no more than two percent body fat. Only a few animals such as hedgehogs just prior to hibernation or marine mammals have fat levels even approaching those of humans (Pond, 1987).

In 1969 it was discovered that fat cells (adipocytes) in experimental rats are fully developed at approximately three months. Overfeeding caused fat cells to increase in size later in life, but the number of fat cells remained constant. The original assumption that the same phenomenon would be confirmed in primates, including humans, proved false. Heavy apes have more fat cells than thin specimens, while the size of individual cells remains the same. It seems that the ability to form new fat cells persists in primates, especially in humans. Humans, who tend to have small fat cells, have ten times as many of them as the average mammal of similar size. The persistence of this ability can be considered a Type I characteristic. Conversely, the great quantity of human fatty tissue can also be seen as a Type II characteristic if we consider that fatty tissue begins to play a significant role only in the later phases of fetal development, as evidenced by the extreme thinness of premature infants.

Pointing to a human bodily idiosyncrasy already mentioned in this chapter, Pond (1991) offers still another reason for the relatively large amount of fat in the human body: "Compared to rodents, carnivores and ungulates, the humerus and femur of humans are relatively long, and the carpals, tarsals, and digits of the hands and feet are relatively short." It turns out that the most important fat deposits are concentrated in the same spots in humans and other mammals, namely, in the proximal portions of the limbs. In humans, these parts— the thigh and upper arm— are much longer in proportion to total body size and therefore have larger fat deposits, while the relatively fat-poor peripheral portions of the limbs are shorter. The end result is that fat deposits underlie a larger percentage of the skin in humans. If the relatively greater development of proximal parts of the limbs is a retardation phenomenon (hypermorphosis), so is the expansion of the corresponding fat deposits.

There are other aspects, however, to the distribution of fatty tissue in humans. Pond continues:

> Some features of adipose tissue in women, notably its presence in the greatly enlarged and rounded breasts and buttocks, cannot be explained simply as the consequence of its greater abundance. These features are not prominent until puberty and there are no corresponding sex differences in other primates. In typical mammals, most of the adipocytes (in small, fast growing rodents, possibly all) form during suckling and enlarge whenever nutritional status promotes it. Pregnancy may lead to adipose tissue expansion, but there are few detectable changes at sexual maturity. However, in humans puberty is accompanied by changes in the relative growth of adipose tissue in both sexes, and in girls these changes are particularly extensive and conspicuous. Thus both in the timing of its expansion and in its distribution, the adipose tissue of women deviates further from other primates than does that of men (1991).

Darwinian selection mechanisms such as sexual selection are usually cited to explain the fat deposits that develop in human females during puberty (Caro, 1987; Pond, 1991). But perhaps we should look in a different direction. In the human fetus as a whole, the development of fatty tissue follows a cephalocaudal growth pattern. "The process of fat formation starts in the head and neck, and rapidly progresses to the trunk, and then to the limbs" (Poissonet et al., 1984). Fat appears on the maxilla and cheeks around week 17 of gestation, on the neck in week 18, on the ribcage in week 20, on the abdomen and arms between weeks 20 and 21, and on the legs between weeks 22 and 23. Furthermore, an inverse, disto-proximal sequence is evident in the limbs. Fat is present in the hands and feet between weeks 19 and 20, and the hands and feet of newborns still appear to have more fat than the other parts of the limbs. Fatty tissue in women's breasts and hips is associated with the distoproximal or caudocephalic developmental pattern of the limbs and appears around puberty, when the limbs' distoproximal growth pulse reaches the torso (see *Figure 39*). In this regard, it should be pointed out that the breasts develop from the mammary ridges, which are directly connected to the arms in the human embryo (see *Figure 51*, p. 108). It is interesting to note that although breast size does not increase in the lower catarrhines, it is clearly evident (although associated with milk production rather than with the beginning of puberty) in the more retarded anthropoid apes (Grzimek, 1972-1975). Gender-related and species-specific differences are presumably associated with differences in growth rhythms around puberty (which begins earlier in human females, for example) and present opportunities for further research.

The Digestive Tract

During the first month of gestation, the digestive tract appears as the result of a process of curvature and involution (see *Figures 40* and *41*). The ends begin to develop earlier than the central portion; that is, the mouth area and the large intestine, which are located at the two ends, have a developmental head start over the other parts of the digestive tract. In the more retarded animals, therefore, we will expect to find that the mouth zones and large intestines remain relatively small in comparison to the more centrally located small intestines, which develop later.

In fact, the oral cavity of human beings is small, which is very important for the development of speech. And even in adult humans, the large intestine is small in comparison to the small intestine. Snipes and Kriete (1991) provide some numbers for purposes of comparison:

Percentage of the total volume of the intestines

	Small intestine	*Appendix*	*Large intestine*
mouse:	58.3	21.3	20.5
rat:	49.6	38.7	11.7
guinea pig:	24.0	62.5	13.5
rabbit:	12.3	79.9	7.8
dog:	70.7	6.5	22.7
sheep:	58.9	11.1	30.0
goat:	71.1	8.4	20.4
pig:	50.2	6.4	43.4
cow:	72.9	4.3	22.9
horse:	25.1	30.0	44.9
human:	95.8	1.2	3.0

(The figures for human beings seem extreme because the volumes listed are for a flat-walled digestive tract. In humans, the wall of the adult small intestine is highly convoluted and thus has a very large surface area.)

The different proportions among parts of the digestive tract in different mammals can be explained in terms of specialization. (In certain herbivores, for example, the caecum [appendix] is large and houses bacteria capable of breaking down cellulose.) But since humans are unspecialized omnivores, the extreme proportions found in human beings are difficult to explain as the result of specialization and apparently result from hypermorphosis of a growth pattern common to all mammals, namely, the fact that the large intestine begins to develop earlier than the small intestine, and the latter subsequently accelerates in growth. The relatively long small intestines of the large anthropoid apes support this interpretation (Grzimek, 1972-1975).

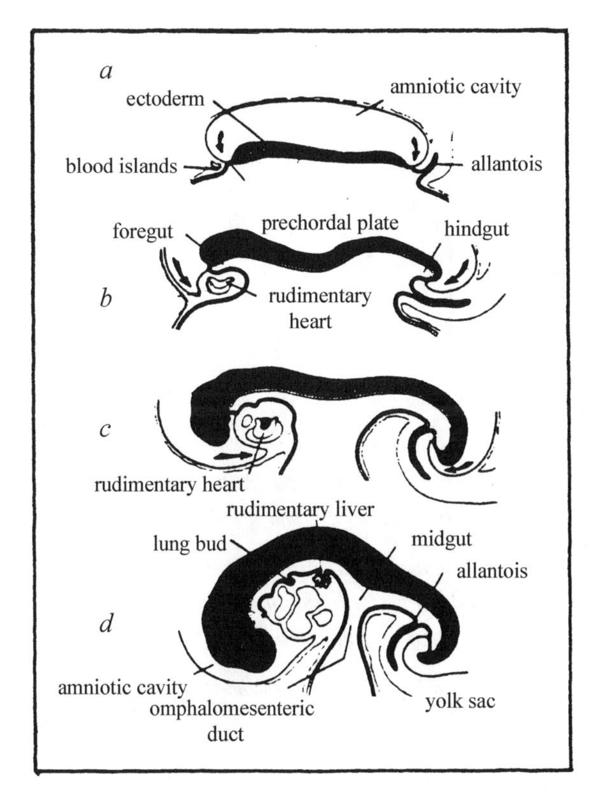

Fig. 40: Vertical (sagittal) section through a human embryo in four successive early stages. In Stage (a), the actual embryo (dark) still has the shape of an elongated disk. In the stages that follow, both ends of this disk gradually curve inward, forming first the foregut (mouth area) and hindgut and only later (in Stage (d), at the end of the first month) the midgut. Thus the mouth and the large intestine begin to develop earlier than the actual intestinal tract, whose primordium (the omphalomesenteric duct) is still connected to the yolk sac. The rudimentary heart is originally located at the head end of the embryonic disc and moves into the interior as the curvature progresses (Langman, 1966).

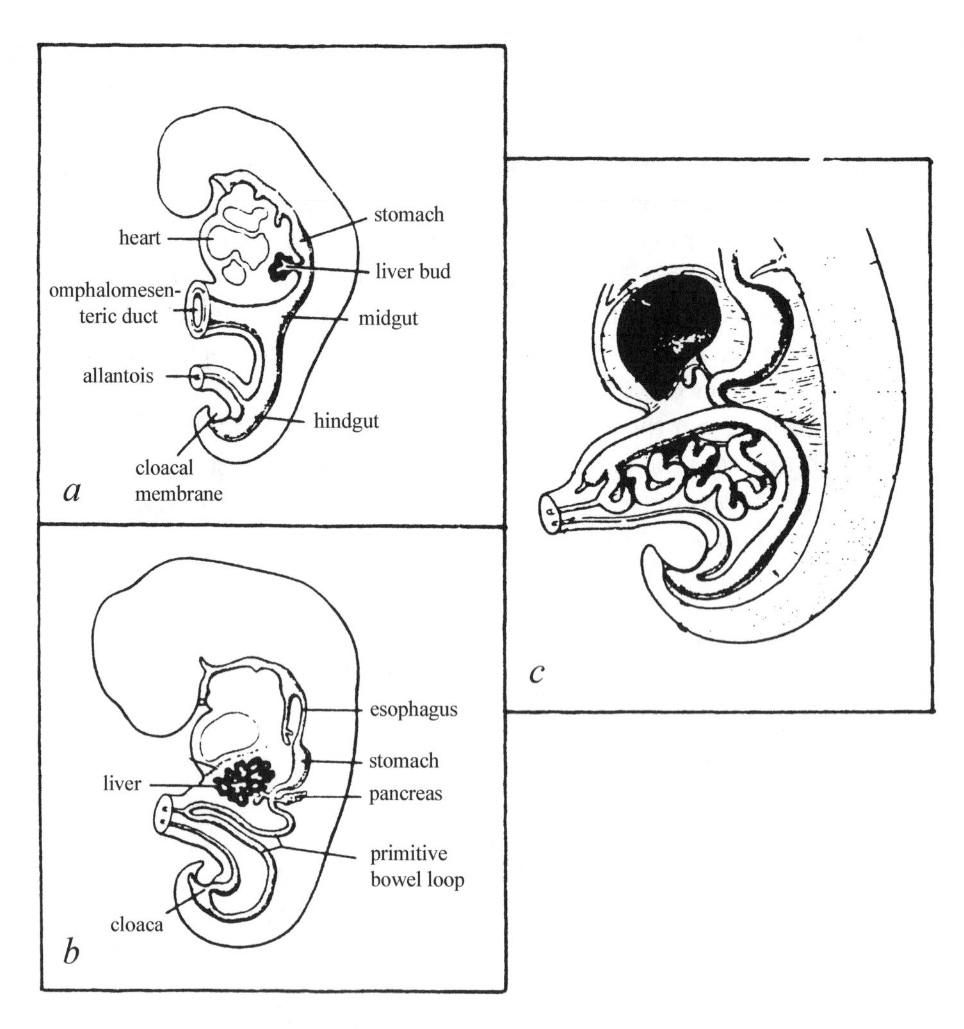

Fig. 41 a-c: Development of the digestive tract.

(a): embryo at 25 days (3 mm). The development of the hindgut is relatively advanced; the stomach is still shapeless and its location (behind the heart) is higher than it will be later.

(b): embryo at 32 days (5 mm). The hindgut is still longer than the rest of the intestines. The stomach has moved downward. The lungs are beginning to develop between the esophagus and the heart, and the diaphragm is forming between the heart and the liver.

(c): embryo at 8 weeks (30 mm). The intestines have a developmental head start on the rest of the pelvic area and form an outward curve at the navel, which disappears around the end of the third month (Langman, 1966).

Hill demonstrates that the large intestine is still quite long in children but later decreases in length relative to the small intestine. Hill's observation is especially true of the lowest portion of the large intestine, the descending colon, as we would expect for purely theoretical reasons (the parts of the intestine that begin to develop during earlier embryonic stages ultimately lag behind the younger portions). Hill discovered the opposite phenomenon in large anthropoid apes, whose descending colon elongates and changes shape. Humans, in contrast, retain the shape of the large intestine that is common to human children and juvenile anthropoid apes. Here, too, the anthropoid apes deviate from the common point of departure: "In the arrangement of the colon the great apes, as in many other features of their anatomy, appear . . . to have attained a gerontomorphic status as compared with Man" (Groves, 1986).

The Breastbone (Sternum)

The human sternum consists of three parts. The manubrium is the handle-shaped uppermost portion; below that is the body and below that the xiphoid process. The manubrium has jointed connections to the collarbones and the first pair of ribs. The second pair of ribs is attached at the transition between the manubrium and the body, and the remaining five pairs are attached to the body (*Figure 42*).

The sternums of humans and anthropoid apes are very different from those of lower primates. For one thing, higher primate sternums are very broad, a phenomenon undoubtedly related to the flattened and broadened shape of the ribcage and thus also to upright posture. Slijper's bipedal goat with its abnormally broad sternum (Slijper, 1942) also exemplifies this phenomenon (see *Figure 43*). Presumably the potential to develop a broader sternum is a general characteristic of mammals, a potential that is amplified in higher primates, where it actually appears. Keith (1923) relates the broadening of the sternum to increased lateral tension on the front of the ribcage in brachiators. In all likelihood, this factor does contribute to the additional broadening of the breastbone in gorillas and orangutans, but upright posture must remain the primary cause, since Slijper's bipedal goat had only stumps for forelimbs and lateral tension on its ribcage was undoubtedly minimal.

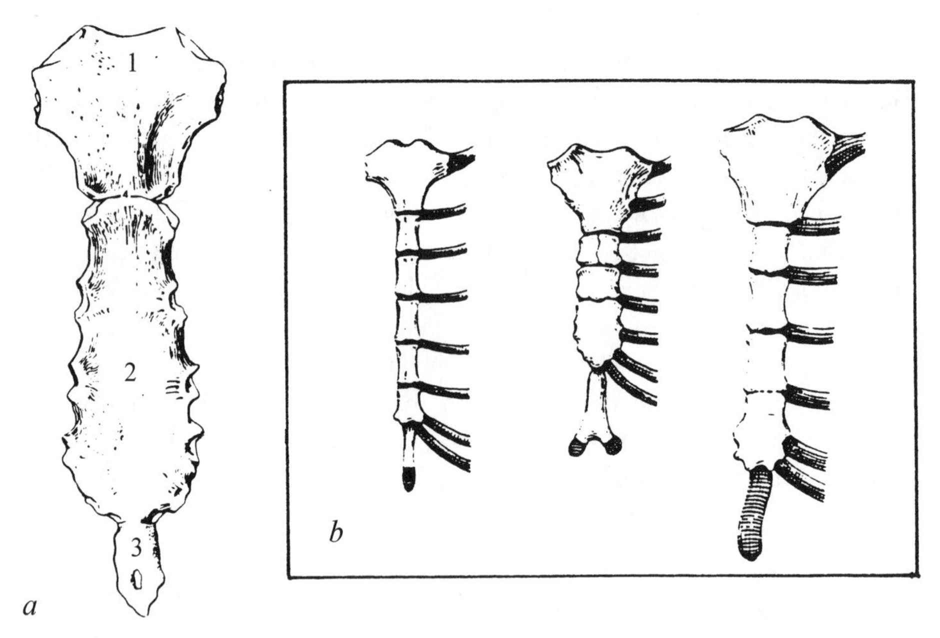

Fig. 42: (a): Human sternum. (1): manubrium; (2): body; (3): xiphoid process. (b): Left to right: macaque, gibbon, and chimpanzee sternums. In the latter two, the sternum is much broader than in lower primates. Note also the partial fusion of the sternebrae (ossification centers) in anthropoid apes and its caudocephalic progression (Schultz, 1961).

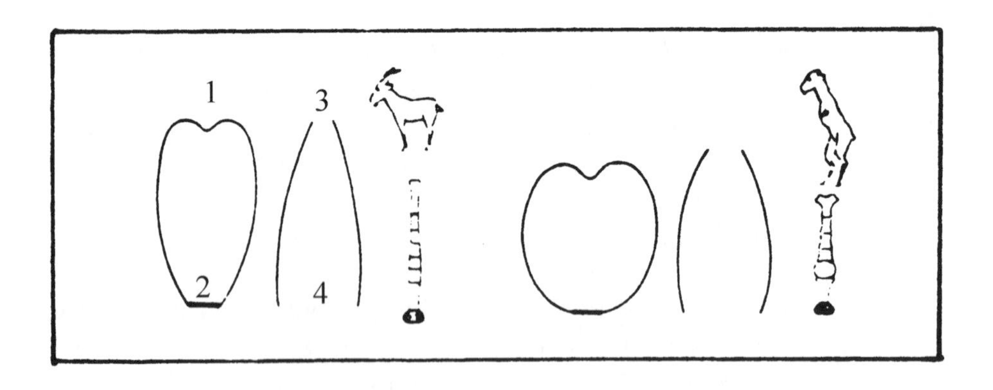

Fig. 43: Horizontal (left), and vertical (center) profiles of the ribcages and sternums (right) of a normal goat and of a goat that was born without forelegs and learned to walk upright on two legs. In the bipedal animal, the ribcage and sternum became quite broad. (1): back (dorsal) side; (2): front (ventral); (3): head (cranial) end; (4): tail (caudal) end (Slijper, 1942).

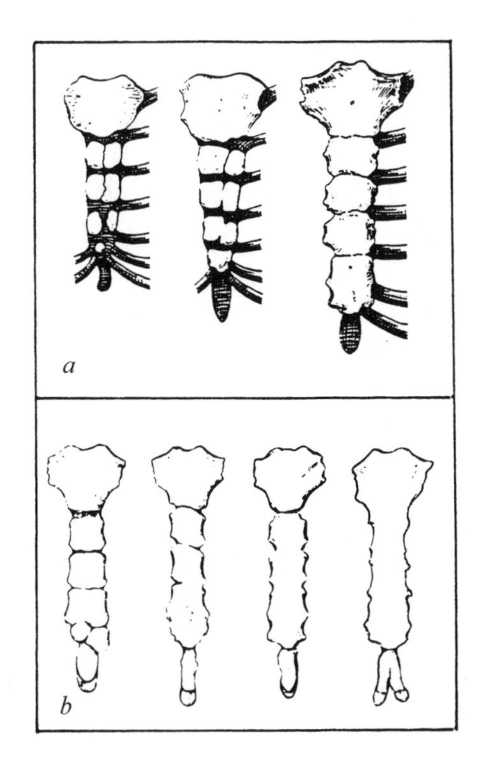

Fig. 44:

(a): breastbone of the gorilla. Left to right: newborn, juvenile, adult. The growth and ossification of the individual sternebrae (each of which develops out of two ossification centers) follow a cephalo-caudal pattern.

(b): breastbone of the human being at 11, 16, 22, and 68 years of age (left to right). Fusion of the sternebrae follows a caudocephalic pattern (Schultz, 1961).

Initially, the body consists of a number of individual sternebrae, which develop in cephalocaudal succession from ossification centers (*Figure 44*; see also Kranich, 1995). Furthermore, the manubrium is larger in proportion to the body during the fetal phase. Thus the breastbone begins to develop according to the cephalocaudal growth pattern (Schultz, 1961).

In most human beings, the body broadens and becomes blunt toward the bottom. This characteristic is absent in anthropoid apes, whose body either remains approximately the same width over its entire length or narrows toward the bottom (Montagu, 1955). The greater breadth of the lower portion of the human sternum can be seen as an example of hypermorphosis; i.e., the lower zone begins developing last and therefore, in a retarded pattern of growth, becomes relatively large. (Incidentally, it is interesting to note that this typically human broadening toward the bottom and the blunt end of the body is reflected in a seemingly nonfunctional but very popular article of masculine clothing, the necktie. The breastbone is clearly longer and narrower in men than in women [Kahle et al., 1975; Schultz, 1961]. Both of these male traits are reflected and exaggerated in the necktie, which can therefore be seen as a representation of a typically human and supermasculine breastbone.)

Another curious phenomenon is the tendency toward ossification of the sternum in the highest primates. The ossification centers in the corpus follow a cephalocaudal pattern in developing into complete sternebrae. When this process is complete in higher primates, an opposite (caudocephalic) wave of ossification begins closing the spaces between the sternebrae (see *Figure 44*). This second ossification process begins in the lower corpus and moves upward, so that the junction of the corpus and the manubrium ossifies last. In humans, this caudocephalic ossification begins around puberty, that is, when the caudocephalic growth tendency reaches the torso. This same process also occurs in anthropoid apes, where it is painfully slow and never completed. In lower primates, it does not occur at all, and all of the sternebrae remain separate for the entire life of the animal.

The more rapid and complete ossification of the human sternum can therefore be considered a Type II characteristic. It occurs to a much lesser extent in animals because the later phase of their development (in which the caudocephalic growth tendency reaches the torso) is selectively suppressed through ontogenetic compression. Presumably, sternum ossification in humans is functionally related to upright posture. According to von Hayek (1970), the upper human ribs that are connected to the sternum exert an upward force of 1.8 kg, which is balanced by the equivalent downward force of the lower ribs.

Goethean and Steinerian Synthesis

We have considered various phenomena that reveal conceptual connections between human and animal evolution. Let's review these connections to clarify what has been presented thus far. We saw that both hypermorphosis (or further development) and fetalization are characteristic of humans. We called fetalized characteristics (such as the limited pattern of hair growth in humans, the domed human forehead, or the foramen magnum's central location on the underside of the skull) *Type I characteristics* and hypermorphic characteristics (such as the length of human legs, thighs, and upper arms, or the width of the ribcage) *Type II characteristics*.

It is interesting to note that Goethe already clearly recognized the existence and significance of these characteristics and summed them up in a kind of synthesis. In 1795 he wrote, "In the human being the animal element is enhanced to serve higher purposes; as far as both the eye and the intellect are concerned, it has been overshadowed." The element of "enhancement" applies to the emergence of the human form, which results in Type II (i.e., specifically human) characteristics that allow typically human capabilities (such as language and the use of tools) to appear. On the other hand, the "overshadowing" element addressed by Goethe refers to what Bolk calls "fetalization"—that is, the avoidance of one-sided animal specializations—and is expressed in Type I characteristics.

In the view of evolution that we have now developed, two factors play a role. One is the tendency toward anthropogenesis. It is present as a potential in every animal but is overwhelmed by a second tendency, the impulse to become specialized in one-sided ways. This second tendency is what drives animals into the domain of Darwinian natural selection mechanisms. As evolution progresses, the anthropogenetic tendency breaks through to a greater extent and specialization becomes less dramatic. In higher animals, the human gestalt is expressed to a considerable extent, especially during fetal development, until ultimately the anthropogenetic tendency emerges at its strongest in human beings.

Figure 45 provides a graphic overview and summary of these concepts. The arrows represent the biographies or ontogenies of different species. All species share more or less the same beginning; their development begins by moving in a common, universal direction (vertical line in the diagram). Later, ontogeny may continue in one of two different ways:

1. It may become more specialized. In the illustration, the horizontal axis represents this component. The more the arrow is deflected to the right, the more strongly ontogeny is impacted by species-specific specialization. This specialization occurs

primarily during later developmental phases, so the greater the distance from the common origin, the more the arrow is deflected to the right. In human beings, animal specialization is "overshadowed;" that is, the arrow representing human ontogeny is deflected only slightly to the right.

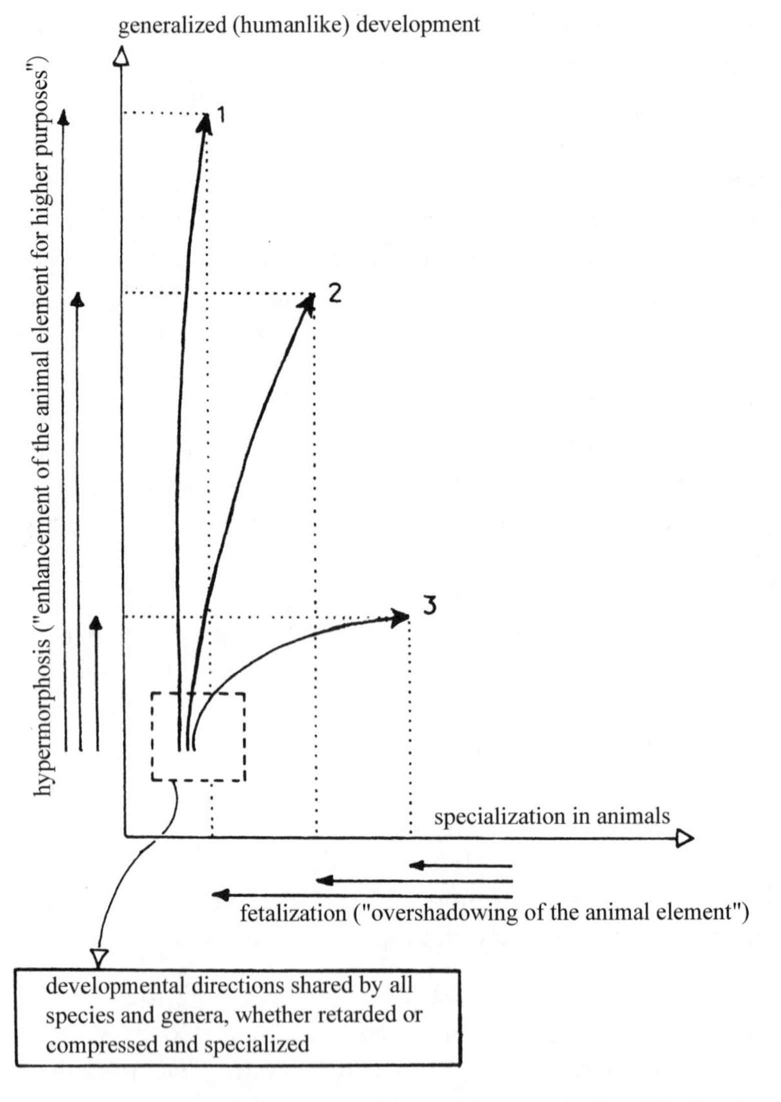

Fig. 45: Graph of the evolutionary tendencies of hypermorphosis and fetalization.

2. On the other hand, a less specialized ontogeny may continue in its original direction. The original developmental tendencies and growth differences persist, and the resulting mature organism (whose position on the chart is marked by the head of the arrow) is characterized by hypermorphic phenomena. In humans, the original tendencies shared by all animals are most persistent, and the latent potentials of the universal vertebrate body structure (upright posture, speech, etc.) are realized; "the animal element is enhanced to serve higher purposes."

The less specialized its ontogeny, the greater the fetalization of the adult organism and the longer its life span. In the diagram, this Bolkian connection between fetalization and life span is represented by the longer arrows of more highly fetalized ontogenies.

Proterogenesis (Schindewolf's Theory)

Schindewolf (1928, 1972) observed that typically human characteristics (such as a high forehead or a flat facial skull) are most evident in young people and later become less pronounced. Among apes, too, humanlike potentials are initially clearly evident but are later overwhelmed by animal developments. Schindewolf, like Bolk, noted that this fact is the exact opposite of what we are led to expect by Haeckel's biogenetic law, which predicts that new characteristics will be incorporated into the *last* ontogenetic stages. In reality, however, humanlike traits in apes (and also in hominids such as *Australopithecus*, etc.) appear during *early* ontogenetic phases and later recede. Schindewolf calls this process "proterogenesis" and describes it as follows:

> A trait or complex of traits appearing in phylogeny for the first time is acquired directly by juvenile stages, whereas adult stages of the ontogenies in question still demonstrate the ancestral, inherited morphology. New traits acquired in early ontogeny expand to successively older stages of phylogenetically subsequent types and increasingly displace the ancestral remnant, whose duration becomes ever more limited (1928).

To summarize these thoughts, a primal humanlike developmental tendency persists to an ever-increasing extent in higher primates, while the corresponding animal tendency recedes. This relationship is summarized in *Figures 23* (p. 55) and *45* (p. 96). Thus Schindewolf is among the thinkers who independently rediscovered Goethe's key thought.

Fetalization and Hypermorphosis: Steiner's Contribution

In animals, as we saw, elementary processes of growth or development are eliminated from the generalized, humanlike developmental pattern and are replaced by a smaller number of specialized and species-specific processes. The number of processes affected in humans is much smaller, and the human life span is correspondingly longer. We can compare this process to a tower (the Eiffel tower, for example) which was originally designed for maximum height. Alterations to the design after construction begins violate the original coherence of the design (which aimed for maximum height) and can therefore be accomplished only at the expense of height. The altered tower will be shorter, just as the life span of an altered ontogeny is shorter.

We have already observed the fact that human limbs are ontogenetic latecomers in comparison to the head and the torso, while the head has a "head start" in comparison to the limbs. Hence Type I characteristics (fetalization) will be most evident in the head while Type II characteristics (hypermorphosis, or additional development along generalized lines) will manifest primarily in the limb system. Schad has presented plausible elaborations on this principle in several publications (1992, 1993).

In 1918— in the same year when Bolk's theory of fetalization was published— Rudolf Steiner first formulated the following elaboration on Goethe's synthesis:

> The course of human development as such is retrograde, and in the human body retrograde development is strongest in everything related to the instruments of our senses and to the head. If I had time for forty lectures instead of one, it would be very tempting to recount all the discoveries of modern science that could be presented as proof that an ascending development manifests in the

> human organism but restrains itself in the head, where retrograde development prevails . . . The principle underlying the human extremities is not subject to retrograde development or devolution but is unique in that it exceeds the standard set by the evolution of the rest of the organism (with the exception of the head). The extremities are subject to overdevelopment; they go beyond the point achieved by the head and the rest of the organism. When growth in the rest of the organism tapers off, the extremities develop the ability to persist, as it were, and transcend the usual standard (1992).

Thus Steiner elaborates on Goethe's synthesis by showing more precisely where the animal element is "overshadowed" in the human organism (fetalization or retardation) and where it is "enhanced" (hypermorphosis or overdevelopment). Seen in combination with Bolk's contribution (the connection between retardation and increased life span), Steiner's elaboration gives rise to a coherent image that we will continue to develop and document in subsequent chapters.

4

THE ANIMAL ELEMENT IN HUMANS

Atavisms or Evolutionary Potentials?

In 1902, Bolk published a report of an autopsy that he had performed on the body of a deceased agricultural worker. The corpse revealed a number of idiosyncrasies, all of which Bolk attributed to exaggerated retention of fetal characteristics. For example, the body had no hair on the torso and limbs. The brain, intestines, ribs, etc. revealed many characteristics that normally appear only fleetingly in specific phases of human fetal development. In many respects, Bolk's subject resembled an overgrown fetus.

Bolk also observed another curious fact. The shape of the man's stomach was very similar to the compound stomachs of certain primate species such as gibbons and langurs (see *Figure 46*). Bolk returned to this issue nearly twenty years later (1920). Why, he asked, did this man, who showed so many signs of exaggerated fetalization, have an unusually complex-looking stomach?

In some primate species the fetal stomach is clearly compound, and traces of the same division are occasionally seen in the stomach of a human fetus. In human beings, it is not uncommon to find poorly defined, transitory (or, rarely, persistent) morphological traits— such as the triple stomach in this case — that are both clearly defined and permanent in some animal species. We generally conclude that the ancestors of human beings once possessed these traits in fully defined form and that their fleeting appearance in the human fetus is a Haeckelian recapitulation of the ancestral form. Similarly, certain organs present in all humans— such as the vermiform appendix or the small muscles behind the earlobe—are (incorrectly) viewed as nonfunctional vestigial organs (von Eggeling, 1920; Seiler, 1974).

Such Haeckelian recapitulations clearly do occur in certain animal genera. Whales, for example, have no hind limbs, but the potential for them is clearly visible in whale embryos. Occasionally we see a whale with rudimentary hind legs—externally visible, superfluous appendages located where the animal's ancestors once had limbs (Hall, 1984). Another classic example of such phenomena is found in the horse, which normally has a single, very highly developed "finger" or "toe" on each of its fore- and hind legs (see Chapter 5). Horses, however, are descended from multi-toed ancestors, and we occasionally encounter horses with two extra toes, each of which may be equipped with an individual hoof (see *Figure 47*). Such isolated and incomplete recurrences of ancestral traits are called atavisms (from the Latin *atavus*, forefather or ancestor).

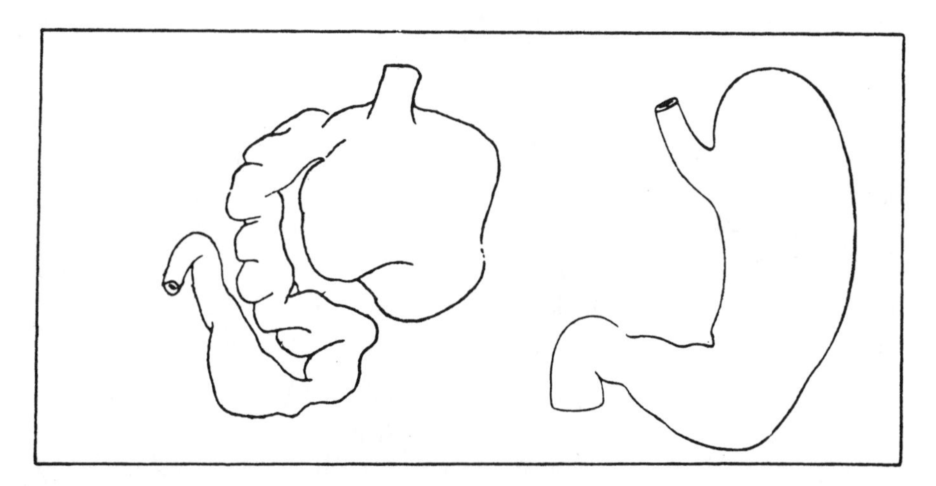

Fig. 46: Left: The compound stomach in a full-term langur fetus. The esophagus empties into the first compartment, which is linked to a second compartment by an intestine-like section. Right: The simple human stomach (Bolk, 1920).

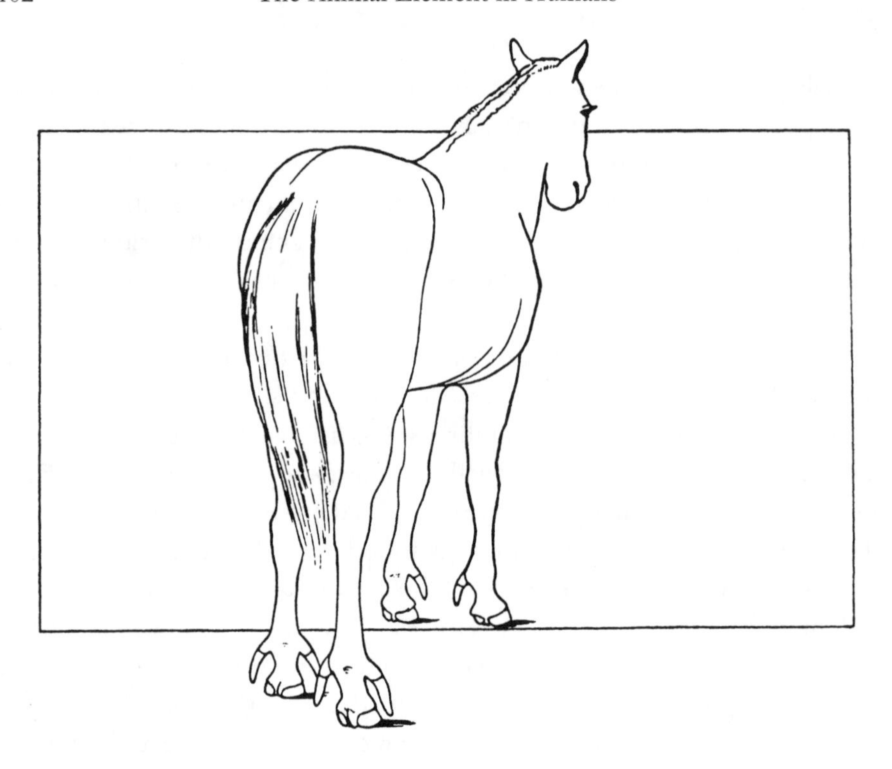

Fig. 47: Horse with extra, atavistic toes on its fore- and hind legs (Marsh, 1892).

In the man Bolk studied in 1902, the vaguely defined triple structure that is occasionally evident in the fetal human stomach persisted into adulthood, as sometimes also happens with the embryonic hind legs of whales. Was what Bolk observed an atavism? In other words, did any ancestors of human beings ever have triple stomachs resembling those of gibbons or langurs? Bolk was inclined to answer this question in the negative:

> In my opinion, this phenomenon should not lead us to conclude that the adult stage of any human ancestor ever had a compound stomach as highly developed as that of a langur . . . I am more drawn to the following viewpoint. Imagine that in the evolution of the primate group as a whole, the tendency to develop a compound stomach emerged, or perhaps it would be more correct to say that a

> phylogenetic possibility became reality. In langur species, this tendency or possibility was realized to a very great extent, . . . while in humans, by contrast, . . . its effect with regard to the stomach is observed to very different extents in different individuals and primarily in the fetal organs. Persistent traces of a triple human stomach, therefore, should not be seen as a recurrence of a formerly pronounced morphological trait but rather as the manifestation of an evolutionary potential that is realized to a much greater extent in other primate species . . . I fully acknowledge that this perspective deprives most so-called atavistic phenomena of their significance as markers in constructing a phylogenetic tree; however, it does justify (with regard to phylogeny) a principle that in our view plays a significant role in ontogeny, namely, that the expression of a morphological element is due not only to the activity of one group of formative factors but also to the dormancy of a second group. It is hard to tell which of these factors is more significant (1920).

In humans, according to Bolk, animal morphological traits are present as potentials but remain latent. (This thought corresponds to Goethe's view that the animal element is "overshadowed" in human beings.) The intrinsic evolutionary factors that produce animal morphologies remain latent in humans. In animals, these factors are fully expressed, but in humans they manifest either not at all or only superficially. Bolk clearly distinguishes between two groups of evolutionary factors: ones that are consistently realized in humans and others that are expressed not at all or only minimally in human beings but unfold fully in certain animals. The sporadic appearance of the latter in humans produces the phenomenon of human "atavisms" (cf. Verhulst, 1996).

Bolk excludes the possibility that the ancestors of humans ever had a compound stomach and that the human species returned to a simpler form during later stages of its evolution. The "atavism" he observed in 1902 simply indicates that evolutionary tendencies that unfold fully in animals are also marginally present in humans.

This phenomenon is illustrated by morphological changes in the skulls of chimpanzees and humans during the period of active growth (*Figure 48*). In the fetal stage, chimpanzee and human skulls are very similar. The adult human skull remains more similar to the fetal form than the ape skull does, but it, too, deviates from the fetal form in ways that parallel the changes in chimps. The forehead becomes less pronounced and the jaw becomes heavier. Some other animal tendencies, however, such as protrusion of the teeth and compression of the rear of the skull, remain almost completely undeveloped in the human skull.

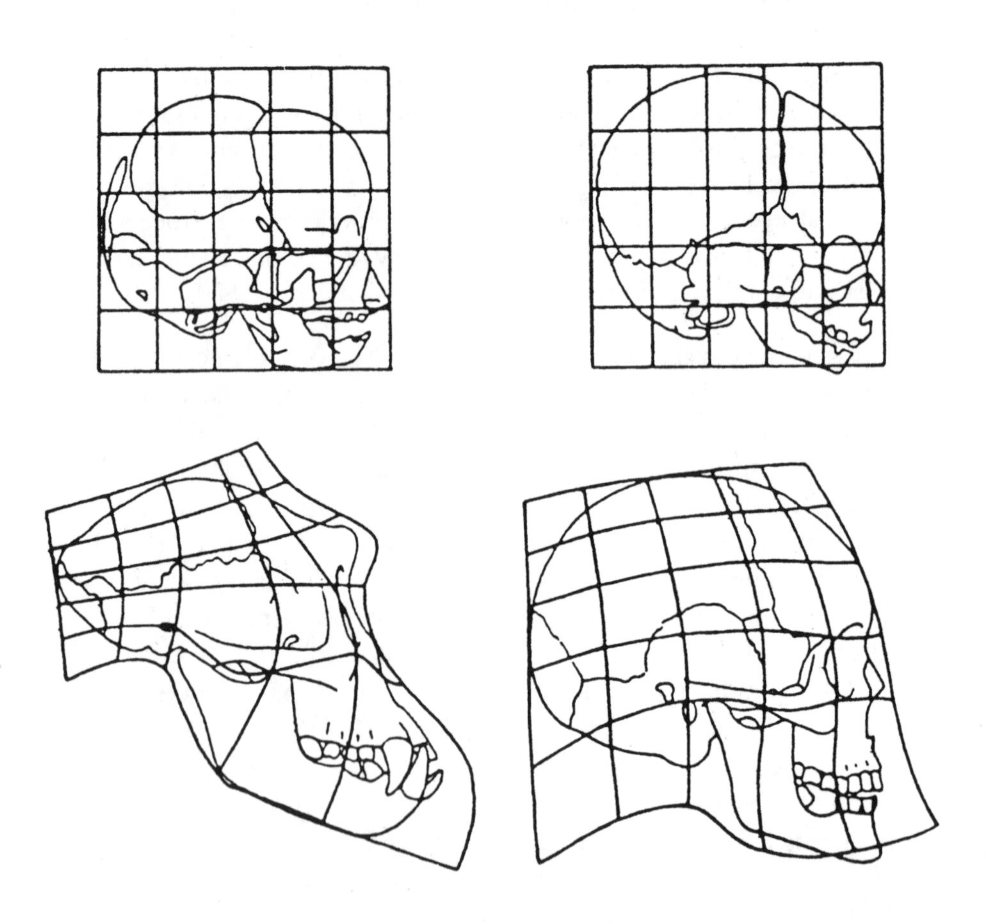

Fig. 48: Fetal (above) and adult skulls (below) of chimpanzee (left) and human (right). The tendencies governing the transition from the juvenile to the adult form are similar in chimps and humans, but in chimps deviation from the original form is much more pronounced (Stark and Kummer, 1962).

We are already familiar with Bolk's study of the foramen magnum, the large opening on the underside of the skull (see *Figure 16*, p. 36). In the fetus of humans and apes, the foramen magnum is located in the center of the underside of the skull. In humans, the foramen magnum remains in this position, but in apes it shifts toward the rear as growth progresses. In the course of his very detailed study, Bolk discovered a curious fact: between the ages of seven and twelve years, the human foramen magnum shifts very slightly toward the rear (1915). That is, this shift occurs in both animals and humans but to a much lesser extent in humans. Clearly, we are not dealing with an atavistic effect in this case. In the fetal skull form common to humans and anthropoid apes, the foramen magnum is not located near the rear of the skull; if it were, its central location in juvenile apes would be incomprehensible. The common gestalt from which humans and anthropoid apes are descended must have been humanlike in character. It has been largely overwhelmed by animal evolutionary tendencies in anthropoid apes but only minimally so in humans.

Bolk also warned against applying a Haeckelian interpretation to certain organs (such as the notochord or the mesonephros) that appear only fleetingly in human embryonic development but persist in some lower animals. In Bolk's opinion, these organs are not atavistic structures; rather, they are integral and functional components of the embryonic organism, which is quite different in character from the adult. We must not make the mistake of regarding forms that occur earlier in human development as incomplete versions of adult forms. "The neonate is not a less complete organism than the adult, and completeness does not begin at a certain point in development. It is present from the very beginning of life, that is, from the moment of conception" (1926a).

The transition from fertilized ovum to adult organism is an incredibly complex process. Interpreting a few impermanent organs as atavistic structures assumes that their transitory presence is superfluous and that a "shorter" or "more direct" path of development must exist between the ovum and the adult organism. This assumption, however, is not based on observation; it is purely hypothetical. From Bolk's perspective, the persistence of some impermanent organs in lower animals signifies simply that these animals abandoned the general humanlike developmental pattern at an early stage and set out on a very different developmental path based on embryonic structures; as a result, the fate of these embryonic organs is different.

To clarify what we have discussed thus far, let's take a closer look at several so-called atavisms in humans.

Darwin's Tubercle

One famous example of an "atavism" in humans is mentioned by Darwin himself in *The Descent of Man* (1871). Some people have a slight projection (called Darwin's tubercle or the auricular tubercle of Darwin) on the external rim (helix) of the ear. Darwin considered it a vestige of our ancestors' pointed mammalian ears (see *Figure 49*). This view initially met with great approval but was contested by Schultz (1968), who noted that such ears are found in catarrhines only in macaques and baboons and are not found in the early fetal stages of any anthropoid apes. Darwin believed that he had seen an orangutan ear that proved the existence of auricular tubercles in anthropoid apes, but in fact the ear he observed was a gibbon ear, and its "tubercle" was the result of poor preservation techniques.

This does not mean that Darwin was incorrect in assuming a relationship between the pointed ears of many mammals and the sporadic appearance of auricular tubercles in humans. The relationship, however, cannot be explained as an atavism. From the Bolkian perspective, the phenomenon of pointed ears is a tendency that manifests only very weakly in humans but very strongly in certain animal species.

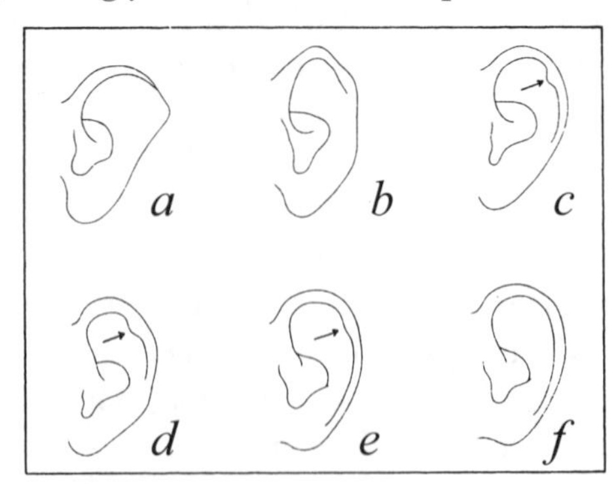

Fig. 49: Hypothetical evolution of Darwin's tubercle. (a): presumed ancestral stage; the rim of the ear comes to a point and is not yet rolled. (b): presumed stage with a partially rolled rim. (c, d, e): human ears with highly, moderately, and weakly developed auricular tubercles, respectively. (f): human ear with no auricular tubercle.

Zygodactyly

In some people (approximately 1 in 2000), the second and third toes are partially grown together (see *Figure 50*). The potential for this phenomenon, called zygodactyly, is present in all human embryos. Thin membranes are present between the bases of the toes for a brief period after the toes separate. The membrane between the second and third toes is somewhat thicker than the others and persists longer (Schultz, 1922).

The potential for zygodactyly is present in all mammals. Webbed second and third toes are found sporadically in many ape species, and zygodactyly is a consistent trait in siamangs, which are close relatives of the gibbons. Incidentally, the siamang owes its Latin name (*Hylobates syndactylus*) to this trait.

Since the siamang is an evolutionary latecomer and a specialized climber, it seems unlikely that this species would have retained an ancestral characteristic abandoned both by other catarrhines and by humans. Gibbon species are presumed to have separated only recently from the line of descent leading to the siamang. In the ancestors of gibbons, therefore, the trait of zygodactyly would have had to persist into quite recent times before ultimately being lost. A Bolkian interpretation seems more reasonable, namely, that the potential for zygodactyly is present in all mammals, including primates; among the catarrhines, however, it unfolds fully only in the siamang. From this perspective, human zygodactyly is not an atavism but simply the occasional expression of a latent potential that emerges to a much greater extent in the siamang (cf. p. 199).

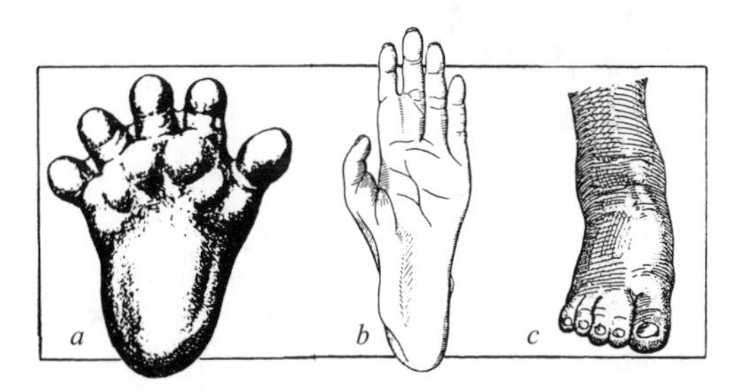

Fig. 50: Zygodactyly. (a): foot of a human fetus at the stage in which the toes are nearly separated. The second and third toes are closer together. (b): foot of a siamang (zygodactyly always occurs in this animal species). (c): a case of pronounced zygodactyly in a human foot (Schultz, 1922, 1925, 1968).

Supernumerary Nipples

Humans (and primates in general) typically have a single pair of mammary glands located on the upper torso, while many species of lower mammals have several pairs. Occasionally, however, the phenomenon of supernumerary nipples also occurs in humans. Because this trait is common in lower mammals, its appearance in humans seems atavistic. I suspect, however, that the actual fact of the matter is different: the embryonic development of lower mammals passes through but then exceeds the potentially human stage, resulting in the development of several pairs of nipples, which may be located on the lower as well as the upper torso.

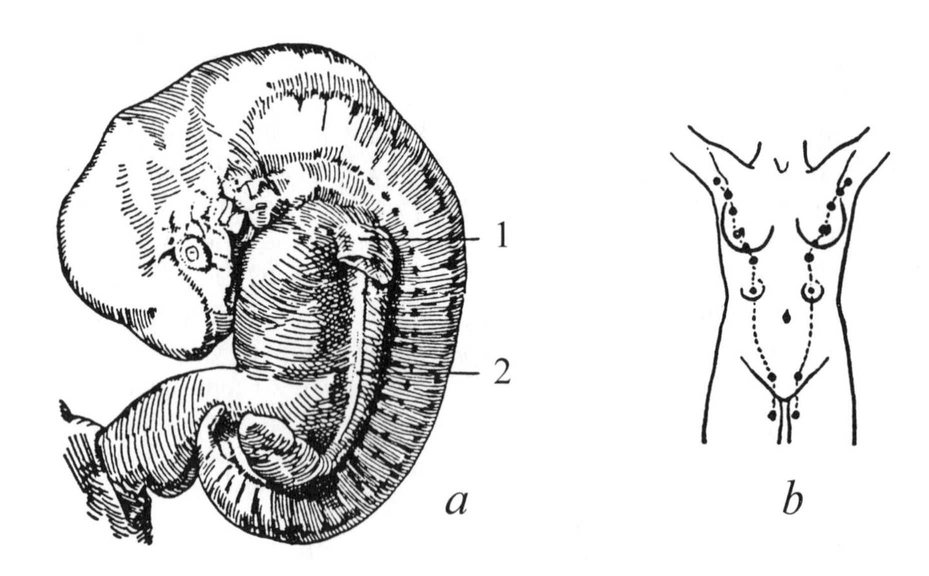

Fig. 51: The development of supernumerary nipples.

(a): A 13-mm-long embryo with excessive development of the mammary ridge. The rudimentary upper limb has been cut away (1). In this example, the mammary ridge (2) runs from the base of the arm to the base of the leg. Normally, only the upper third of the mammary ridge develops in human embryos.

(b): Supernumerary mammary glands or nipples typically develop along the mammary ridge (Arey, 1974).

Mammary glands develop out of the so-called mammary ridge, which runs from the base of the arm toward the base of the leg in the embryo. In the embryos of mammals such as pigs, which have rows of nipples running the entire length of the body, the mammary ridges run the entire length of the torso, from the forelimbs to the hindlimbs (Arey, 1974). In humans, the mammary ridge (and hence the development of mammary glands) is generally restricted to the chest.

Since the mammary ridge runs from the upper to the lower limbs and the upper limbs develop first, it seems likely (although it has not been conclusively proved) that the mammary ridge also develops in a cephalocaudal direction. Hence, restriction of mammary ridge development to the chest in humans can be considered a Type I characteristic. Because the mammary ridge emerges very early in ontogeny, its development is similar in humans and primates. In other mammals, this aspect of development is "propulsive," to use Bolk's terminology, and the mammary ridge continues to develop; as a result, mammary glands later develop over the entire length of the torso. This trait, which is common to all lower mammals, remains latent in primates.

The Tail

Around the sixth week of gestation, when the human embryo is approximately 14 to 16 mm long, it has a tail of sorts. During weeks 7 and 8, this formation is overgrown by the lower torso (*Figure 52a*).

In humans, as in all mammals, the coccyx and the rest of the spinal canal emerge at an early stage of embryonic development. But we also know that the torso develops in a cephalocaudal direction (see p. 67). At an earlier stage, the coccyx is externally visible because the torso below the navel is still completely undeveloped, as evidenced by the tiny lower limbs and the low position of the umbilical cord. As the lower torso grows, the coccyx moves into the interior (*Figure 52c*).

Occasionally a human baby is born with a tail-like appendage at the height of the coccyx or lumbar spine (*Figure 52d*). In the nineteenth century, such appendages were often interpreted as atavisms. Darwin presents this phenomenon as proof that human beings are descended from animals. He writes:

By considering the embryological structure of man—the homologies which he presents with the lower animals—the rudiments which he retains—and the reversions to which he is liable, we can partly recall in imagination the former condition of our early progenitors and can approximately place them in their proper place in the zoological series. We thus learn that man is descended from a hairy quadruped furnished with a tail (1871).

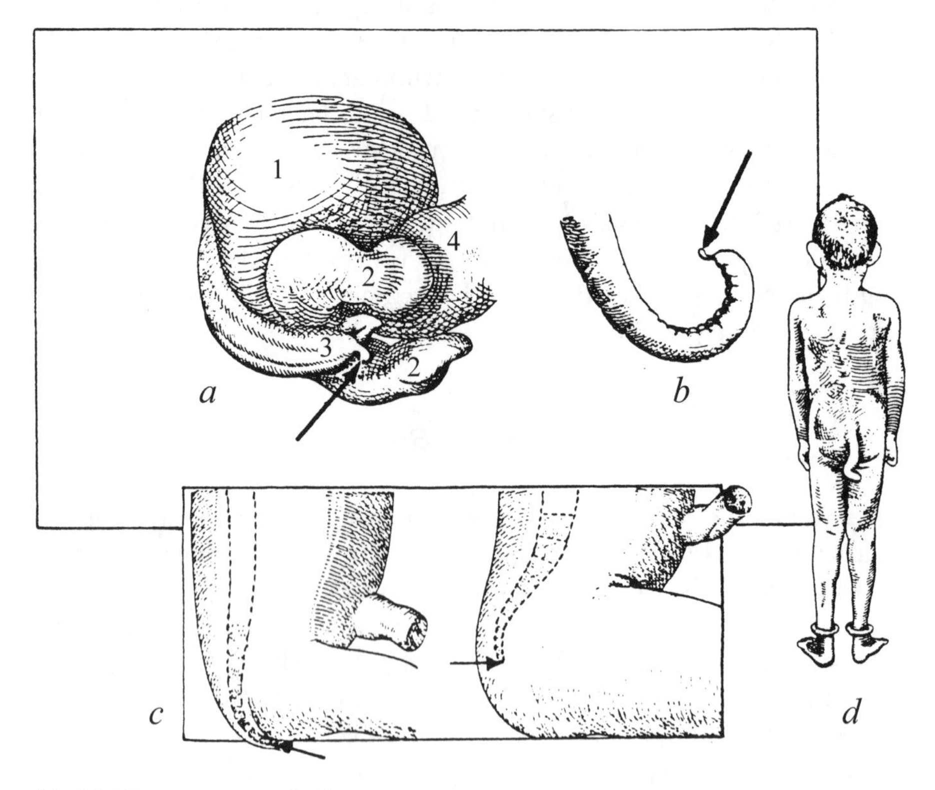

Fig. 52: Human coccyx and tail.

(a): The posterior of an approximately 14-mm-long human embryo. (1): torso; (2): rudimentary foot; (3): coccyx; (4): umbilical cord. The arrow points to the caudal appendage.

(b): Tail of a 107-mm-long capuchin monkey fetus. The arrow indicates the caudal appendage.

(c): Lower torso of a human fetus at 10 weeks (left) and newborn (right). In the interim, the lower torso grows significantly, the coccyx (arrow) shifts upward, the legs become more massive, and the umbilical cord moves upward.

(d): Twelve-year-old boy with a 23-cm-long tail (Schultz, 1924).

Tail-like structures in human beings, however, are fundamentally different from true animal tails. The most significant difference is that human "tails" never contain bones (tail vertebrae) or even cartilage (Dao and Netsky, 1984; Ledley, 1982). Except in extremely pathological cases, the lower spinal column develops normally and has the usual number of vertebrae. (When "tails" occur in humans, they may be located quite high on the back or to the right or left of center.) In contrast, mammal tails invariably have vertebrae, with the exception of the Barbary ape (see below) and one specific mutation in mice. This mutation, however, lacks not only the tail vertebrae but also the coccygeal, sacral, and lumbar vertebrae of normal mice. In Homo caudatus—humans with tail-like appendages— the number and shape of the vertebrae are normal.

Schultz (1924 and 1925) agrees with Harrison, who sees tail development in humans as abnormal growth of the so-called "caudal appendage" found at the end of the embryonic coccyx (see *Figure 52a*). The caudal appendage was formerly considered a vestige of the actual tail presumed to exist in the ancestors of the human being. According to Schultz, this argument is untenable because the caudal appendage also appears at the end of the true tail in ape fetuses (*Figure 52b*). The famous tailless apes of Gibraltar (the Barbary ape, *Macaca sylvanus*) actually have very short tails, which— as Schultz discovered— contain neither vertebrae, cartilage, nor striated muscle tissue (1961). Schulz concluded that the tail of this species is probably a pseudotail, which develops out of the caudal appendage, rather than a true tail. This anomaly is universal in Barbary apes but is the exception rather than the rule in humans. Incidentally, Schultz also ascertained that chimpanzees quite often have pseudotails.

The occasional appearance of a tail-like structure in humans, therefore, cannot be considered an atavism. It is a trait found now and then in tailless apes as well as in humans; in the Barbary ape, it has evolved into a permanent attribute, like zygodactyly in the siamang. Normal, bony tails such as those of most primates are never seen in humans. Thus the potential to develop a pseudotail can be considered a latent endowment that is common to all primates, manifests very seldom in humans and most apes, and has become a fixed attribute of one species, the Barbary ape. These phenomena do not permit us to draw conclusions about phylogenetic relationships; that is, the fact that pseudotails occasionally appear in humans does not mean that a common ancestor of humans and Barbary apes must have had a pseudotail. Nonetheless, many scientists follow Darwin in describing the appearance of tail-like formations in humans as an atavism that purportedly proves that human beings descended from earlier forms of animals (Patten, 1968).

The Future of the Human Species

Klaatsch describes Goethe as being totally unsurprised by reports of a human tail, which he thought quite natural, since he believed the human being to be "structured in a way…that unites so many attributes and natures, therefore [serving], even in physical form, as a microcosm, a representative of all other animal genera" (Goethe, 1796; Klaatsch, 1920).

In the Goethean or Bolkian view, so-called atavisms, while they are not proof for or against the idea that human beings are descended from animals, do suggest that the intrinsic evolutionary factors that produce species-specific, one-sided specialization in animals are also present in weakened form in humans. The central theme of anthropogenesis is expressed in its purest form in humans and inadequately in animals, where it is overwhelmed by tendencies to specialize in one direction or the other. In both humans and animals, the central humanlike gestalt struggles for supremacy over animal specialization tendencies. In animals, the tendency toward specialization wins, while in humans the central gestalt emerges victorious.

A quick look at *Figure 48* (p. 104) suggests that enhancement of and liberation from the animal element has not yet peaked in modern human beings. Human children still seem to prefigure a more highly evolved form that lies concealed in the future (cf. *Figure 53*). The philosopher Arthur Schopenhauer (1788-1860) believed, for example, that genius in human beings could be explained as an unusual degree of liberation from the animal element, revealing the pure, childlike individual (Gans, 1923). In a naïve or inspired way, *Figure 1* (p. 1) expresses the same idea, namely, that in the future the human species will be supplanted by a completely liberated angel-like race of beings. This thought touches on the central mystery of human existence.

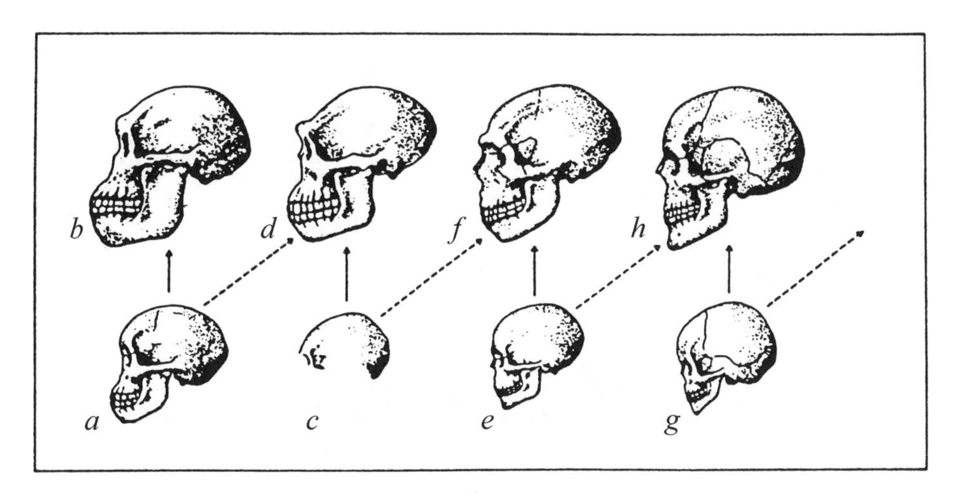

Fig. 53: Adult (above) and infant skulls (below) of (left to right) Australopithecus *(a, b);* Homo erectus *(Modjokerto skull) (c, d); Neanderthal (e, f) and modern human (g, h). In each case, the juvenile form of a prehistoric hominid seems to prefigure the adult form of later, less animallike types. There is no indication that this process will stop, since comparable differences still exist between infant and adult skull forms in modern human beings (Kipp, 1980) On this phenomenon, Montagu (1955) comments, "A fetalized pithecanthropine, to judge from the juvenile Modjokerto skull, or a fetalized australopithecine, to judge form the juvenile* Australopithecus africanus, *would more closely resemble modern man than these fossil juvenile forms would the adult members of their own type . . . In almost all the traits in which the juvenile members of these fossil forms differ from the living apes and their own adult form they most closely resemble modern man, for example in the comparatively globular form of the skull, the thinness of the skull bones, the absence of brow ridges, the absence of crests, the form of the teeth, and the relative size and form of the brain."*

5

THE UPPER LIMBS

The Hand

Although some early terrestrial vertebrates (tetrapods) had more than five fingers (see *Figure 101*, p. 190), the five-digit hand soon emerged as the basic pattern (Gee, 1989). *Figure 54* compares the human hand skeleton to that of a mammal-like reptile that lived during the heyday of the dinosaurs. It is interesting to note that digit I (the "thumb") of this primitive reptilian foot is already different from the other digits in that it has one fewer phalanx. This structural difference is ancient, dating from a time when the "hand" did not yet serve any grasping function and was strictly an organ of locomotion, with all fingers playing more or less equivalent roles. The many ancient handprints reproduced by Klaatsch (1920) have clearly separate thumbs and look almost like crude human hands. They prefigure, as it were, the special significance of the thumb in primates and especially in human beings. In the basic mammalian pattern, each finger has three phalanges, but the thumb has only two.

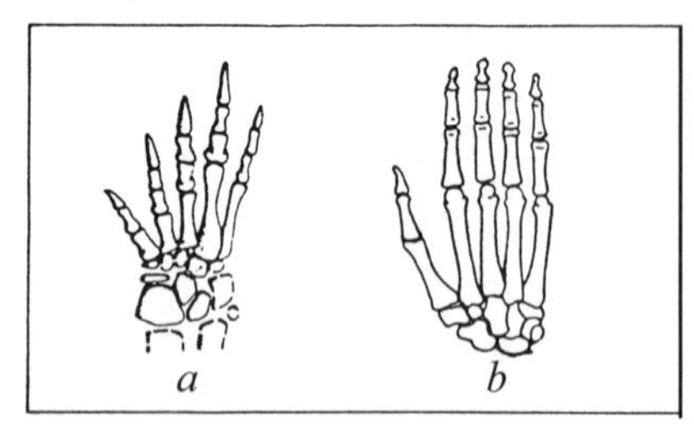

Fig. 54: The five-digit hand of:

(a) a reptile from the late Paleozoic period;

(b) a human.

In both cases, the thumb has one phalanx fewer than the other fingers (Romer, 1962).

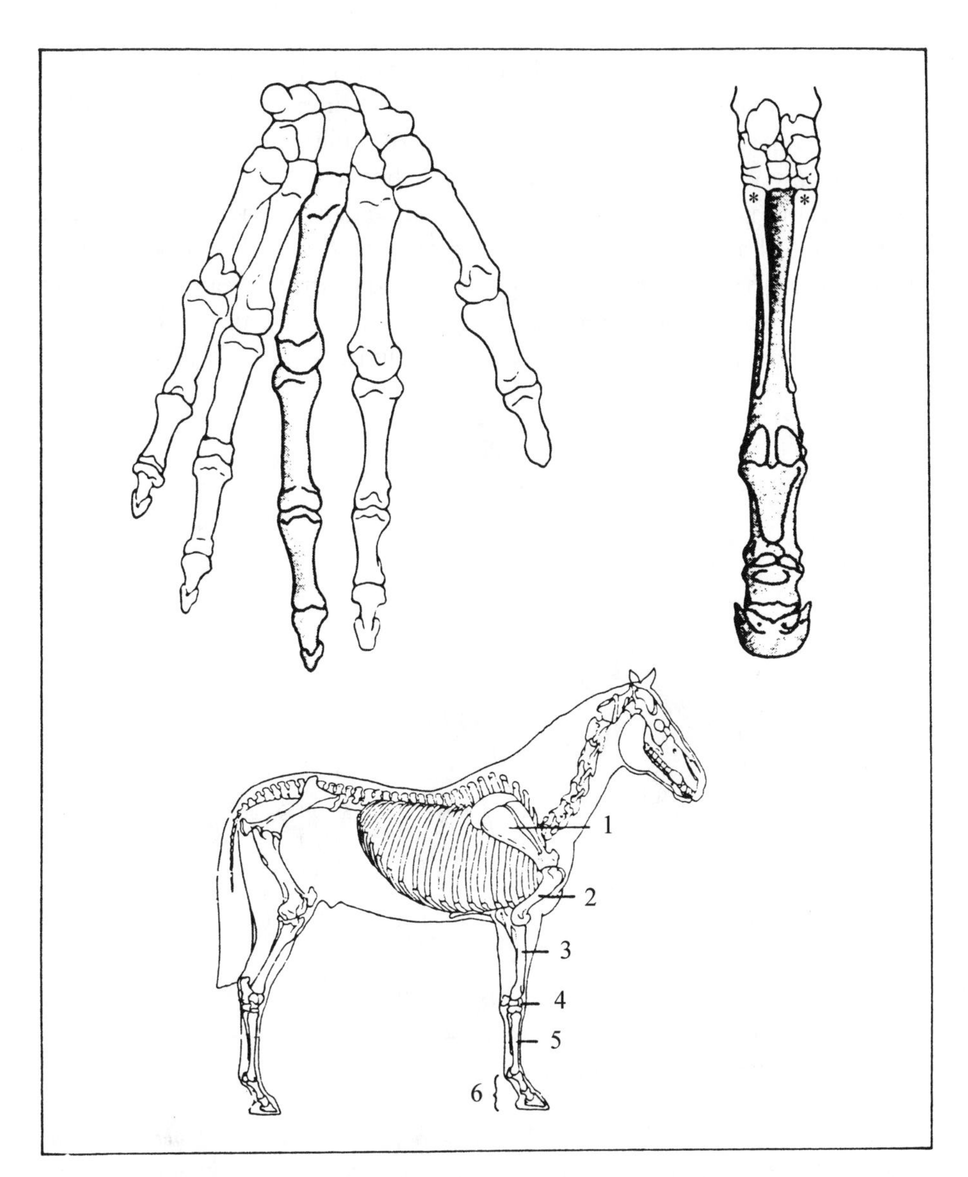

Fig. 55: Skeleton of the horse. Above: Human hand and horse "hand." In the horse, the middle finger has become huge. Two other fingers are normally reduced to the so-called splint bones (*) but become larger in exceptional cases (see Fig. 47, *p. 102 for this atavism).*

Below: The "hand" of the horse constitutes a major portion of the leg. The forelimb includes: (1) shoulder blade, (2) humerus, (3) radius, (4) carpal bones, (5) metacarpus, (6) phalanges. Other specializations in the horse include the absence of the clavicle and ulna (Latour, 1958).

In some animal species, such as salamanders, the primitive five-digit foot has been fairly well preserved throughout their evolution, while in others it has undergone extreme specialization. A familiar example is the horse (*Figure 55*), whose foot has been completely transformed to serve the functions of running and jumping. Only the middle digit of each foot has been retained and dramatically elongated. Digit I (the "thumb") and digit V (the "little finger") have completely disappeared, and of the two other digits only the so-called splint bones (reduced metacarpal bones) remain. The elongation of the "hand" is even more pronounced in the giraffe (*Figure 56*).

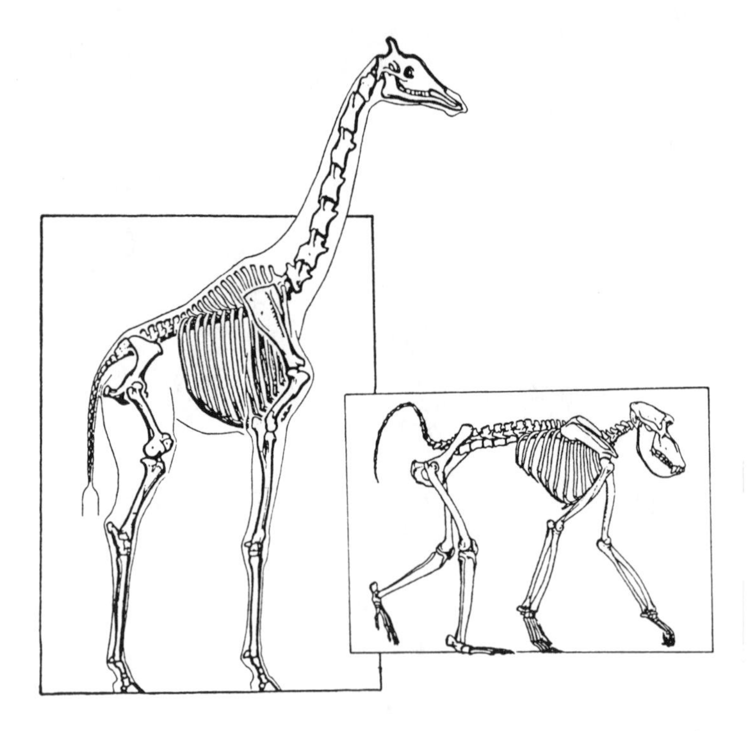

Fig. 56: Skeletons of the giraffe (by Tank, reproduced from Kranich, 1999) and the baboon. Both animals are true ground-dwelling quadrupeds. The giraffe has an extremely specialized "hand;" extreme growth of the metacarpal and metatarsal bones results in greatly elongated limbs (cf. Fig. 55*). In contrast, the hand of the baboon has remained relatively unspecialized. In the giraffe, the leg itself (radius plus humerus or tibia plus femur) is surprisingly short. In the baboon, the limbs are considerably longer relative to the torso, and the hands and feet are significantly shorter in proportion to the total length of the limbs. These phenomena must be attributed to the more retarded character of this animal (hypermorphosis).*

Other examples of advanced specialization are found among the flying vertebrates such as bats, birds, and pterodactyls (*Figure 57*). In these animals, some fingers have been greatly elongated, others shortened or eliminated.

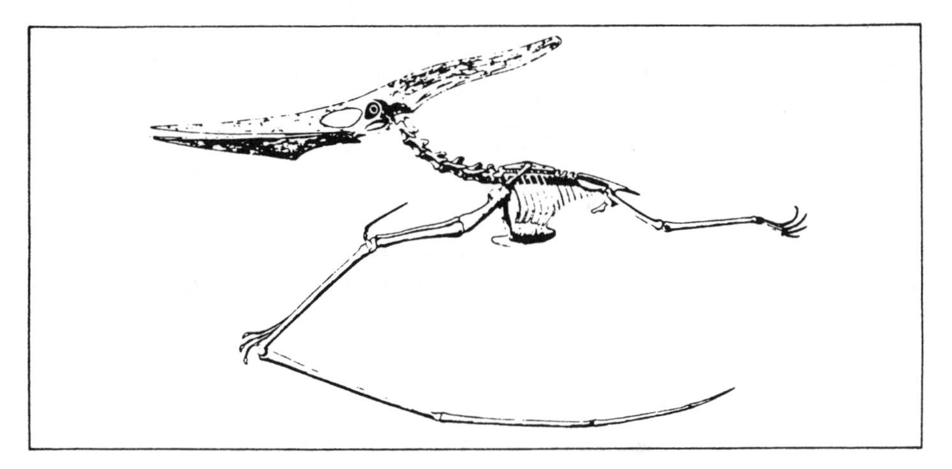

Fig. 57: Pteranodon*, a flying reptile from the Cretaceous period. With its four-meter wingspan it offers an example of extreme specialization of the hand (Eaton, 1910).*

Basic Hand Movements

The human thumb represents the culmination of an evolution that can be traced far back into the animal kingdom, to the earliest primates (see *Figure 58*). The tree shrew (*Tupaia*) still has a very primitive hand; its thumb is not yet capable of any specialized movements. In tarsiers, the thumb can already rotate around the joint that connects it to the metacarpal bone. The capuchin monkey, a platyrrhine, has evolved the additional ability to move the metacarpal bone of the thumb back and forth independently. In the gorilla (and in anthropoid apes in general), this bone also rotates. As we see, the thumb becomes increasingly mobile as we ascend through the primates, achieving maximum mobility in humans and anthropoid apes. But as we will also see, further specialization of the thumb in anthropoid apes entails the loss of possibilities resulting from the evolution we have just sketched. In contrast, only the human hand incorporates the universal structure. Landsmeer (1986) writes:

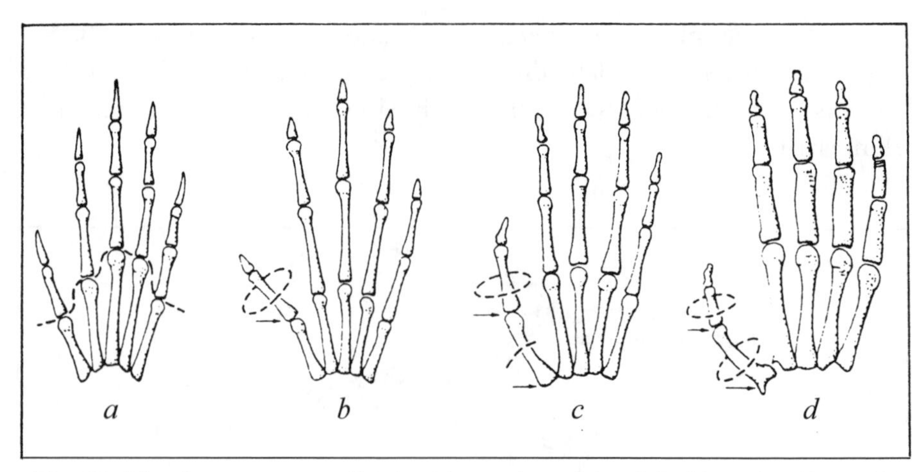

*Fig. 58: Thumb movements possible in: (a) tree shrew (*Tupaia*), (b) tarsier, (c) capuchin monkey, (d) gorilla (Napier, 1962).*

> The all-important point is that in the human hand it is not possible any longer to distinguish, by its structural setting, one function that would be prevalent . . . The full-fledged thumb makes the hand into a functional structure not skewed in any way whatsoever. The result is that the human hand is endowed with a functional continuum, a universal domain, and each particular function of the hand is basically designed or created within this domain.

Napier (1962) distinguishes four basic hand movements (see *Figures 59* and *60*):

- *Divergence*, or the weight-bearing position: In this primitive gesture (a universal mammalian capability related to the hand's support and locomotion functions), the fingers are splayed. Human beings use this hand position when pushing a large, heavy object (such as a car that will not start), for example.

- *Convergence*, or the hook grip: All fingers are bent. Two convergent forepaws form a grasping organ. Many mammals (such as squirrels) use this grip for holding food. When the fingers are bent still more, the hand forms a type of hook. Humans use the hook grip to carry a handbag or open a drawer. In apes, this grip can play a major role in climbing.

- *Prehensility*, or the power grip: The four fingers are wrapped around an object that is pressed against the palm of the hand, and the thumb augments the pressure exerted by the fingers. Human beings use the power grip to hold a hammer or unscrew a jar lid. This grip, a universal capability in higher primates, is contingent on having an opposable thumb separate from the other fingers.
- *Opposition*, or the precision grip: The object is held between the fleshy inner pads of the tips of the thumb and one or more fingers. The precision grip is a typically human gesture; apes either cannot use it at all or do so very clumsily. It is used, for example, to hold a writing implement or thread a needle. This grip is both strong and extremely precise.

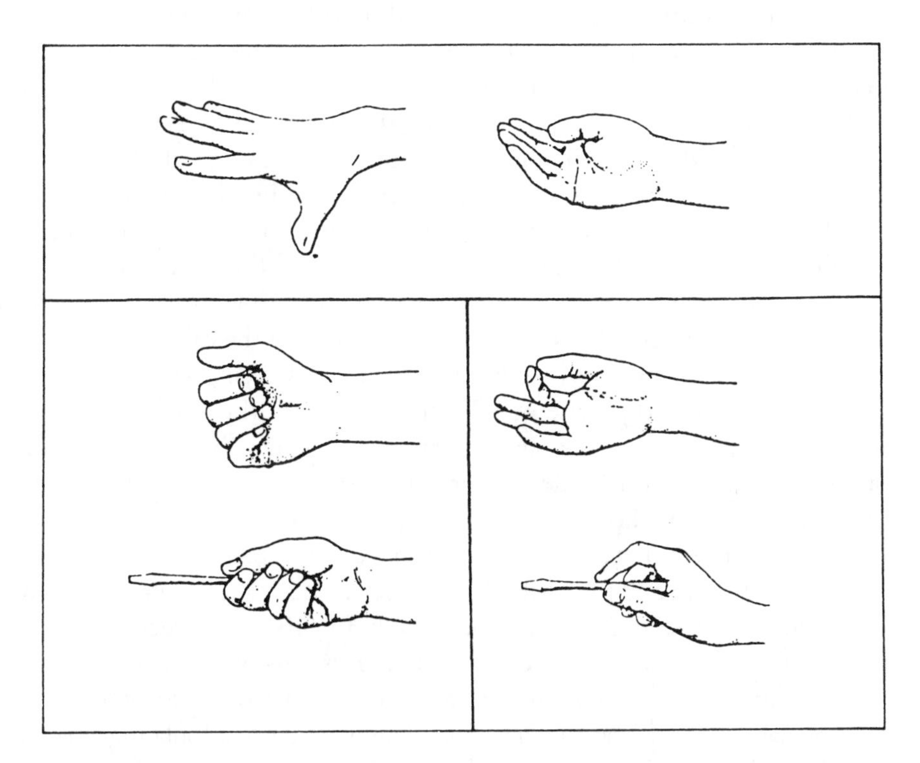

Fig. 59: Basic hand positions. Above: the two primitive positions, divergence (left) and convergence (right). Below left: prehensility/power grip. Below right: opposition/precision grip (Napier, 1962).

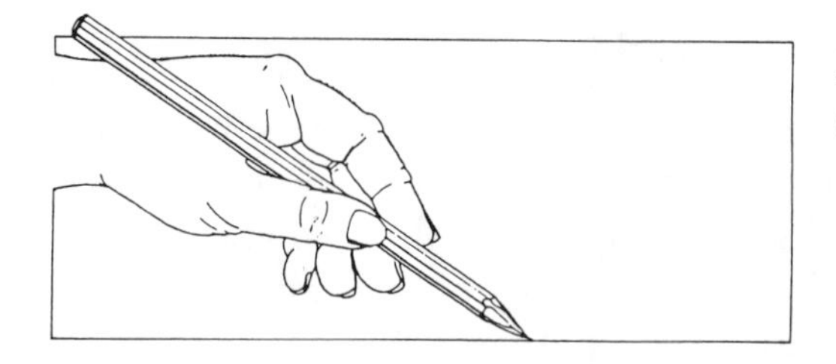

Fig. 60: Use of the precision grip in a typically human activity.

Hand Specialization in Anthropoid Apes

In principle, the thumb of anthropoid apes is just as mobile as that of humans, and yet these animals cannot perform the precision grip.

Figure 61 shows that the thumb is disproportionally shorter in anthropoid apes than in humans. This shortening is especially apparent in orangutans, whose thumbs often have only one phalanx. A shorter thumb is a specialization that adapts the hand to using the hook grip, especially for climbing. The palm elongates, while the thumb becomes short and weak and moves downward toward the wrist. The elongated palm and strong fingers form a hook, but the thumb is not involved. In the heavy anthropoid apes (but not in gibbons), the digital flexor muscles are shortened, so the fingers naturally tend to bend when the hand is extended. As a result, using the hook grip takes less energy when the animal is hanging from a branch by its arms. This specialization is most pronounced in the orangutan, the most one-sided climber of the three large anthropoid apes.

A shorter thumb makes the precision grip impossible. Anthropoid apes cannot bring the inner sides of the upper phalanges of their thumbs and fingers together. Only the very ends of these fingers meet, which does not allow for a secure and precise grip (Krantz, 1965). Anthropoid apes generally use a type of pseudo-precision grip, holding the object between the thumb and the side of the index finger in the grip people use to hold playing cards. All anthropoid apes hold objects either between the thumb and the ends of the fingers, between the fingertips and the palm of the hand, or between the thumb and the back of the fingers. The "cigarette grip"—holding an object between the index and middle fingers—is observed only in the African anthropoid apes. None of these grips is as precise and secure as the human precision grip. Krantz notes that chimpanzees and orangutans, in particular, use their mouths like third hands when manipulating objects. In anthropoid apes, the hand is used not only for manipulation but also for locomotion, and the mouth is still a prehensile organ.

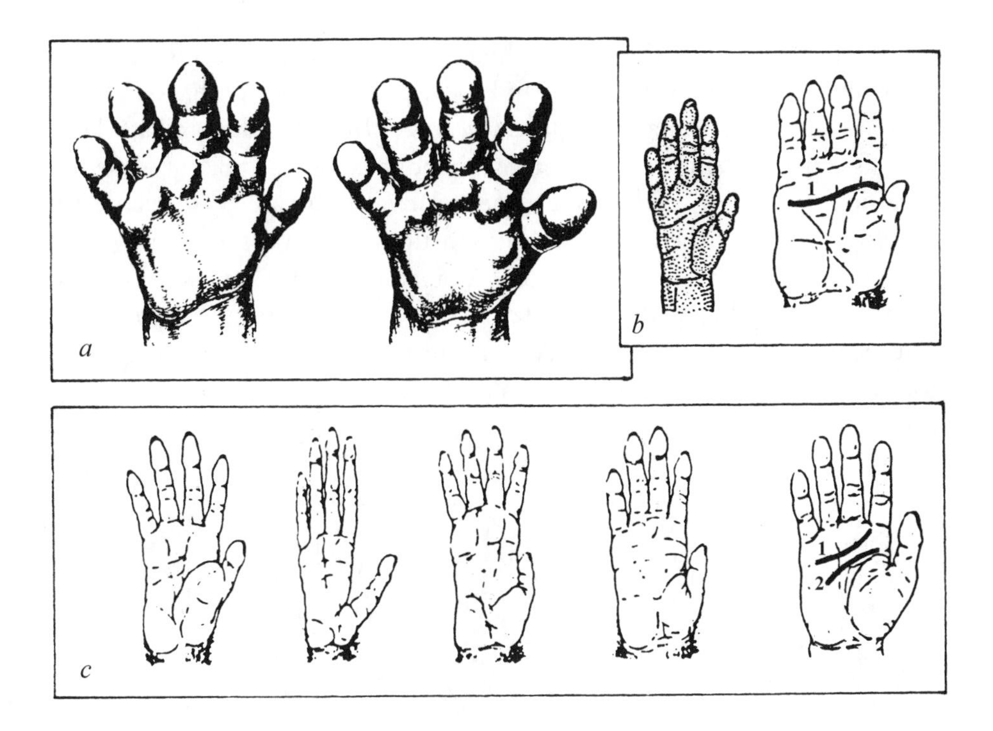

Fig. 61: The primate hand.

(a) The fetal hand of a macaque (left) and human (right). The fingers are not yet completely separated (torso length of both fetuses is 24 mm). At this stage, all primate hands still have the same rather broad shape. The base of the thumb is quite high and the thumb reaches above the base of the index finger (Schultz, 1968).

(b) Gorilla hand at a later fetal stage (four months, left) and in the adult (right). The adult hand has regained its original broad shape. This regression is typical of the gorilla (Schultz, 1931 and 1968). 1: The heart and head lines merge into one, as is typical of the hands of apes.

(c) Left to right: hands of the adult macaque, gibbon, orangutan, chimpanzee, and human. None of the adult catarrhines have thumbs that reach above the base of the index finger; consequently, although the thumb rotates, it is too short to perform the precision grip. Note the remarkable specialization that occurs in the gibbon – the separation between the thumb and the other fingers extends into the metacarpus. The orangutan's thumb is greatly reduced in size (Schultz, 1968). The human palm has both a heart line (1) and a head line (2).

In chimpanzees and gorillas, which are both less active climbers than orangutans, the thumb is longer and located higher on the hand. This is exhibited in lower primates too (see *Figure 62*). But even the gorilla, whose hand is especially similar to the human hand, has lost the ability to use the precision grip. The fairly broad and relatively humanlike gorilla hand develops by way of a strange detour. In the early fetal stage, the gorilla hand is broad, as it is in all primates. At a later stage it elongates, as is the case in other anthropoid apes. Later still, it widens again (see *Figure 61b*). The gorilla's development includes a number of such detours—in the foot and the penis, for example—as if the animal were trying to "retract" certain specializations.

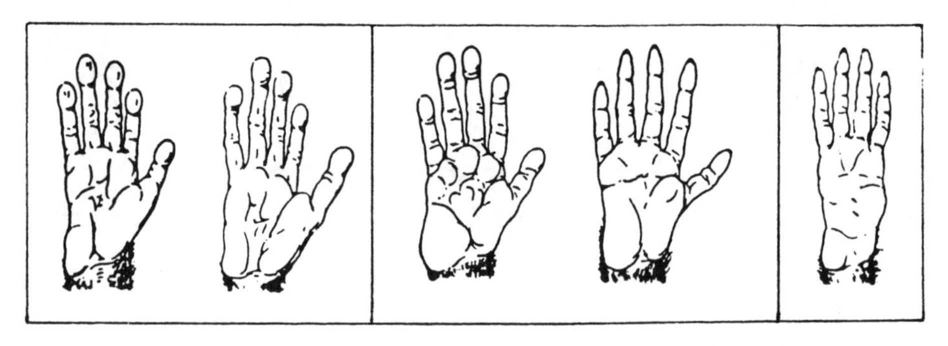

*Fig. 62: The thumb in lower primates. Hands of (left to right): two prosimians (lemur, bush baby) and three platyrrhines (night monkey, capuchin monkey, spider monkey). Although least closely related to humans, these primates all have a long thumb that reaches above the base of the index finger, a trait also found in humans but never in the more closely related catarrhines. Nonetheless, the prosimians and platyrrhines cannot implement the precision grip because their thumbs rotate either not at all or very little. In the most highly evolved platyrrhine, the spider monkey (*Ateles spp.*), the thumb is completely absent (Schultz, 1968).*

An additional specialization serving locomotion, however, is evident in the hands of African anthropoid apes. These animals, called "knuckle walkers," support the weight of the upper body and arms on the middle phalanges of both hands (see *Figure 63*), with the hands in the hook-grip position, which is also very important for climbing. The muscles and bones of the arms and hands of chimpanzees and gorillas reveal a number of adaptations related to this way of walking. Both of these apes have special "walking pads" on the middle joints of the fingers. Certain muscles and tendons have been shortened and details of the wrist joint altered. As a result, the capacity for certain movements has been lost in African anthropoid apes. For example, they cannot straighten both their fingers and their palms at the same time; when the palm is aligned with the underarm, the fingers bend by themselves. Even with the fingers bent, the wrist joints of chimpanzees or gorillas are only slightly hyperextensible; that is, the hands cannot bend more than slightly backward beyond the straight position. These adaptations stabilize the hook grip and are therefore advantageous not only for knuckle walking but also for climbing. Essentially, knuckle walking makes sense only as the mode of ground-level locomotion of apes that are already specialized climbers of the brachiator type. In such animals, certain aspects of the hands' fine motor coordination have been sacrificed for the benefit of locomotion.

Fig. 63: Orangutan (left) and gorilla (right). The gorilla, like the chimpanzee, is a knuckle walker; that is, the weight of the upper body is supported on the middle phalanges. In contrast, the orangutan's weight usually rests on the side of the hand (Schwartz, 1987).

No traces of these specializations, which are typical of the hands of African anthropoid apes, are found either in humans or in prehistoric hominids. The hands and wrists of juvenile anthropoid apes, however, are nearly as mobile as those of humans. Straus (1940) discovered that two of a newborn chimpanzee's fingers can still be fully extended when the hand is aligned with the forearm. This mobility persists for the first few months of life but then disappears quickly. Hyperextensibility of the wrist joints is also much greater in very young chimpanzees than in older ones. Tuttle (1969) concludes that humans and African anthropoid apes must be descended from a common ancestor whose hands were more humanlike than apelike. The greater mobility of the human hand in comparison to that of anthropoid apes is a Type I characteristic.

The Primitive Shape of the Human Hand

Napier (1962) quite rightly observes that:

> In the case of the hand, at least, evolution has been incremental. Although the precision grip represents the ultimate refinement in prehensility, this does not mean that more primitive capacities have been lost. The human hand remains capable of the postures and movements of the primate foot-hand and even of the paw of the fully quadrupedal mammal, and it retains many of the anatomical structures that go with them. From one stage in evolution to the next, the later capability is added to the earlier.

Thus there can be no talk of specialization in the case of the human hand. In the course of anthropogenesis, the hand never loses capabilities but always gains new ones. In this respect, human evolution follows a fundamentally different course from that of the horse or the bat, for example, who relinquish great potentials for the sake of highly specialized abilities. The human hand "specializes in nonspecialization." It is an excellent example of enhanced primitiveness in the Goethean sense (see *Figure 45*, p. 96). "The human hand displays an exceptional degree of primitiveness – an astonishing conclusion if we consider that it is capable of specialized movements and exceptional sensitivity, precision, subtlety, and expressiveness" (Napier, 1993). It is *the* universal organ of manipulation. The very shape of the human hand hints at tools, calls for tools, is incomprehensible without tools. For this reason, every form of technical education ought to begin with a study of the hand.

The universal, five-digit form of the human hand is a Type I characteristic. As we see, this ancient organ has undergone the most extreme and varied specializations in countless animals over the course of evolution. It has been modified into paws, flippers, wings, hooves, etc. Although the thumb has always been differentiated from the other fingers, no animal has ever solved the riddle it presents because only the human line of evolution remained true to the original form. The higher primates begin to solve the riddle of the thumb. In primates in general, the thumb becomes more mobile, the hand's ability to manipulate objects is refined, etc. In comparison to that of the lower primates, the movable and opposable human thumb is a Type II characteristic. It appears to an ever-increasing extent in the higher primates but takes on meaning only in the final stage, when the precision grip appears. The effect of this development is explosive. In the appearance of technology (the actual use of tools), the full significance of all prior evolution faithful to the original gestalt becomes evident.

The mobile human thumb is a Type I characteristic, however, in comparison to the thumbs of anthropoid apes. These apes have moveable, humanlike thumbs but are unable to take advantage of thumb mobility, which becomes meaningless in certain respects because their hands have become specialized to serve locomotion. In both the most highly evolved platyrrhines and the catarrhines (including the anthropoid apes), the degeneration of the thumb is strikingly obvious (see *Figures 61* and *62*). This tendency, however, is absent in the lower primates and in human beings.

The Hand Index

The hand index is the width of the hand as a percentage of its total length. In the fetus, the hand is relatively short and thus the hand index is high (see *Figure 61a*). According to Schultz, during the earliest phase of embryonic development, the hand index is high and roughly the same in both humans and apes. After that, the human hand remains relatively short while the hand index of apes decreases rapidly. In general, adult apes have long, narrow hands. Schultz (1926a) gives these examples of the hand index in adults of various species:

humans:	ca. 45%
gibbons:	21-25%
orangutans:	22-28%
capuchin monkeys:	31-34%

In Chapter 3 we saw that the short, light human hand fits well into the total picture of the retarded, caudocephalic growth pattern of the limbs. This pattern, however, does not explain the longer hands of apes, which are a specialization related to these animals' mode of locomotion. Again, in comparison to ape hands, the human hand, which remains short, is a Type I characteristic.

The Thumb Index

We humans have relatively highly developed thumbs in contrast to the catarrhine apes, which are supposedly our nearest relatives. In the later phases of their growth, the thumbs of apes become consistently shorter in relationship to the hand. In chimpanzees, a clear decrease is noticeable in the thumb index (the length of the thumb as a percentage of the length of the hand) in the course of individual development. Schultz (1940) provides the following values for chimpanzee thumb indices:

fetus:	55-62%
newborn:	49-56%
juvenile:	46-56%
adult:	44-52%

A decrease in the thumb index is also observable in the individual development of the orangutan (Napier, 1967):

fetus:	45-49%
newborn:	44-53%
juvenile:	40-48%
adult:	37-47% (male)
	40-46% (female)

Schultz discovered that the thumb index in macaques decreases rapidly in the earliest fetal stage but then stabilizes (1937). He observed that this decline in the thumb index is common to all primates except humans (1926a, 1937). As we will explain in detail in Chapter 6, the fingers and toes do not emerge all at once but follow a specific sequence, namely: ring finger, middle finger and little finger, index finger, thumb. Thus the thumb is an ontogenetic latecomer in comparison to the other fingers. As a result, retardation and hypermorphosis should tend to produce a relatively large thumb, which is exactly what we observe in humans. The human thumb is initially small in comparison to the rest of the fingers but later catches up. Montagu (1931) discovered that the total length of the thumb phalanges is not quite 48 percent of the length of the middle finger's phalanges in a human fetus between the ages of five and seven months. In an adult, this number increases to nearly 57 percent.

In apes, however, the decrease in the thumb index results from elongation of the other digits, including their metacarpal bones. This process cannot be explained by retardation and hypermorphosis and must therefore be considered a specialization.

Like the thumb, the index finger is also a latecomer in comparison to the ring finger and middle finger. In human adults, there is considerable individual variation in the relative lengths of the second (index) and fourth (ring) fingers, but the metacarpal bone of the second digit is consistently longer than that of the fourth digit. In apes, these proportions are reversed (Jouffroy et al., 1993); that is, the fourth digit (phalanges plus metacarpal bone) is longer than the second. In prosimians and other terrestrial vertebrates, the fourth digit is usually very long (see, for example, *Figure 62*, p. 122, and *Figure 54a*, p. 114).

The precision grip, a typically human capability, is a direct result of enhanced development of the thumb and index fingers in humans. Another curious consequence of the emancipation of the index finger is the fact that human palms almost always have two lines (see *Figure 61*, p. 121), known to palm readers as the "head line" and the "heart line," while the palms of nonhuman primates have only one. "The difference…reflects the fact that the human index finger has a great deal of independent mobility, while, in non-human primates, the index finger tends to work in harness with the remainder" (Napier, 1993).

The Digital Flexor Muscles

The difference between the human hand and the anthropoid ape hand is clearly apparent from the configuration of the deep digital flexors — muscles that are noticeable on the underside of the arm when the fist is clenched. In humans, we find two such muscles (see *Figure 64*):

- The flexor pollicis longis (long thumb flexor), which is attached to the radius. Its tendon extends to the base of the last phalanx of the thumb.
- The flexor digitorum profundus (deep flexor muscle of the fingers), which is attached to the ulna. The belly of this muscle ends in four tendons that are attached to the final phalanges of the four fingers. This muscle moves the joints of the wrist, metacarpus, and fingers.

This configuration, in which one muscle moves the thumb and another the remaining fingers, reflects ancient differences in form and function that set the thumb apart from the other four fingers (see *Figure 54,* p. 114).

In anthropoid apes, the arrangement of the digital flexors is different. In chimpanzees, gorillas, and orangutans, the thumb and index finger share a flexor muscle, and the thumb's flexor tendon is either poorly developed, degenerated, or altogether absent. Even when the tendon is present, it is almost always very weak. Hence, in the majority of anthropoid apes, the thumb will not bend independently of the other fingers. Straus also observed a general weakening tendency among the other muscles that move the thumb in anthropoid apes. In gibbons, the thumb degenerates to a lesser extent but reveals a number of remarkable specializations adapted to the climbing habits of this genus. In contrast, the human thumb, according to Straus, has "a generalized structure, completely lacking not only the defective qualities found in the pollices [flexor muscles] of the great apes and most other brachiators but also the peculiar specializations relating to the thumbs of the Hylobatidae" (1942).

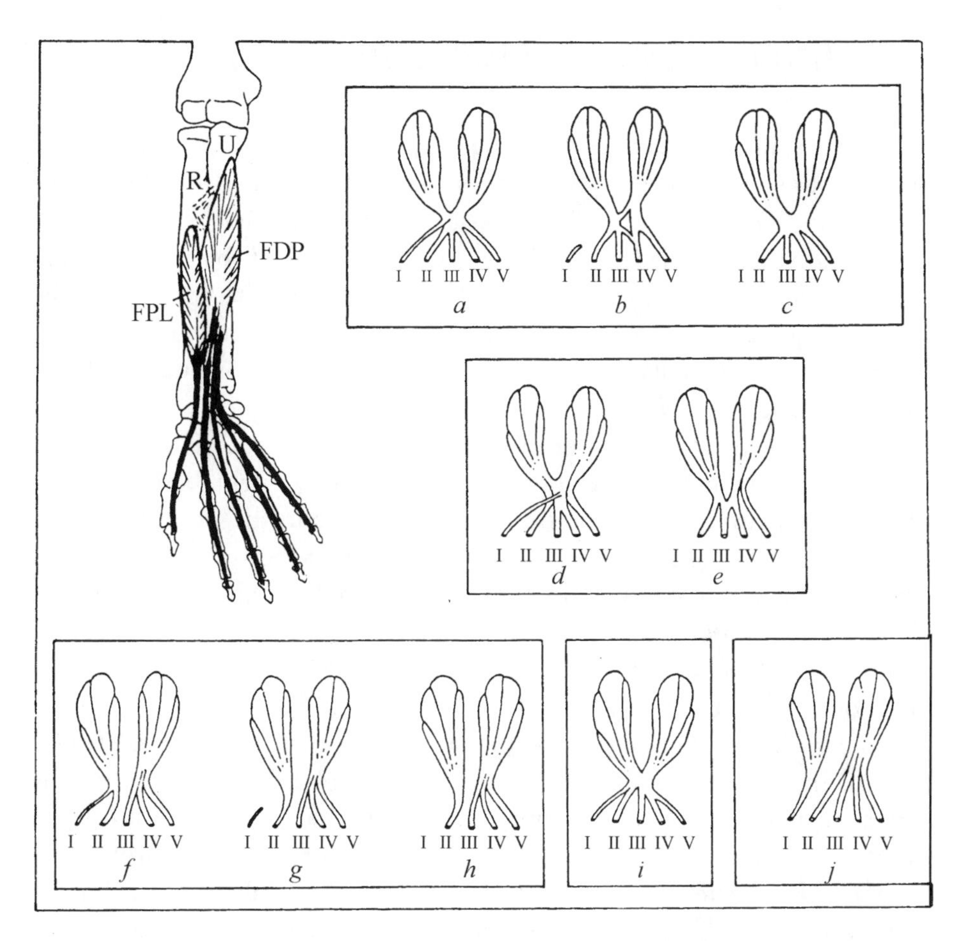

Fig. 64: The deep digital flexor muscles. Above left: outline of the locations of the long thumb flexor (flexor pollicis longus: FPL) and deep digital flexor (flexor digitorum profundum: FDP) in the human forearm. U: ulna, R: radius. From the bellies of these muscles, long tendons (black) extend to the base of the last phalanges of the fingers. The skeleton of the hand is shown from the palm side.

The illustration also gives schematic representations of the corresponding muscles and tendons in a number of primates. The fingers are numbered from the thumb (I) to the little finger (V). (a) capuchin; (b, c) variants in spider monkeys (thumb tendon degenerated or absent); (d) macaque; (e) colobus monkey (a catarrhine with no thumb and therefore no thumb tendon); (f, g, h) three variants found in gorillas, chimpanzees and orangutans (the thumb tendon is either weak, degenerated, or absent); (i) gibbon; (j) human (Straus, 1942). Only in humans is the special status of the thumb reflected in the configuration of the deep digital flexors.

Arm Location and Ribcage Configuration

In mammalian quadrupeds, the shoulder blades are located at the sides of the torso. This location places the forelimbs close together below the torso, which makes walking easier. In humans (and in anthropoid apes) the shoulder blades are located on the back, and the forelimbs are attached at both sides of the torso. In humans, the line connecting the shoulder joints passes through or very close to the spine, while in most mammals this line falls slightly in front of (or below) the spine (see *Figure 22*, p. 44, and *Figure 56*, p. 116).

The lateral location of the arms has a remarkable and self-reinforcing consequence. First, this position is favorable for upright body posture. Because the arms and the spine fall more or less in the same plane, the arms do not create forces that act on the joints of the spine. Such forces are almost inevitable in mammals whose legs are located below the torso. The lateral position of human arms helps free them from the function of locomotion. Second, this position greatly enhances the arms' range of motion. Human beings can extend their arms in opposite directions; reaching objects standing off to the side does not require any particular effort. In contrast, in animals whose shoulder blades are located on the sides of the torso, forelimb movement is restricted to the area directly in front of (or under) the torso. The great mobility of the human arm represents a logical extension of the highly refined motor abilities of the human hand. The hand's opportunities would be severely restricted if it were attached to a less mobile arm. In humans, the universal character of the hand, the lateral location of the arm, and upright posture of the body are three closely related traits.

The configuration of the ribcage plays a central role in this complex relationship. In quadrupeds, the ribcage (thorax) is deep, in stark contrast to the flattened human ribcage. A broad thorax is an advantage in acquiring upright posture because its weight remains closer to the weight-bearing spine and thus exerts a lesser force. In contrast, a narrower, deeper ribcage has advantages for quadrupedal locomotion because it places the shoulder blades on the flanks and points the shoulder joints downward, in the same line as the forelimbs.

The carnivores serve as a good illustration of the role of ribcage configuration in upright posture. The bear has a wide ribcage with proportions similar to those of a human child (which may have something to do with the popularity of bears as stuffed toys) and is also quite good at walking upright. For example, Kranich (1999) says of bears, "These animals have traits that reflect in crude and unfinished form, so to speak, what will appear in perfected form in humans." As quadrupeds, however, bears are less agile than cats, which have deeper, less humanlike ribcages. Cats, on the other hand, are not as good at walking on their hind legs.

The thoracic index (the ratio of thorax width to thorax depth at specific points) is a measure of the breadth of the ribcage. The higher its value, the broader the ribcage. According to Davenport (1934), the ribcage is often quite broad and flat in primitive mammals but very narrow in large quadrupeds that are specialized runners (ungulates, elephants, etc.). The lower primates occupy an intermediate position, while in anthropoid apes the ribcage is very flat and approaches human proportions. Specializations in some other animals (such as moles and other diggers) also result in a very flat ribcage.

The flat configuration of the human ribcage is often considered a specialization because it favors upright posture and increases the range of motion of both the arms and the hands by shifting the arms to a lateral (and thus more mobile) position. This view is incorrect. The high thoracic index in humans is a logical continuation of a universal mammalian tendency. The ribcages of all mammals are very deep during early embryonic development (*Figures 65* and *66*) but then gradually become flatter, although only until a certain phase of development is reached. In humans, the thoracic index normally continues to increase throughout the growth period. In animals such as dogs, a reversal occurs at a certain point in time; that is, the thoracic index peaks and then declines. This reversal has even been observed in many quadruped primates such as baboons, macaques, capuchins, etc. (Schultz, 1926a).

For this reason, Schultz was critical of researchers who saw the narrow ribcage of the human embryo as a Haeckelian recapitulation of an ancestral (quadruped) condition. For example, a dog fetus has a thoracic index that initially rises to 100, but after that the ribcage narrows again. Schultz comments:

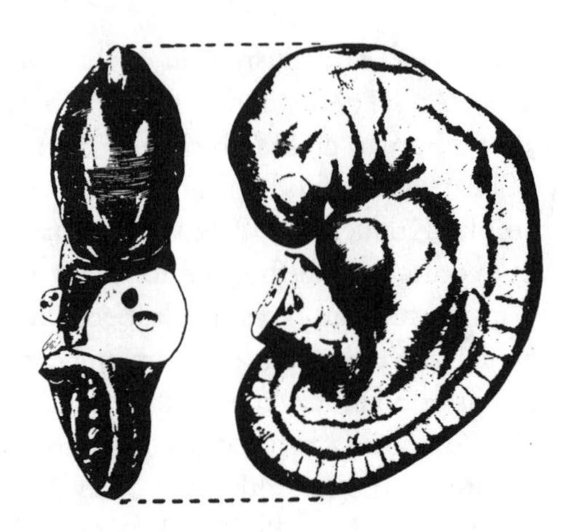

Fig. 65: Front and side views of a 4.2-mm-long human embryo. Note the laterally flattened thoracic area. To compare this to later developmental phases, see Figure 22, *p. 44 (Blechschmidt, 1961).*

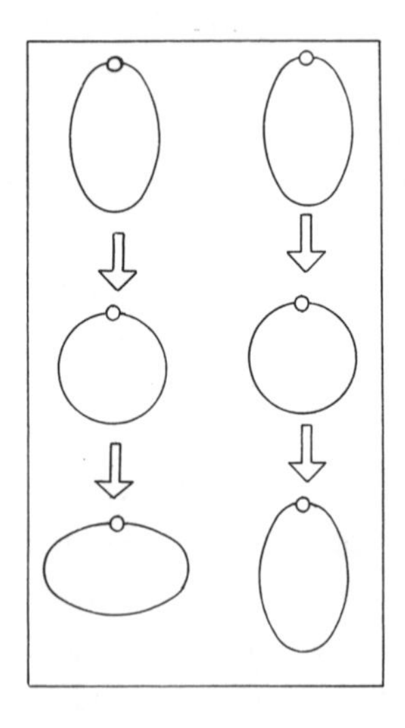

Fig. 66: Schematic representation of changing chest proportions in (left) humans and (right) animals (cf. Figure 22a, *upper left); the small circle indicates the location of the spine. In humans (and also in anthropoid apes), the development of these proportions proceeds consistently in the same direction (i.e., the ribcage always grows broader). In lower apes and most other mammals, however, a reversal of the original broadening tendency occurs, and the ribcage becomes narrower as the animal matures.*

> In order to be consistent, one would have to make the absurd assumption that the ancestors of the dog had an upright posture . . . The chest shape in early embryonic life cannot be regarded as the ontogenetic reappearance of some phylogenetic condition, but is directly determined by causes of a purely topographic nature, especially by the growth of the heart, lungs, and liver" (1926a).

The high thoracic index in adult humans is a clear example of a Type II characteristic. It appears quite late in human development, its structural significance depends on a typically human condition (upright posture), and it is clearly present already in the most retarded primates, the anthropoid apes. The broad, flat human ribcage emerges as a result of consistent prolongation or hypermorphosis of an earlier developmental tendency that is present in all mammal embryos. Any mammal with a deep ribcage deviates from this universal pattern; that is, its thorax narrows again after a certain stage in development. This deviation from the universal pattern produces a body structure that is specialized to serve quadrupedal locomotion. In contrast, humans remain true to the original developmental pattern, which permits the development of a bodily configuration that favors both the acquisition of an upright stance and the liberation of the hand from the demands of locomotion.

The Shoulder and Other Examples of the Progressive Emergence of the Human Gestalt

The generalized character and optimized mobility of the human shoulder are directly and functionally related to the universal character of the human hand. Oxnard (1969) states, "There is increasing mobility of the shoulder in passing from prosimians through monkeys and then apes . . . The human shoulder is most mobile of all as befits a forelimb not used for locomotion but possessing important tactile, prehensile, manipulatory, communicative, and agonistic functions."

The Shape of the Shoulder Blade

In several respects, the structure of the human shoulder represents the culmination of an evolution that can be traced far back into the animal kingdom. Mobility of the forelimbs is intimately related to the shape of the shoulder blade. "In general a proportionally wide scapula permits a smaller range of movement of the upper limbs than a relatively narrow scapula" (Schultz, 1930). Primates generally have relatively taller and narrower shoulder blades than other mammals (see *Figure 56*, p. 116, and *Figure 67*). *Figure 67* shows that a systematic decrease in the width of the shoulder blade occurs in the phylogenetic sequence of primates from prosimians to lower apes, anthropoid apes, and finally humans. It is interesting to note that although the highly evolved platyrrhine *Ateles* (spider monkey) is much less closely related to humans than are the lower catarrhines, the shoulder blade in *Ateles* is much more similar to the human shoulder blade.

Humeral Torsion

Not only in humans but also in specialized quadrupeds such as horses, the elbow joint rotates around an axis that is roughly perpendicular to the body's direction of movement. In humans, however, the shoulder blade and shoulder joint are located on the back. Furthermore, the head of the humerus faces the rear in horses but is laterally located in humans. Consequently, the relationship between the structure of the hinge joint of the elbow and the head of the humerus is not the same in humans as it is in quadrupeds (see *Figure 68*). This different relationship results in a rotation (torsion) of the human humerus.

Humeral torsion begins at an early stage of fetal development (Martin and Saller, 1957-1959). It parallels the increase in the thoracic index. Thus the shoulder blades gradually shift toward the back of the body. In primates, torsion is most pronounced in humans, followed by the African anthropoid apes, then gibbons, orangutans, higher platyrrhines, and finally catarrhines (see *Figure 68*; Evans and Krahl, 1945; and Knussman, 1967). Torsion contributes to the human hand's ability to manipulate objects. The unique location of the human scapulohumeral (shoulder) joints in relationship to each other makes it easy for the hands to come together in front of the body. Thus humeral torsion can be seen as a Type II characteristic.

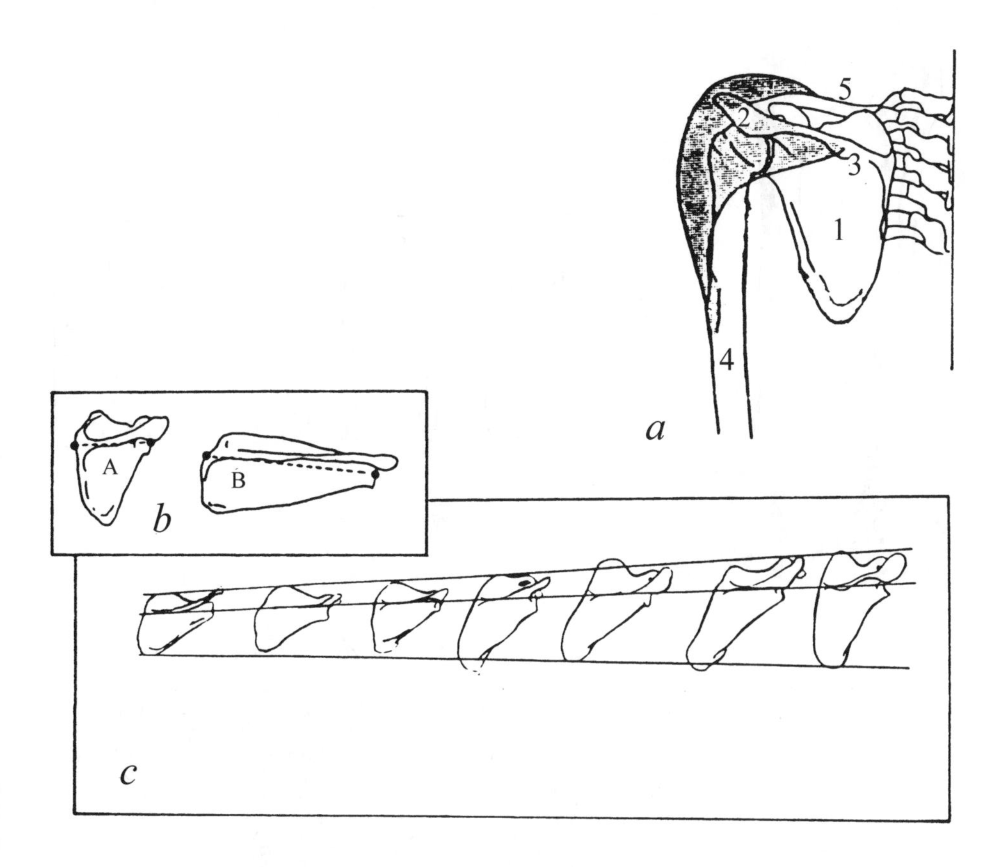

Fig. 67: Shoulder and shoulder blade.

a: Human shoulder (rear view). The location of the deltoid muscle is indicated in gray. (1): scapula (shoulder blade); (2): acromion; (3): scapular spine; (4): humerus; (5): clavicle or collar bone (Campbell, 1974).

b: Human scapula (A) compared to that of an average mammal (B) (Shea, 1986). The dotted line indicates the longitudinal direction. The human shoulder blade is relatively tall and narrow.

*c: Scapulae of (left to right): a prosimian (*Indri*), two catarrhines (macaque and baboon), a platyrrhine (*Ateles*), gorilla, chimpanzee, and human. The scapular index (the width of the shoulder blade divided by its height) decreases consistently over this series of primates. The extreme narrowness of the human shoulder blade is related to optimum shoulder mobility (Inman, Saunders, and Abbot, 1944).*

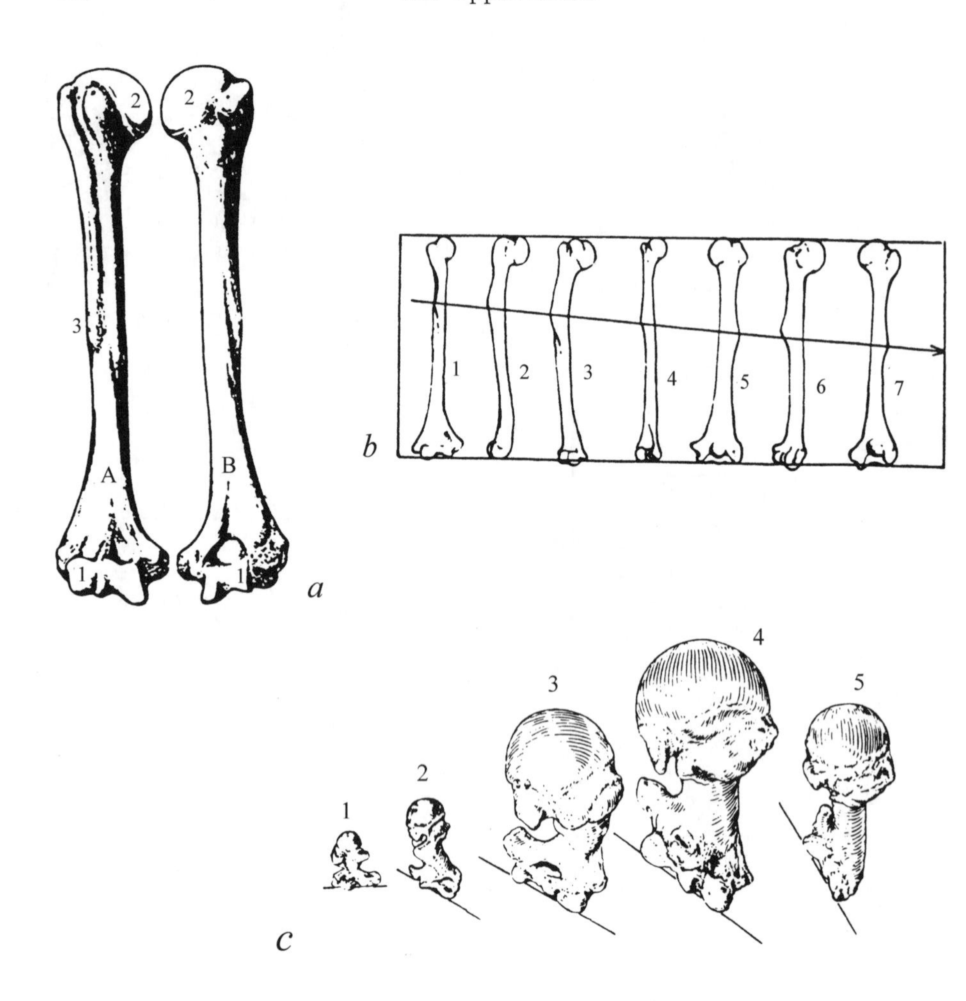

Fig. 68: The humerus.

(a): Right humerus of a human being, seen from in front (A) and behind (B). (1): humeral trochlea (connects to the elbow joint); (2): head of the humerus (connects to the shoulder joint); (3): deltoid tuberosity (insertion site of the deltoid muscle).

(b): Location of the deltoid tuberosity in: (1): indri (prosimian); (2): baboon; (3): chimpanzee; (4): spider monkey; (5): gorilla; (6): chimpanzee; (7): human.

(c): Humeral torsion in: (1): indri; (2): baboon; (3): chimpanzee; (4): gorilla; (5): human. View from diagonally above the shoulder joint. The straight line is perpendicular to the direction of the humerus shaft. Note that in humans and anthropoid apes, the head of the humerus is greatly enlarged in comparison to the shaft (Kahle et al., 1975).

The Deltoid Muscle and Its Insertion Sites

Humans are better able than animals to extend their forelimbs to the side. The deltoid muscle, which plays a major role in this gesture, is also highly developed in humans. This muscle is like a cap covering the lateral bony elements of the shoulder girdle and part of the humerus. The deltoid muscle constitutes almost 40 percent of the total muscle mass of the shoulder and upper arm in humans, 37 to 39 percent in anthropoid apes, and not quite 20 percent in other mammals, including the lower apes (Inman et al., 1944). (In specialized quadrupeds such as horses, however, this muscle has completely disappeared.) The increasing importance of the deltoid muscle is associated with the emergence of a more substantial acromion (*Figure 67*). The acromion serves as one of the origins of the deltoid muscle (Inman, Saunders, and Abbot, 1944).

The activity of the deltoid muscle is enhanced not only by its increasing mass but also by an increasingly efficient lever structure. The muscle is attached from below to a specific point on the humerus (the deltoid tuberosity, see *Figure 68*). This insertion site shifts ever lower as we move from prosimians through lower apes and anthropoid apes to human beings. In humans, leverage is improved not only by the low insertion site but also by the greater length of the humerus relative to the arm as a whole (see p. 76 above).

In the more retarded primates, the great degree of development of the deltoid muscle and acromion and the low position of the deltoid tuberosity (a result of the increased length of the proximal portion of the humerus) can be considered examples of hypermorphosis. That is, retardation (prolongation) of limb growth results in greater development of the parts of the limbs closer to the torso. The same can be said about the head of the humerus in humans and anthropoid apes, which becomes very large in comparison to the humeral shaft (*Figure 68*). The proximal portion of the humerus develops late in comparison to the distal portion — that is, the part farther from the torso (see *Figure 69*). Thus the head of the humerus will be larger in a more retarded organism. The enlargement of the head of the humerus, the acromion, and the deltoid muscle increases the shoulder joint's range of lateral and upward movement (Corruccini and Ciochon, 1976). The resulting increase in shoulder mobility among most primates is directly related to the universal possibilities of the unspecialized and retarded human hand. Although further research is essential, we can safely say at this point that the most striking features of the human upper arm and shoulder seem to be Type II characteristics. These traits, which are already clearly apparent in the more retarded primates, are directly related to the fact that the human hand has been totally freed from serving the function of locomotion and has evolved into a mobile, universal organ of manipulation. Furthermore, most of these characteristics make sense as direct consequences of hypermorphosis; that is, they came about through retardation (prolongation) of the distoproximal growth of the limbs.

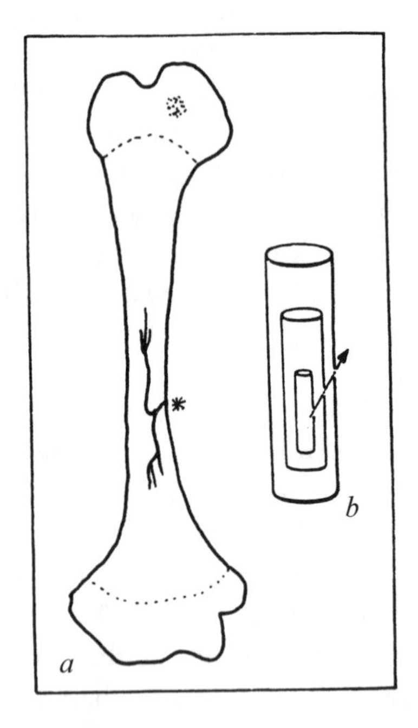

Fig. 69:

(a): Humerus of a human newborn, showing a portion of the bone arteries. The arteries pass through the canalis nutriens, which runs diagonally upward (*).

(b): The diagram shows three different stages in bone growth, drawn inside each other. Their relative positions are determined by the fact that the proximal portion grows faster than the distal portion. The arrow indicates the direction of the developing canal (Braus, 1929).

Nails and Claws

The human fingernail consists of a hardened corneal plate (the unguis) that continues growing throughout the life of the individual, pushing outward from root to tip to cover the underlying nailbed. The fertile proximal portion of the nailbed, which constantly produces new keratinous substance, is called the matrix. The crescent-shaped spot (the lunula) that is visible through the unguis at the base of the nail is part of the matrix. The distal portions of the nailbed are sterile and do not produce new substance, although even there the unguis and the nailbed are very firmly connected.

Nail and skin meet in an acute angle under the tip of the nail. At this location, the skin is covered with a special kind of keratinization called the subunguis. The subunguis is very small in humans but much more extensive in many animals (see *Figure 70*), assuming huge proportions in horses, for example.

Many mammals (such as rats, cats, and hedgehogs) have claws, which differ from nails in several respects:

- The unguis curves sharply to the left and right, so that the lateral edges are close together and almost completely surround the distal phalanx (see *Figures 70* and *71*). However, "The longitudinal curvature of the unguis varies considerably. While sometimes no greater than in humans or apes, it is often much greater" (Boas, 1894).
- The distal phalanx also assumes a clawlike shape. Under a nail, the same bone remains generalized in shape and does not adapt to the shape of the unguis (see *Figures 70* and *71*).
- The matrix, the fertile portion of the nailbed, is more extensive in clawed animals. "The distal (forward) thrust …of the basal matrix is almost always greatest in the middle and recedes gradually at the edges. In humans, for example, the distal tip of the basal matrix is a slightly curved line, while in other species it thrusts forward much more or even (as in carnivores) develops a long, pointed, median process that may even reach the terminal

matrix (as in the dog), dividing the sterile portion of the nailbed into two separate halves . . . Whenever such a median process is present, the unguis becomes much thicker in the center than on the edges, which is very important for maintaining the sharp tip of the claw because the thinner lateral portions of the unguis abrade faster than the central portion" (Boas, 1894).

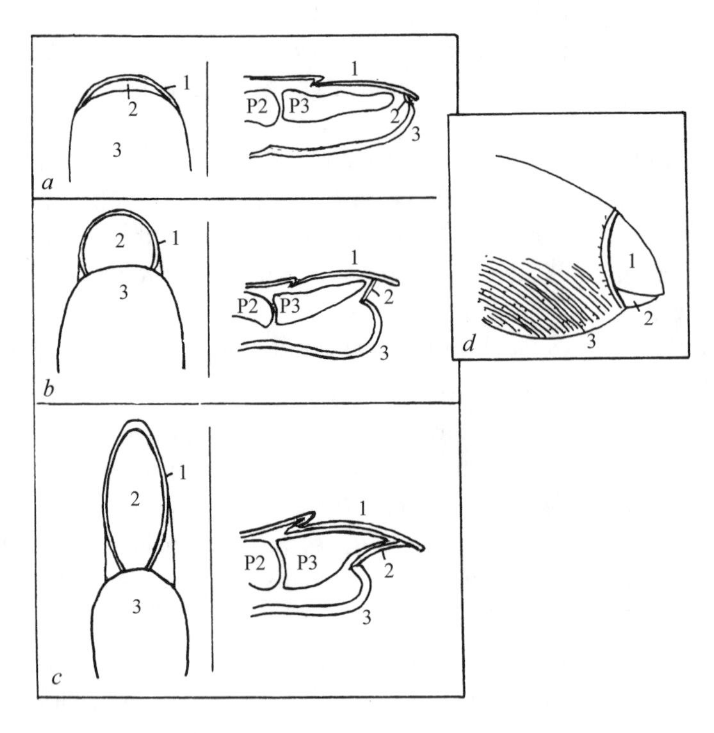

Fig. 70: Schematic representation of the tip of the finger in (a) humans, (b) apes, and (c) clawed mammals.

(left: view from below; right: longitudinal section).

(1): nail plate or unguis; (2): sole plate or subunguis; (3): the pad or fleshy ball of the fingertip; (P2): middle phalanx; (P3): distal phalanx.

The tegula of apes (b) (see p. 143) represents a transition form between the human nail and the mammalian claw (c), which surrounds the distal phalanx, giving it a clawlike shape.

(d): fingertip of the chimpanzee. The subunguis is much more highly developed than in humans.

The ontogeny of the claw passes through a nail-like stage; see *Figures 71* and *72* (Clark, 1936; Panzer, 1932). During the earlier stages of development, the unguis is limited to the dorsal side (back) of the fingers. Later it becomes claw-shaped, growing to the left and right and wrapping around the phalanx. The nail is a retarded claw that has remained juvenile. Thus the human nail must be considered a Type I or fetalized characteristic. This increased paedomorphosis, observed even in the claws of reptiles, climaxes in human beings.

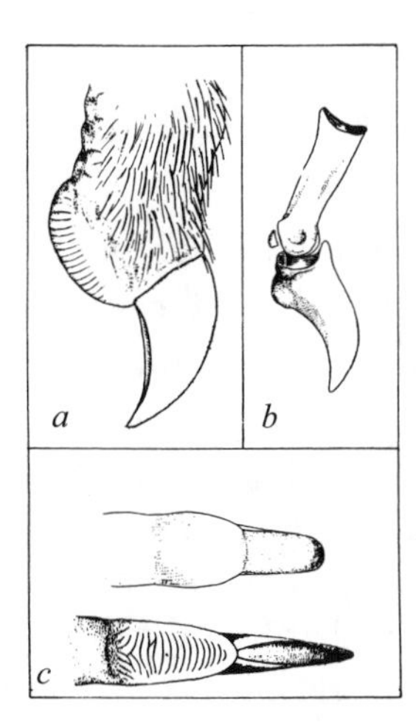

*Figure 71: The second toe of a tree shrew (*Tupaia*).*

(a): side view;

(b): side view of the skeleton;

(c): view from below (above: fetal stage; below: adult stage).

The distal phalanx of the adult toe has assumed a distinctly claw-like shape. Note also that in the fetus, the unguis has not yet grown around the tip of the toe; that is, the claw is still in the juvenile, nail-like stage (Panzer, 1932).

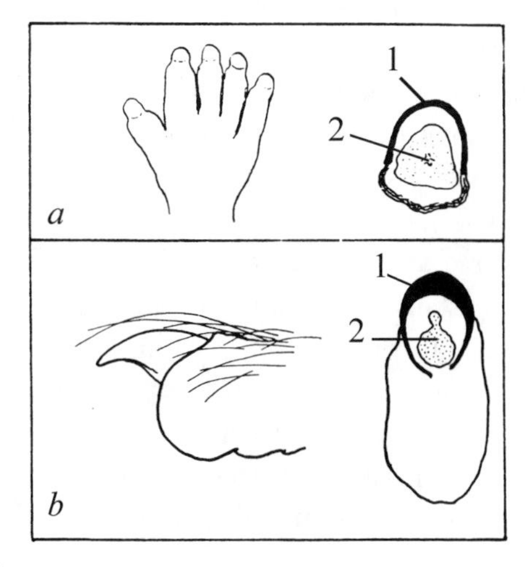

Fig. 72:

(a): Foot of a mouse fetus; right, cross-section through one digit. Note that the nail plate (1) (unguis) has not yet grown around the phalanx (2).

(b): side view of the tip of the digit of an adult mouse; right, cross-section. By now a true claw has formed; that is, the nail plate extends all the way around the phalanx (Clark, 1936).

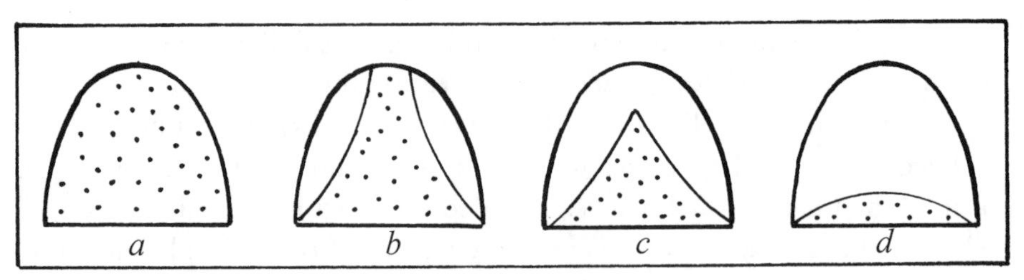

Fig. 73: Highly schematic representation of the distribution patterns of fertile nailbed or matrix (dotted) and sterile (white) nailbed in: (a): birds, crocodiles (b) dogs, Varanus *spp.; (c):* Tupaia *spp., lemurs; (d): humans (see also Clark, 1936). The fertile matrix from which the nail plate grows gradually retreats toward the base of the nail in animal genera whose development is more retarded.*

The first important juvenilizing tendency appears in the basal matrix, the fertile portion of the nailbed, which extends less and less far toward the tip of the finger in higher vertebrates. In most reptiles (crocodiles, turtles, etc.) and in birds, the entire nailbed is fertile, but in some reptiles (such as the genus *Varanus*, which includes the Komodo dragon), we find "a great similarity to the claws of mammals, especially carnivores" (Boas, 1894). In these species, as in mammals, the distal area is sterile and the proximal, fertile matrix has a long median process like that found in dogs. Boas writes, "I must admit that I was very surprised by the great similarity between saurian and mammalian claws. I had begun my investigation expecting to encounter the pattern familiar from crocodile or tortoise claws. The difference is all the more surprising because the saurians show no other signs of being closely related to mammals. It would never occur to anyone that the saurians, of all the reptile species alive today, are the closest relatives of mammals" (see *Figure 73*).

As we have already seen (p. 27 above and *Figures 13*, p. 29 and *15,* p. 32.), in terms of their morphology, certain platyrrhines are unexpectedly similar to humans (more so than are anthropoid apes), although from the phylogenetic perspective they are quite distant relatives of humans (see *Figure 12*, p. 21). The reason for this is that the human body remains general and unspecialized. Whenever an animal species retains any less specialized characteristics, its appearance is more humanlike, regardless of the degree of relationship. The same principle applies here on a larger scale. Human beings are mammals. Whenever the claws of a reptile fail to become specialized—that is, when they remain fetalized—they more closely resemble the claws of mammals, regardless of the degree of relationship. Fetalization in this case means earlier cessation of the growth of the rudimentary fertile matrix, which develops in a proximodistal direction (that is, away from the torso) during the fetal period.

In the ascending evolutionary sequence of primates, the nails show an increasing tendency to remain fetalized. Although in apes most fingers and toes still have claws (although the big toe always has a nail), the regression is obvious in comparison to typical mammalian claws. In platyrrhines, the shape of the nail is narrower and more compressed on the sides than it is in Old World catarrhines. This intermediary form between a claw and a nail is called a tegula (Bruhns, 1910). Old World primates, and anthropoid apes in particular, have the flattest nails, although chimpanzee nails are still significantly different from human nails (Vigener, 1896). The less pronounced the development of the claw, the less the subunguis develops, leaving more of the fleshy pad of the fingertip free—an important factor in the precision grip, for example.

The transition from claws to nails, therefore, is a good example of the connection that Goethe observed between fetalization ("overshadowing of the animal element"—in this case the claws) and the expression of humanlike potentials that lie dormant in the unspecialized pattern of vertebrate development ("enhancing the animal element for higher purposes"—in this case the precision grip).

6

UPRIGHTNESS

Bipedalism

Unlike apes, human beings walk upright (see *Figure 74*). Upright walking is prefigured even at a very early stage of evolution. Schultz points out a pronounced tendency toward bipedalism in primates (1950a). Many prosimians are already able to run on their hind legs with neither hands nor tails touching the ground. Most lower catarrhines can *stand* upright, but anthropoid apes can also *walk* upright, although their attempts are short-lived and fairly clumsy, with knees bent and hands held ready to provide additional support. Some individual anthropoid apes, however, are capable of impressive accomplishments with regard to upright walking. The weight of the body rests primarily on the forelimbs in most mammals but on the hind limbs in primates (see *chart*, page 146).

In primates, shifting the center of gravity toward the rear prefigures the freeing of the forelimbs that is realized in humans. In spider monkeys (*Ateles*), which have several humanlike traits (see Chapter 1), this shift is especially pronounced.

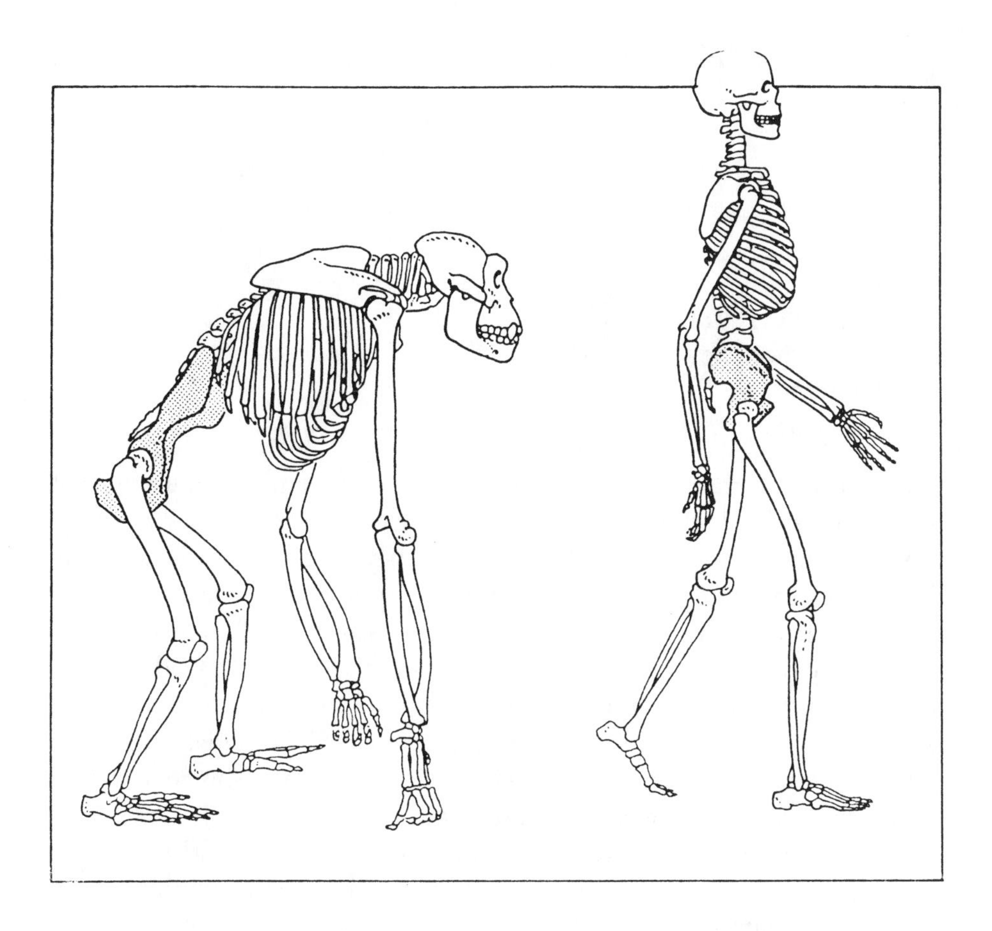

Fig. 74: Skeleton of: gorilla (left), human (right), in typical walking postures. The gorilla walks on all fours. Its hands touch the ground with the second, specially adapted phalanx of each digit ("knuckle walking"). The gorilla skeleton differs from the human skeleton in several ways. Its arms are longer and its legs shorter (the intermembral index—the ratio of the length of the forelimbs to that of the hind limbs—is much greater in anthropoid apes than in humans); the cervical vertebrae have large processes (insertion sites for powerful neck muscles); the skull is carried in front of the body rather than balanced on top of the spine; the ribcage is suspended below the spine; the orientation of the shoulder blades is horizontal; the pelvic bones almost touch the ribs; the sacrum tilts less to the rear; the knees are bent; the big toes are short.

Percentage of body weight carried by:

	Forelegs	*Hind legs*
dog	63	37
horse	55	45
cow	56	44
camel	66	34
elephant	55	45
spider monkey	29	71
baboon	44	56
macaque	46	54

The above chart (Martin, 1990) shows a gradual shift of the center of gravity toward the hind limbs. This shift should be interpreted as a retardation phenomenon. Consider, for example, a newborn rat, which is initially barely able to move at all. It first learns to move its head; later, the head and torso begin to move relative to the pelvis and the still immobile hind limbs. Later still, the young rat learns to "pirouette" its entire body around one hind paw that is still immobile (Golani, 1992). As this example shows, a young mammal's ability to move follows a cephalocaudal pattern of development. The point above which movement occurs (shoulder, pelvis, hind paw) gradually moves downward, with the parts of the body that are closer to the head becoming mobile first. This cephalocaudal pattern of development is very common and can be observed even in amphibians.

Young mammals, such as puppies at play, have a strong tendency to rest their body weight on their hind limbs. (Golani gives several astonishing examples.) Older animals, however, again reveal the opposite tendency, transferring most of their weight to their forelimbs. At this stage, therefore, the original cephalocaudal pattern is reversed.

This reversal does not occur in humans. Thom (1992) comments, "It has often been stated that 'hominization' of primates occurred through a neoteny process, a return to an ontogenetically primitive state. This fits astonishingly well with Golani's observation. It seems that bipedality in man occurred through a restriction of the hind legs to their basic function, that of carrying." Human beings, therefore, do not undergo the ontogenetic reversal that we can observe in quadrupeds. In this respect, upright posture is a Type I characteristic.

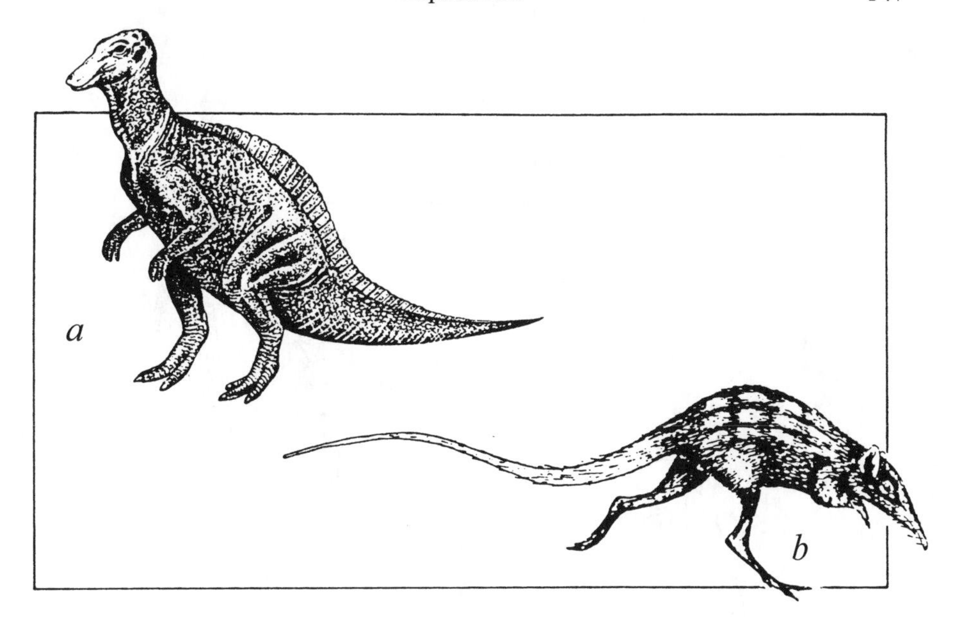

Fig. 75: Hypothetical reconstruction of prehistoric examples of bipedalism.

(a): Ouranosaurus*, one of the many dinosaurs that walked on two legs. Fossil remains of this species were discovered in Africa. Length ca. 7 meters (Charig, 1985).*

(b): Leptictidium nasutum*, a mammalian insectivore that lived in Europe during the early Tertiary period. Length ca. 75 cm (Benton, 1991). In 1986, this species was introduced in the press as a mammal that "walked like a human" as early as fifty million years ago (*New Scientist*, 24 July, 1986). Basically, however, the species has nothing in common with humans other than bipedalism.*

The original tendency (shifting the point of reference for body movement in the direction of the hind limbs), which is realized to a considerable extent in primates and less so in other mammals, is taken to its logical conclusion in humans. In this respect, upright posture is an enhancement of general mammalian development and therefore a Type II characteristic.

Bipedalism of sorts is found not only among primates but also in various other animals, such as birds, kangaroos, some rodents, and extinct theropod dinosaurs (see *Figure 75*). And anthropoid apes, even when they walk on all fours, are fairly upright because their arms are relatively long (*Figure 74*).

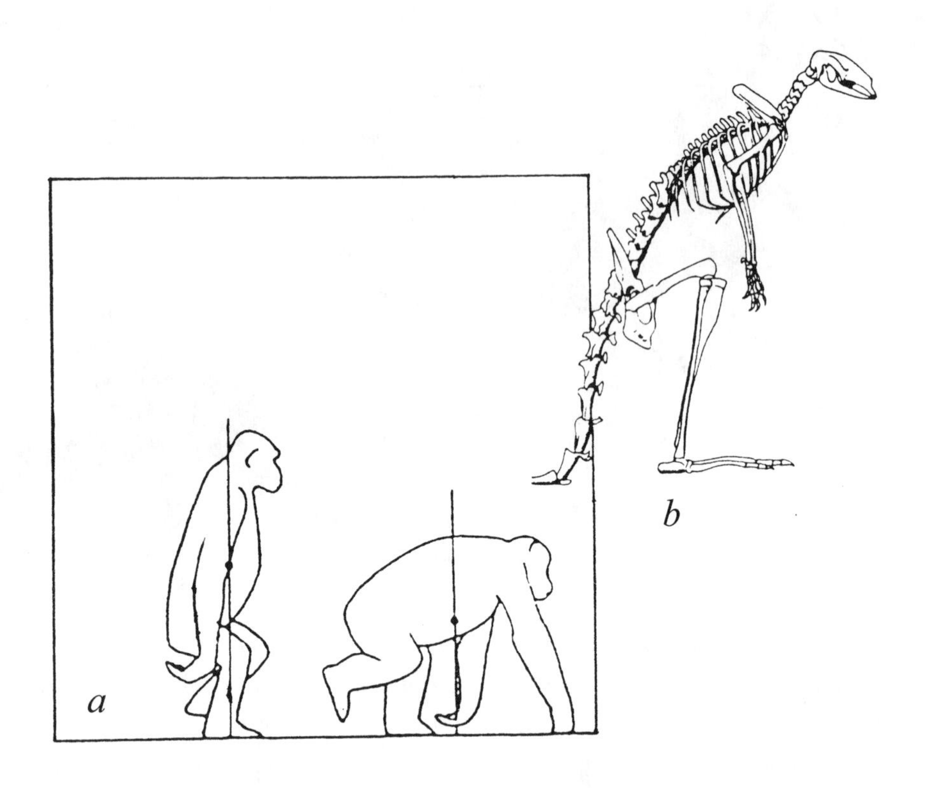

Fig. 76: Animal bipedalism.

When a chimpanzee walks upright (a), it experiences quite strong torques about the most important joints—the neck, the axis of the hips, and the knees—because its center of gravity is not in a line with these joints. Even when a chimp walks on all fours, the head still causes a significant torque about the atlas joint (Robinson et al., 1972).

In a kangaroo, as represented in the skeleton (b), the mechanics of balance are not at all the same as in human upright posture (cf. Figure 77*).*

Bipedalism, however, is not the only factor in human upright walking. As we will see, the unique biomechanical equilibrium that characterizes human body posture is not even approximated by any other species. The forces that human beings must overcome in order to walk upright are minimal, and staying upright takes little or no effort. A person who is standing consumes only seven percent more oxygen than someone who is lying down. In quadrupeds such as dogs, the difference in oxygen consumption between lying and standing is much greater because constant muscle exertion is required to keep the neck, head, and bent limbs extended, since in most standing mammals the limbs are not straight (Abitbol, 1988). Thinking in terms of natural selection and adaptation, we would have to conclude that "adult humans are better adapted to upright posture than dogs are to walking on all fours" (Abitbol, 1988). This conclusion, however, is almost impossible to reconcile with the idea that upright walking is a relatively recent variation on the ancient four-legged body plan. Quite the contrary, the extreme energy advantages of completely upright human posture support the view that the human body represents the original vertebrate structural pattern. All animal species have modified this pattern to meet specialized goals; invariably, the consequence has been drastically reduced efficiency. Human upright posture is fundamentally different from the body postures of higher animals in that it is freed from the constraints of gravity (see *Figure 76*).

The process of learning to stand is also different in humans than it is in animals. In animals, the ability to walk is inborn and often well developed at birth. This is not true of human beings.

> The mastering of the erect bipedal type of locomotion is a relatively prolonged affair and appears to be a learned process, not the result of inborn reflexes. The congenitally blind child never spontaneously attempts to stand or walk but must be carefully taught. No one can watch the struggles of an infant as it first attempts to stand, holding onto the edge of a chair or tightly grasping in its hands the supporting fingers of a parent, without feeling that this is pure experimentation rather than the maturation of an inborn reflex (Inman, Ralston, and Todd, 1981).

The same authors also cite findings that suggest that " . . . the characteristic patterns of walking seen in the adult are not achieved until the child reaches the age of 7-9 years. Apparently, before this age, the child is experimenting with the neuromusculoskeletal system, modifying the displacements that occur in various segments with changes in bodily proportions, and developing improved neural controls."

In other words, human beings acquire uprightness throughout the first seven years of life and do so through imitation rather than instinct.

Balance in Human Posture

The unique aspects of human posture were the subject of detailed research as early as the nineteenth century. In 1890 Braune and Fischer published a very thorough study (conducted for military purposes) in which they attempted to determine the exact locations of both the entire body's center of gravity and the subordinate centers of gravity of separate body parts (head, torso, upper arm, forearm, hand, thigh, lower leg, foot). Their work, referred to below, produced some noteworthy results.

First, Braune and Fischer established quite exact relationships between the locations of certain joints and specific subordinate centers of gravity. The center of gravity of the upper arm, for example, always lies exactly on the line connecting the shoulder and elbow joints, while that of the lower leg lies on the line between the knee and the ankle joint, and so on. The center of gravity of the torso (not including limbs and head) lies on the line connecting the atlas joint (on which the head rests) with the center of the axis of the hip joints.

When a person stands in a position in which the shoulder, elbow, wrist, atlas, hip, knee, and ankle joints all lie in a single plane (the "normal plane"), all subordinate centers of gravity except those of the feet also lie in the same plane. The researchers discovered that the resulting posture could be considered a natural position, which they called the "normal position" (see *Figure 77*).

In the normal position, the center of gravity of the entire body (excluding the feet) lies in the normal plane which runs through the heels. The centers of gravity of the feet lie a few centimeters in front of the normal plane, which of course has no bearing on the balance of the body as a whole, since the feet rest directly on the ground. All of the other joints and subordinate centers of gravity mentioned above fall in the normal plane, which also runs through the body's overall center of gravity and is perpendicular to the line of vision.

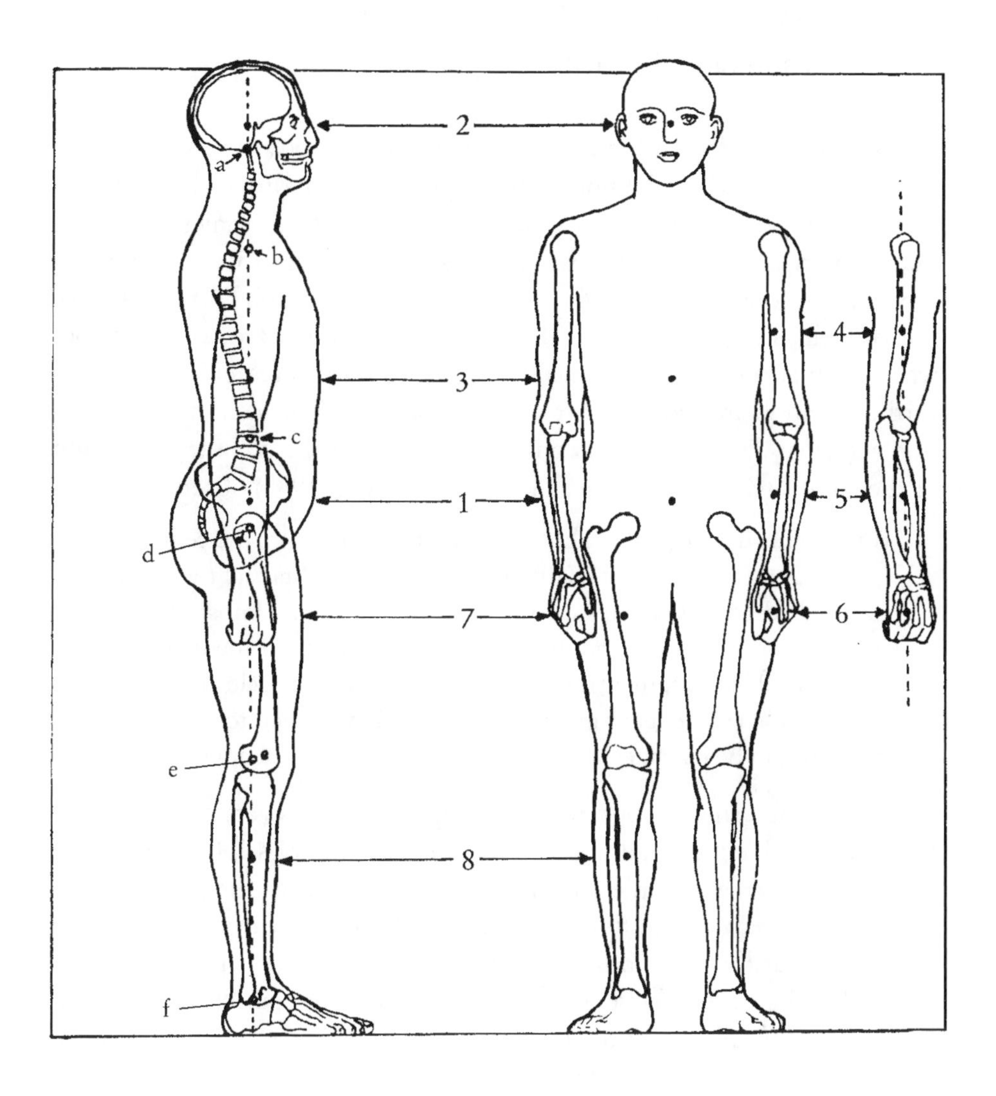

Fig. 77: The normal position. The broken line indicates the normal plane, which runs through the body's center of gravity (1). All of the important subordinate centers of gravity also lie in this plane: head (2), torso (3); upper arm (4), forearm (5), hand (6), thigh (7), lower leg (8). The major joints also lie in this plane: atlas (a), shoulder (b), elbow (c), hip (d), knee (e), ankle (f) (Braune and Fischer, 1890).

The Torso's Center of Gravity

Because its center of gravity is located directly above the axis running through the hip joints, the human torso produces no torque about the hip axis; that is, in the normal position, no force is exerted (unless it is produced by muscle action) that would cause the torso to rotate around the hip joints. Consequently, only minimal muscle exertion is required for human beings to stand upright. If the torso is tilted forward, as it is in apes walking on two legs, a great deal of muscle exertion is required to prevent further rotation around the hip axis (see *Figure 78*). The semi-upright position is a great disadvantage in comparison to either the completely upright or the completely quadrupedal stance. H. Helmuth (1985) observes:

> In the fully upright human body in anatomical position, the center of gravity is located in front of the second sacral vertebra and the gravitational force acts downward through the hip joint, almost directly through the knee joint and through the foot in front of the tibio-tarsal joint. Thus, moment arms are minimal, and muscle forces to counter the moments of gravity are also minimal. Therefore, upright posture can be considered a successful energy-saving adaptation.

The same author also states that upright posture is very advantageous for carrying objects. In addition to making carrying possible because the hands remain free, complete uprightness also produces the mechanically most advantageous conditions for carrying. When we carry something in our hands, our torso inclines so that the combined center of gravity of the torso plus the object remains directly above the hip axis. In the upright position, the torque about the hip axis remains almost zero even when we are carrying something. In contrast, holding something in any other position almost invariably involves increased torque.

Helmuth eliminates the possibility that human uprightness developed gradually in the course of evolution (as neo-Darwinists continue to suggest; see *Figure 78*). No paleontological finds indicate that such transitional forms ever existed. Purely from the perspective of energy expenditure, the semi-upright posture itself is a disadvantage, and carrying something in this position is even more problematic. Hence, although the hands are freed from bearing the body's weight, they are not truly functional.

Fig. 78: A frequently encountered illustration of human evolution inserts a semi-upright hominid between the hypothetical quadrupedal ancestor of human beings (left) and the upright Homo erectus *(right) (Washburn, 1978). Such an intermediary form, however, is very unlikely for biomechanical reasons, and paleontologists have never proved that it existed.*

All accumulated evidence from biomechanical, medical and physiological investigations points convincingly to the fact that a forward flexed posture and the lifting of weights in this position can be disadvantageous or even dangerous, because it places a high and continuous stress on the intervertebral discs, the vertebrae, the muscles and the ligaments alike (Helmuth, 1985).

Knees and Feet

In the normal position, the line connecting the hip joints cuts the verticals through the knee joints. In the normal, upright position in humans, there is no muscle activity at all around the straightened knee joint (Carlsöö, 1972). When the knee is straightened, slight rotation of the tibia and femur "locks" the knee and prevents buckling. (This does not mean, however, that an upright stance is not possible with the knees active and slightly bent; everyone is familiar with the difference between standing at ease and standing at attention.) In contrast, apes run with bent knees, a posture that requires constant muscle exertion to prevent buckling (Campbell, 1974). In the upright stance, maintaining the shape of the arch of the human foot also does not require muscle activity (Aiello and Dean, 1990).

The Resting Position

Although the normal position is a natural posture, it is somewhat different from relaxed upright posture, which Braune and Fischer call the "resting position" (*Figure 79*; see also Woodhull et al., 1985). In the resting position, the lower leg tilts forward slightly (approximately five degrees), positioning the center of gravity above a spot closer to the center of the supporting surface formed by the feet. As a consequence, the calf muscles must do a little work to counteract the torque about the ankle joint. The center of gravity of the body as a whole (with the exception of the lower leg) lies slightly in front of the knee joint; the knee joint is effectively stabilized since tipping forward around this joint is impossible. The hip joint is also stabilized because the ribcage and head are tilted back slightly in relationship to the hip area. The resting position is a variation of the normal position in which the hips and knees are stabilized and the body's center of gravity lies above the center of the supporting surface, an advantage that is achieved at the expense of a slight force exerted by the calf muscles.

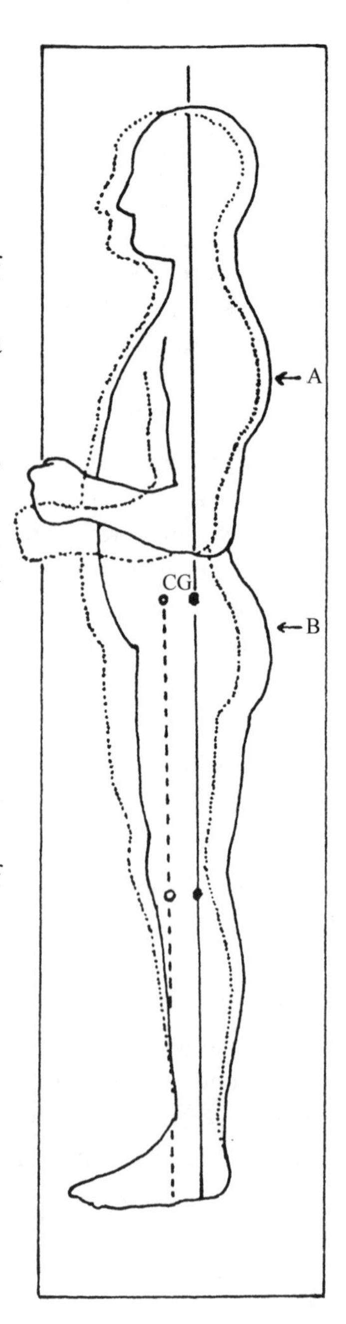

Fig. 79: The resting position. The normal position (solid line) and the resting position (broken line) are shown superimposed, along with the center of gravity (CG) and its vertical projection for both postures. In the resting position, the slight forward angle of the lower leg shifts the center of gravity forward a few centimeters; the knee joint remains directly below the overall center of gravity. In the resting position, the chest (A) shifts slightly backward in relation to the hips (B) (Braune and Fischer, 1890).

Upright Walking

Strictly speaking, upright posture is not static. Even when standing still, most people unconsciously shift their weight from one leg to the other, moving from side to side six to eight times a minute in the sagittal plane that divides the body into right and left halves. These motions, which may achieve amplitudes of several centimeters, are broken into eight to ten tiny, jerky movements per second. As a result of these movements and the activity of various organs such as the heart, lungs, and so forth, the body's center of gravity is in constant motion. Nonetheless, it remains within the same small area of the hips, at a height above ground of approximately 55 to 57 percent of total body height (Carrier, 1984).

During walking, of course, the center of gravity moves even more. In relationship to the ground, it traces a lemniscate-like curve within the frontal plane, which divides the body into front and back halves and is perpendicular to both the sagittal plane and level ground (*Figure 80*). Even then, however, the center of gravity remains in almost the same place relative to the body as a whole (Inman, Ralston, and Todd, 1981).

From an overall perspective, we can say that the normal position is the natural postural point of reference from which the typical motions of standing and walking are derived.

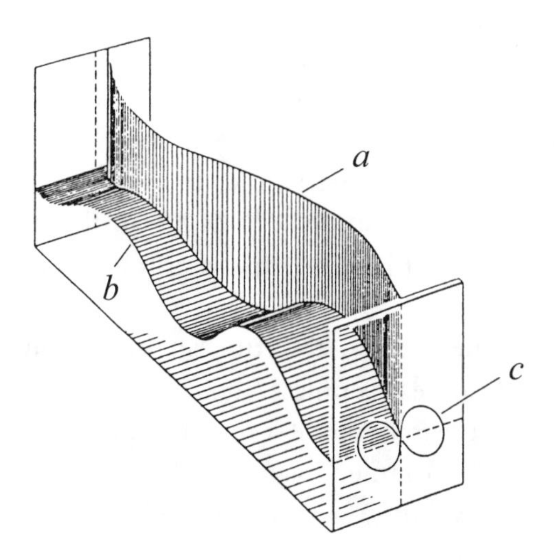

Fig. 80: Schematic representation of the movement of the human body's center of gravity in the course of one step cycle. Side-to-side motion (in the horizontal plane) is indicated by line (a), vertical motion by line (b). As a result of the combination of the two wavelike segments (a) and (b), the center of gravity moves along the lemniscate-like loop (c) within the frontal plane, which is formed by the body and stands perpendicular to the direction of walking (Inman, Ralston, and Todd, 1981).

Upright Posture and the Head

The human head is fundamentally different from that of animals in that it is balanced above the torso. All mammals carry their heads in front of their necks, but in humans the position of the head is essentially in balance.

The atlas joint—the atlas is the uppermost vertebra—connects the skull to the spine. On the skull side, this joint consists of the occipital condyles, two processes located on either side of the foramen magnum (see *Figure 81*). When our line of sight is nearly horizontal (as it is when we look to the horizon, for example), the head is in a labile state of balance and the neck muscles are relaxed. Schultz (1941) observes that the head of a newborn orangutan is nearly as well balanced as that of a human infant. In humans, the head's balance, if it changes at all as growth continues, is merely refined, but in adult anthropoid apes the skull shifts completely out of balance, as evidenced by their powerful neck muscles (*Figure 82*). Many animals also have processes on the vertebrae and on the back of the head (the so-called occipital crest) that serve as insertion sites for the neck muscles. Human neck muscles are relatively insignificant.

The central position of the occipital condyles in humans must be considered a Type I characteristic. In apes, this trait is initially present but later disappears (see *Figure 81*). As we saw earlier, the flat human facial skull is also a Type I characteristic since the typical ape muzzle develops only later (see *Figure 2*, p. 3). On the whole, therefore, the head's position of balance above the torso can be seen as a trait that is prefigured in earlier stages of evolution but becomes meaningful only in the context of human uprightness.

Certain monkeys, such as macaque species, are primarily bipedal in the early juvenile stage, a trait that disappears completely in adolescence (Michejda and Lamey, 1971). Some researchers see a direct connection between this behavioral change and the foramen magnum's shift toward the rear of the head in the course of growth (Masters et al., 1991).

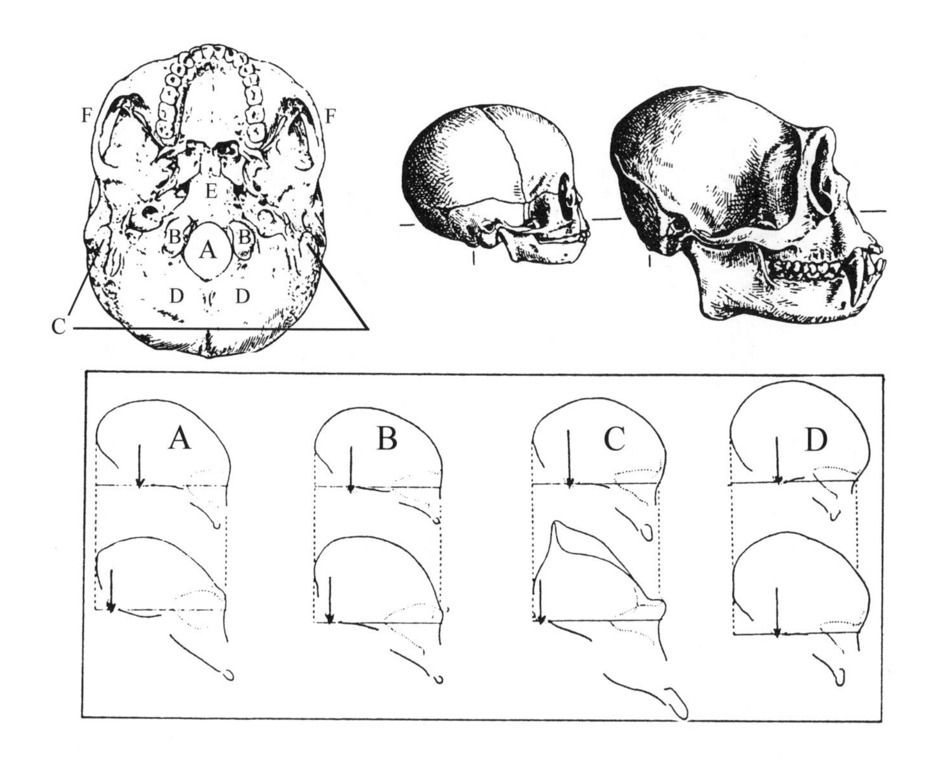

Fig. 81: The balance of the human head (I):

Above left: the underside of the human skull. (A) foramen magnum; (B) occipital condyles; (C) mastoid process; (D) occipital bone; (E) the diagonal "pars basilaris" of the occipital bone; (F) zygomatic bone (cheek bone).

Above right: Skulls of a very young gibbon and an adult specimen, drawn to the same scale. The little vertical lines indicate the position of the occipital condyles, the head's connection to the atlas joint.

Below: The position of the occipital condyles (arrow) and the eye sockets (broken line) in the skulls of juveniles (upper row) and adults (lower row): (A) macaque; (B) gibbon; (C) gorilla; (D) human (Schultz, 1949a, 1968). All skulls are shown tilted forward and without the lower jaw. In the adult ape specimens, the occipital condyles have shifted to the rear, while the adult human skull retains the configuration common to juvenile primates. Schultz (1955) repeatedly expresses his astonishment at this Bolkian idiosyncrasy: "The essential question must still remain unanswered: Why has the joint between the head and spine come to lie so far forward early in development in simian primates, if this specialization is not retained by the great majority of species and is of clear advantage for only one species—the human?"

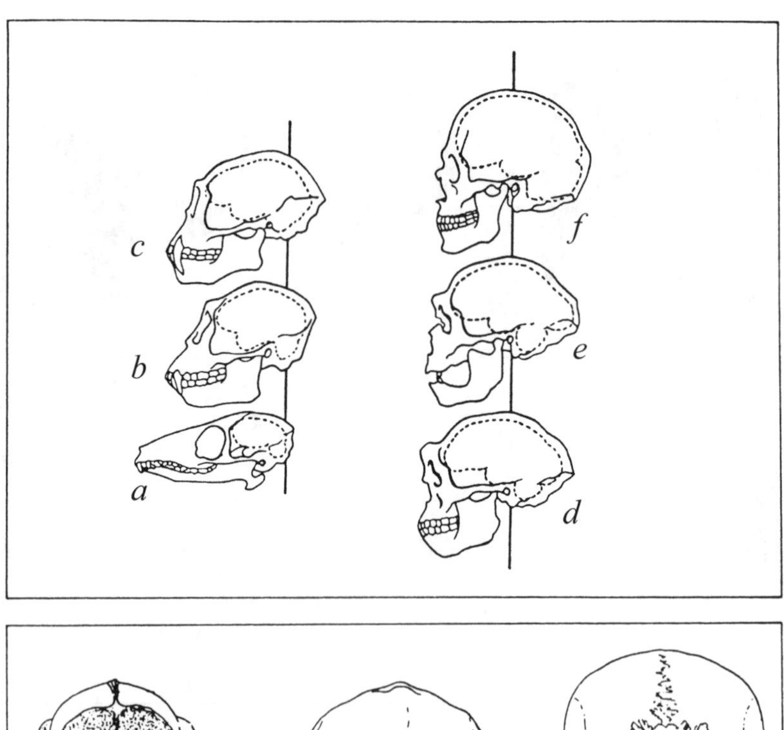

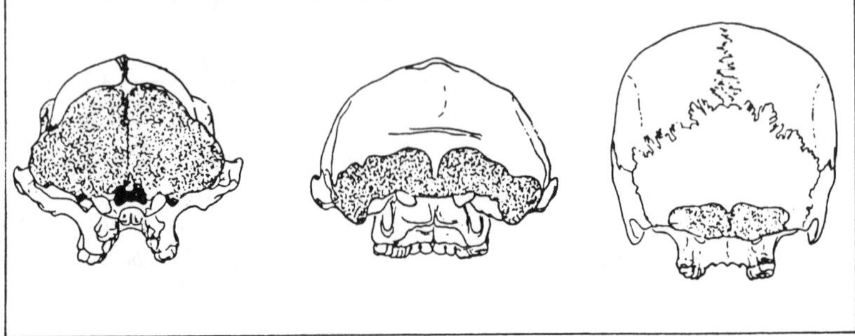

Fig. 82: The balance of the human head (II):

Above: Skulls of: (a) Tupaia *(often considered the nearest relative of the primates); (b)* Cercopithecus*, a catarrhine; (c) gibbon; (d)* Homo erectus*; (e)* Homo sapiens neanderthalensis*; (f) modern human. The vertical line indicates the location of the occipital condyles. In the phylogenetic series of modern primates and fossil hominids, the skull tends to become increasingly balanced.*

Below: Rear view of the skulls of (left to right): gorilla, Homo erectus*, modern human. Gray shading indicates the insertion area of the neck muscles. In the gorilla, this area is very extensive and bounded by the occipital crest, which provides additional attachment possibilities for the gorilla's powerful neck muscles. In the modern human, this area is minimal because the neck muscles no longer support the head (Campbell, 1974).*

The Mastoid Process

Behind the earlobe on either side of the human skull is a remarkable protuberance called the mastoid process. This process is undeveloped in newborns. It is a human characteristic that emerges late in both ontogeny and phylogeny. Even in prehistoric hominids, the mastoid process is usually relatively undeveloped. But as Schultz (1949a) points out, it is also present in anthropoid apes. In humans, the mastoid processes develop quite quickly after birth and are therefore present by the time uprightness is achieved, but in apes (especially chimpanzees and gorillas), they appear only in the course of adulthood and remain underdeveloped in many individual animals (Schultz, 1952).

The mastoid process is the insertion site for a specific muscle, the sternocleidomastoid, which connects the mastoid process with the breastbone and the collarbone (*Figure 83*). It plays a role in various head movements. Contraction of the muscle on one side causes the head to turn in the other direction. Simultaneous contraction of the muscles on both sides of the head tilts the head forward. When the head is tilted backward, contracting these muscles causes the head to return to the upright position. This last function is made possible, or at least easier, by the fact that the muscle is attached to the protruding mastoid process. The process functions as an arm of force around the rotational axis that runs through the occipital condyles (Krantz, 1963); see *Figure 83*.

The ability to return the head from a backward-tilted to an upright position is crucial in humans. In adult anthropoid apes, however, the skull is so heavily weighted toward the front that this issue does not arise, and the mastoid process serves no purpose, as is evident from the delayed appearance (or, in some individuals, total absence) of this feature. The skulls of prehistoric hominids were also heavily weighted toward the front (see *Figure 82*). Hence, the mastoid process is a Type II characteristic. It is prefigured in higher animals but becomes meaningful only in the context of the human constitution.

The Location of the Organs of Balance

When a human being stands in the normal position, the vertical projection of the body's center of gravity runs through both the head's center of gravity and the atlas joint and exactly bisects the line connecting the two sets of organs of balance in the inner ears. Because our feet do not move smoothly over the ground when we walk, the upright body may tilt either forward or backward, resulting in rotation around the fulcrum of the foot joint. Such rotation causes the greatest displacement (and is therefore most perceptible) at head height. Thus the location of the organs of balance in humans—at the top of the body and close to the vertical projection of the center of gravity—is optimal.

It is worth noting that in quadrupeds, the location of the organs of balance—at the extreme end of the body and far from the center of gravity—is not logical in this respect. This location makes sense only in an upright body. Hence, its occurrence in animals can be seen as prefiguring uprightness.

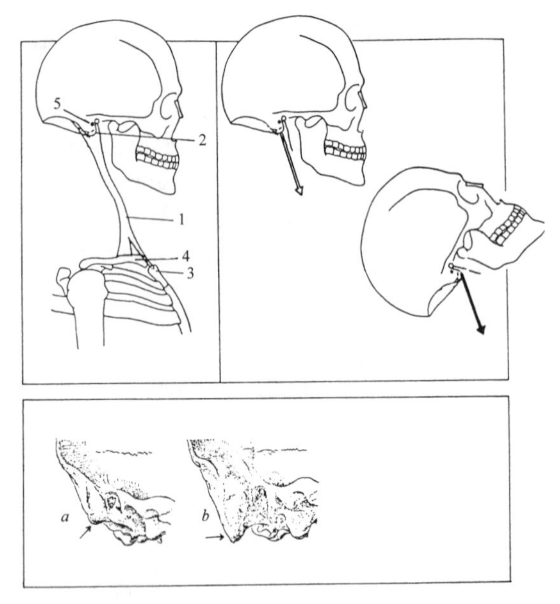

Fig. 83: The mastoid process.

Above, left: The sternocleidomastoid muscle (1) connects the mastoid process (2) with the breastbone (3) and the collar bone (4). The dot (5) indicates the position of the axis of rotation through the joint between the atlas (the uppermost cervical vertebra) and the occipital condyles.

Above, right: The direction of pull of the sternocleidomastoid muscle when the head is in the normal upright position and tilted backward. In the latter instance, contraction of this muscle returns the head to the normal position because the mastoid process functions as the short arm of a lever (Krantz, 1963).

Below: The mastoid process in adult male gorillas. (a) Specimen with a weakly developed process; (b) specimen with a highly developed process (Schultz, 1950b).

The Effect of Upright Walking on the Birth Canal

In humans, the sacrum tilts backward relative to the lumbar vertebrae. The so-called lumbosacral angle (see *Figure 84a*) is a quantitative expression of the degree of tilt. In humans, this angle is normally at least 70° (Abitbol, 1987a). It is consistently somewhat larger in women than in men.

The lumbosacral angle is only 20° in newborns and achieves its final size in five-year-olds. This increase seems to be biomechanically related to the acquisition of uprightness during the first few years of life. In children who learn to walk early, the lumbosacral angle increases rapidly, but in those who learn to walk exceptionally late, the skeleton seems to lose some of its flexibility and the lumbosacral angle remains smaller for life.

In apes, too, the lumbosacral angle increases if upright walking is practiced from an early age (Preuschoft et al., 1988). Even more interesting, perhaps, is Slijper's 1942 study of a goat, born with only stubs of forelimbs, that learned to walk on its hind legs in a semi-upright position (see *Figure 43*, p. 92 above). Its lumbosacral angle was considerably enlarged. Slijper also points out that the lumbosacral angle is extremely wide (almost 56°) in bears. Certain other relatively unspecialized traits in bears, such as a flattened thorax and plantigrade feet, favor sole walking and make upright, two-legged walking easier for them than for most other mammals.

The wide lumbosacral angle in humans is often considered a specialization. Essentially, however, and from a purely physical perspective, the characteristically human trait is not a wide lumbosacral angle as such but simply the fact that humans are born with a flexible pelvis that permits a wide lumbosacral angle to develop as uprightness is acquired. Human uprightness, however, is not inborn but is achieved through individual effort. The potential to develop a wide lumbosacral angle through uprightness also exists in apes and other mammals, as we have seen, but reveals its full significance only in humans.

In fact, the wide lumbosacral angle in humans serves two purposes. On the one hand, it is a biomechanical consequence of uprightness. On the other hand, giving birth to a large-headed baby is possible only if the distance between the pubic bone and the sacrum is great enough (see *Figure 84b*).

The lumbosacral angle is invariably narrower in nonprimates than in humans. In dogs, for example, it varies between 4° and 14°. It is larger in the lower primates, ranging from 20° to 35° in rhesus monkeys, as Abitbol (1987a), among others, discovered. The widest lumbosacral angles are achieved in anthropoid apes—44° in chimpanzees and 60° in gibbons, who display a pronounced talent for bipedalism. The very wide lumbosacral angle (40° or more) of certain baboon species, on the other hand, is probably directly related to the difficulty of giving birth to their relatively large young (Schultz, 1961).

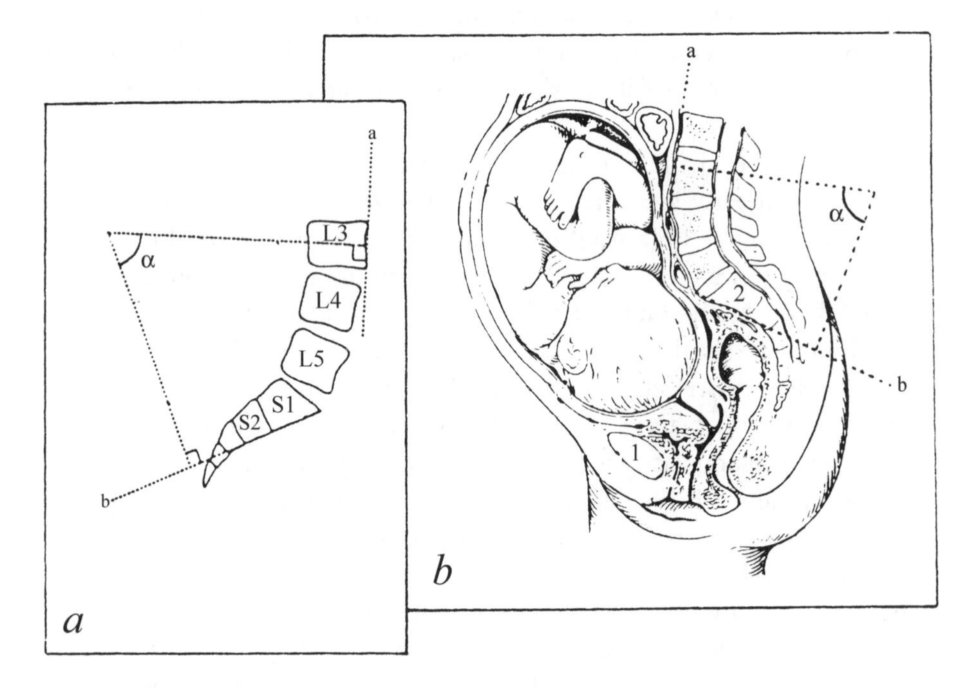

Fig. 84: The lumbosacral angle.

(a): The lumbar line (a) is drawn along the front of the third lumbar vertebra, (L3). The sacral line (b) is drawn along the front of the first two sacral vertebrae, (S1) and (S2). The angle (α) between two lines drawn perpendicular to (a) and (b) is the lumbosacral angle. The wider this angle, the more the sacrum tilts backward.

(b): The strong backward tilt of the sacrum makes human birth possible. The birth canal runs between the pelvic bone (1) and the sacrum (2).

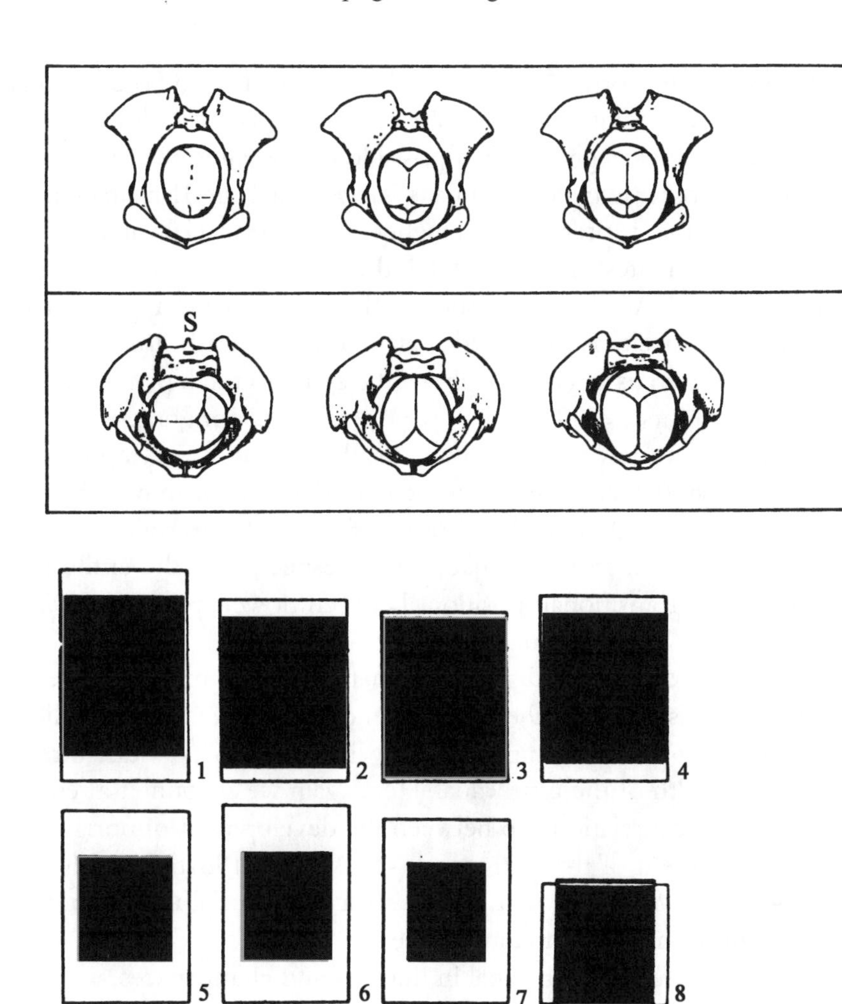

Fig. 85: Above: Passage of the head of the fetus through the birth canal in a chimpanzee (upper row) and in a human (lower row). In anthropoid apes, the birth canal is very roomy, but human birth requires different orientations of the baby's head at the beginning, middle, and end of the birth canal (shown from left to right) (Lovejoy, 1988). Note that the greater width of the human sacrum (S) plays an important part in this process.

*Below: Schematic representation of the relative sizes of the birth canal (white rectangle) and the head of the newborn (black rectangle) in: (1) spider monkey (*Ateles*), (2) proboscis monkey, (3) macaque, (4) gibbon, (5) orangutan, (6) chimpanzee, (7) gorilla, (8) human. The horizontal dimension of the rectangles represents the width and the vertical dimension the depth of the birth canal and the head. In anthropoid apes, the birth canal is abnormally large in comparison to the relatively small size of their offspring (Schultz, 1949b).*

It is especially worth noting that although a wider lumbosacral angle is achieved by acquiring uprightness, it also serves a totally different purpose, namely, it makes giving birth possible by widening the birth canal. The wide lumbosacral angle in humans is clearly a Type II characteristic. It is a tendency that is prefigured in higher animals (especially in primates) but achieves its greatest development and reveals all of its functions only in human beings. As *Figure 85* shows, the lumbosacral angle in large anthropoid ape species is much wider than necessary for birth.

The sacrum transfers the weight of the spine to the hipbones via the two sacroiliac joints (see *Figure 14*, p. 31). Human uprightness places a greater burden on these joints. In humans, the sacrum is correspondingly broad and heavy (which as such can be considered an example of hypermorphosis, see p. 170), and the upper surface of the sacroiliac joints is very large. Schultz discovered that in this respect, too, the anthropoid apes occupy a transitional position between lower primates (whose sacrums are narrow) and humans.

Although the great width of the human sacrum is directly related to uprightness, it also has another consequence. Since the sacrum is like a wedge separating the two hipbones, any increase in its width directly increases the width of the birth canal. Here again we encounter a remarkable biomechanical relationship between the development of uprightness and the enlargement of the birth canal (see Abitbol, 1987b for additional examples). *Figure 85* shows that a wide sacrum is crucial in human birth but irrelevant to the birth of chimpanzees.

If we compare the birth canal in humans and chimpanzees, we make another surprising discovery. In humans, the pelvic passage in women is wider in comparison to the total width of the pelvis than it is in men. This gender difference appears only at puberty, and the same phenomenon can be observed in chimpanzees. But while widening the birth canal makes sense in human females because of the large heads of human infants, the increase seems totally superfluous in female chimpanzees. This increase is another example of a Type II characteristic in that it is prefigured in higher animals but reveals its significance only in humans.

This example illustrates how a trait that appears totally functional may not come about in order to serve that particular function. If we consider the widening of the human birth canal in isolation, it seems to be a convincing example of Darwinian adaptation. But as Schultz (1949b) points out, a significant degree of pelvic widening also occurs in female anthropoid apes, where it is apparently totally superfluous, but not in gibbons, whose birth canal is quite narrow. If we look at the human species in association with anthropoid apes, the idea of functional widening of the birth canal becomes much less convincing. Schultz writes, "If it is assumed that the pelvic sex differences of primates are the necessary result of the disproportionately large size of the head of the newborn, the exceptions to this rule, found in the anthropoid apes, remain to be explained." Attempts have also been made to explain the enlargement of the birth canal of large anthropoid apes in terms of allometry; that is, as disproportional growth in comparison to the growth of the rest of the body (Rosenberg, 1992). This explanation, however, is not totally valid because large catarrhines such as baboons, in comparison to the smallest platyrrhines, show no enlargement of the birth canal.

The Vertebrae

The mammalian spine can be subdivided into five regions (see *Figures 86* and *87*):

- the cervical vertebrae (of which there are seven in humans and most mammals)
- the thoracic or chest vertebrae, which support the ribs
- the lumbar vertebrae
- the sacrum, which consists of a number of fused vertebrae
- the coccygeal vertebrae.

The number of vertebrae in these regions varies among primate species, which makes interspecies comparisons difficult. Lower primates often have three sacral vertebrae and a total of nineteen thoracic and lumbar vertebrae, while humans have five and seventeen, respectively. In more highly retarded primates, the lower lumbar vertebrae tend to "sacralize"; that is, to be integrated into the sacrum. This phenomenon is reminiscent of the caudocephalic pattern of sternum bone fusion and may be a manifestation of increasing opposition between the cephalocaudal and caudocephalic growth tendencies in the torso. It is quite possible that we must also look in this direction to explain the disappearance of the tail in the most highly retarded primates.

The ontogeny of the spinal column reveals a sequence of three general developmental tendencies.

- During the first phase, the thoracic region has a developmental "head start" on the cervical and lumbar regions. The development of the neural tube—and later of the vertebrae—begins in the middle and then moves in both directions, toward the head and toward the tail. This initial centrifugal developmental principle is evident in all vertebrates and is already very apparent in fish species. In fish larvae, the central region initially develops strongly, while the head and tail remain small. In adult fish, the heads and tails are much larger in comparison to the thoracic region, where the rate of growth is much slower (Fuiman, 1983). The same centrifugal tendency is also very apparent in the ossification pattern of the human spinal column (Kjær et al., 1993; O'Rahilly et al., 1980).

- This process is then reshaped by the cephalocaudal growth tendency described earlier in this book. Embryonic development reaches a stage in which the cervical vertebrae are much larger than the more caudally located vertebrae. The sacral vertebrae, the "latecomers" to this developmental trend, are initially relatively small (see *Figure 87*).

- In the third and final process, an antagonism of sorts emerges between the cephalocaudal growth tendency, which governs torso development, and the later-appearing caudocephalic or distoproximal growth tendency, which predominates in the development of the legs. These two trends struggle for supremacy in the pelvic region.

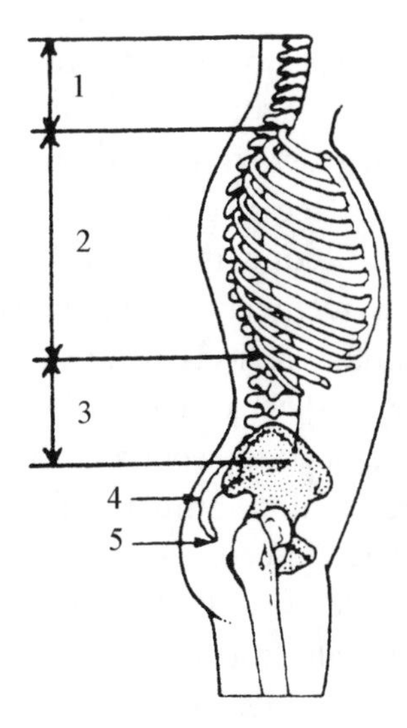

Fig. 86: The division of the spinal column: (1) cervical region, (2) thoracic region, (3) lumbar region, (4) sacrum, (5) coccyx (Aiello and Dean, 1990).

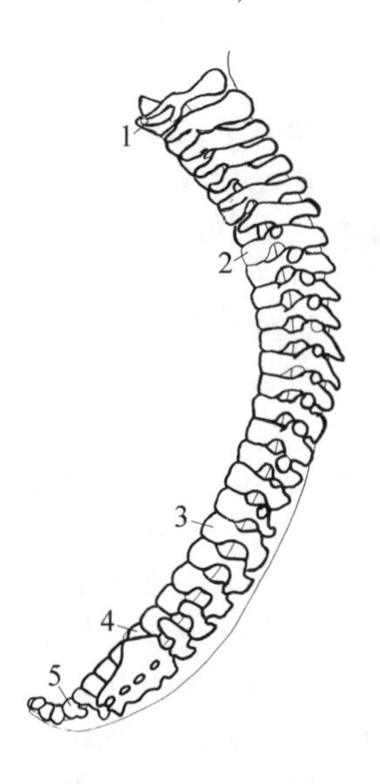

Fig. 87: The spinal column of a 27-mm-long human embryo (O'Rahilly et al., 1980). (1) first cervical vertebra, (2) first thoracic vertebra, (3) first lumbar vertebra, (4) first sacral vertebra (sacrum), (5) first coccygeal vertebra. At this stage of development the cervical vertebrae are very large; they have a developmental "head start" over the lumbar vertebrae and the sacrum, which are relatively small (cf. Fig. 86 *and* 88*).*

What should we expect to find in a retarded vertebrate? Let's recall the central principle: in an organism with retarded growth, organs or body parts that develop later grow larger and are better developed than parts that begin to grow earlier (see *Figures 25* and *26*, p. 62).

If taken to its logical conclusion, the first developmental thrust (see above) will produce a relatively short thoracic region and somewhat longer cervical and lumbar regions. And in fact, in comparison to apes, human beings do have long necks and well-developed lumbar vertebrae.

Retardation of the second, cephalocaudal developmental tendency will cause relatively greater development of the lumbar vertebrae, because the lumbar region is the "latecomer" to this developmental trend. This effect can be observed very well in adult humans. The more caudally located vertebrae, especially the lumbar vertebrae and the sacrum, are exceptionally well developed because they continue growing for a longer period of time. Schultz (1938b) notes that in humans the lumbar region (as a percentage of total torso length) continues to grow after birth, while in primates (such as macaques or chimpanzees, for example) the percentage remains constant.

The third tendency leads to suppression of the tail and more pronounced development of the lower (hind) limbs. In the human embryo, the coccyx is initially exposed but is later incorporated as the legs grow. Tracing this tendency in the evolutionary sequence of vertebrates, we note that fish have poorly developed limbs and powerful tails, while the tail is still highly developed in typical reptiles (such as lizards) but much less so in mammals. The tail of a rat grows slowly until after birth, when it grows much faster than the other parts of the spine, with the fastest rate of growth at the tip (Reiche and Schwarze, 1989). In monkeys, limb development is more pronounced than in other mammals, but postnatal growth of the tail is slower. As a result, young monkeys have disproportionally longer tails than adults do—a phenomenon that is especially pronounced in New World monkeys with prehensile tails (Schultz, 1926c, 1956, 1961). The tail is suppressed completely in some macaque species and in anthropoid apes, and anthropoid apes also have more highly developed lower extremities.

In a certain respect, therefore, tail development and hind limb development are incompatible. Since the legs are ontogenetic latecomers in comparison to the tail, they are more developed in more highly retarded species, while the tail is suppressed to a greater extent. This process reaches its climax in human beings. We might even say that in this respect the snake, whose limbs are completely undeveloped, is the opposite of the human being.

Figure 88 summarizes the three developmental tendencies.

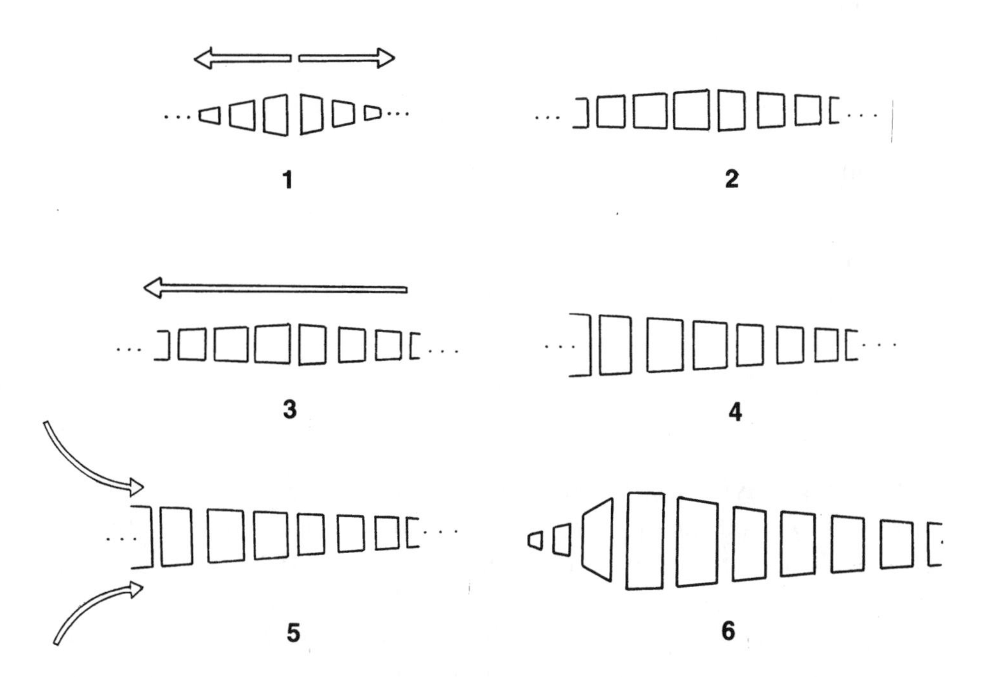

Fig. 88: Schematic representation of the three successive streams of development and growth in the spinal column. The head end is on the right and the tail end on the left.

(1) The initial centrifugal developmental stream begins in the chest and expands in the direction of the head and tail ends. (2) Retardation of this process increases the relative size or length of both the cervical vertebrae and the lumbar vertebrae and sacrum.

(3) The second, cephalocaudal growth stream begins at the head and later expands in the direction of the tail end. (4) Here, retardation increases the size of the lumbar vertebrae and sacrum.

(5) The third, distoproximal growth stream governs limb development. (6) Here, retardation increases the influence of this late-appearing developmental stream as compared to the two earlier streams. The result is suppression of the coccygeal vertebrae in favor of the hind limbs.

The Cervical Vertebrae

In comparison to anthropoid apes, adult humans have conspicuously long necks. The location of the human head above the torso permits it to turn freely, while the ape head is sunk between the shoulders, so to speak (see *Figure 2*, p. 3, and *Figure 30*, p. 70). The human cervical spine is proportionally longer than that of any other primate. Schulz (1949a) gives the following average values for the length of the cervical region as a percentage of torso length:

	Fetus	*Newborn*	*Adult*
capuchin	23.9	21.8	19.2
macaque	22.3	19.1	16.4
gibbon	30.0	26.1	22.3
orangutan	28.9	29.6	23.7
chimpanzee	26.9	25.8	22.5
human	29.9	28.0	25.9

A long cervical spine, however, is clearly no guarantee of an outwardly visible and mobile neck. Although the cervical vertebrae are proportionally longer in the human fetus than in the adult, the adult neck appears longer, as *Figure 30* clearly shows. This fact is related to the sinking of the shoulders in the course of growth, which makes the neck more prominent. This sinking also occurs in lower primates, but in anthropoid apes the opposite tendency may appear. Schultz (1930) points out that the neck of a young gorilla is quite visible because its shoulders are still low and its neck muscles poorly developed. In the adult gorilla, however, the neck disappears completely, surrounded by a forward-tilting head, high shoulders, and powerful neck and shoulder muscles. *Figure 2* (page 3) shows the same development in the chimpanzee.

These data indicate that a long cervical spine (relative to torso length) is a trait that appears as early as the fetal phase of development. The cervical spine is longer in more highly retarded species (lower primates have much shorter necks than anthropoid apes, which in turn have shorter necks than humans) because the initial centrifugal developmental stream produces a longer neck. In this respect, the long human neck is a Type II characteristic.

In fact, the great length of the human neck is directly related to uprightness. A vertically oriented neck exerts no torque about its base. In contrast, when the orientation of the neck is more or less horizontal, a pronounced torque occurs and strong neck and shoulder muscles are required in order to support the head. The adult gorilla, whose cervical spine is quite long, is a good example. The gorilla's cervical vertebrae, which are nearly horizontal in the walking position (see *Figure 63*, p. 123, and *Figure 74*, p. 145), have long processes that serve as insertion sites for very highly developed neck muscles (see *Figure 89*).

A long neck creates unique opportunities for humans. First, it provides space for the descent of the larynx (for details, see Chapter 10). It also permits great mobility of the head and thus also of the line of sight. In anthropoid apes, the cervical spine is also quite long, but head mobility is surprisingly limited (Schultz, 1961). High shoulders effectively limit lateral rotation in anthropoid apes, who cannot raise their heavy lower jaws above shoulder height. The same large, bottom-heavy muzzle also gets in the way of bending the head forward, and the processes of the cervical vertebrae prevent it from tilting back.

Extremely long necks with very heavy cervical vertebrae are also found in some mammals outside the primate group, notably horses and giraffes (see *Figures 55* and *56*, pages 115 and 116). There are also mammals, however—dolphins, for example—that have tiny, short cervical vertebrae (Delmas and Pineau, 1984). (Remarkably, almost all mammals have seven cervical vertebrae; this number seems to be part of the universal mammalian structure.) The exceptional size of the cervical vertebrae in many quadrupeds cannot be explained by the effects of the three growth tendencies discussed above (see *Figure 88*) and must therefore be considered a specialization (see *Figure 45*, p. 96). This specialization is related to the fact that in these animals, the neck functions as a limb to a certain extent. Because the forelimbs are devoted entirely to locomotion, the neck needs to be long enough to reach the ground or foliage hanging from a high limb. The disproportional size of the cervical vertebrae in quadrupeds demonstrates that their walking position, too, is secondary (i.e., a specialization). This means that the original, universal mammalian structure was not suited to quadrupedalism.

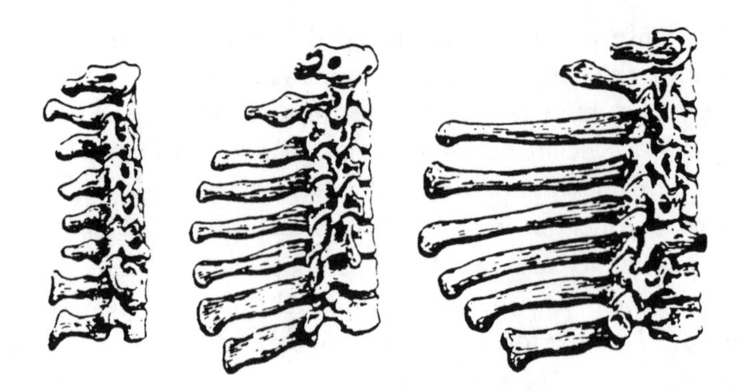

Fig. 89: Processes of the cervical vertebrae of: (left) gibbon, (center) orangutan, (right) gorilla. These processes, which are very large in the large anthropoid apes, serve as insertion sites for the neck muscles, which must be able to support the head as it is carried in front of the body (Schultz, 1961).

In contrast, we find that human vertebrae simply increase steadily in size from the head to the pelvis (see *Figure 90*) and that human vertebral proportions are very well explained by the action of the three successive growth streams, which can be seen as a universal tendency in vertebrates.

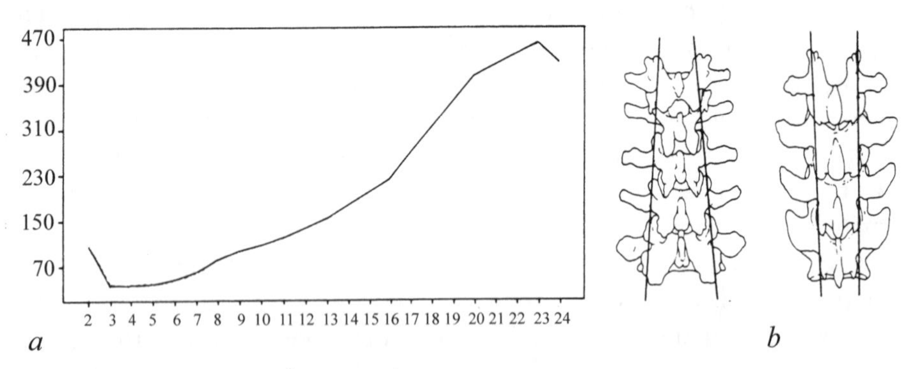

Fig. 90: (a): Volume in mm^3 of human presacral vertebrae as a function of position. The vertebrae are numbered from the neck to the coccyx (#1, the atlas, is not included). In humans, the volume increases systematically from the neck to the sacrum.

(b): The lumbar region in humans (left) and chimpanzees (right), as seen from behind. The lower lumbar vertebrae are wider than the upper in humans, but not in anthropoid apes (Aiello and Dean, 1990).

The Thickness of the Vertebrae

Thick vertebrae are a fetal characteristic; the spine is disproportionally thicker in young primates than in adults. When Schultz (1953) calculated the ratio of the smallest diameter of the middle thoracic vertebra or middle lumbar vertebra to the length of the torso he discovered systematically higher values in juveniles than in adults:

	Thoracic vertebra	*Lumbar vertebra*
gibbon (juvenile)	6.0	9.2
gibbon (adult)	3.9	6.5
orangutan (juvenile)	6.9	10.2
orangutan (adult)	5.0	6.7
chimpanzee (juvenile)	6.1	8.0
chimpanzee (adult)	4.4	6.5
human (juvenile)	6.3	9.4
human (adult)	5.3	8.0

(Average thickness of the middle thoracic and lumbar vertebrae of adult and juvenile primates, expressed as percentages of torso length.)

As we see, the numbers for human children are not strikingly different from the values for higher apes, but the decrease in thickness is less in humans. In adulthood, humans have disproportionally thicker vertebrae than other primates. Humans remain closer to the fetal condition than apes do. In this respect, the massiveness of human vertebrae is a Type I characteristic.

Vertebral thickness is not purely a function of body size. According to Schultz, the gorilla, with a weight of 165 kg, has lumbar vertebrae no larger than those of a small human being, and the relatively small gibbon has thicker vertebrae than the lower catarrhines. In humans, the lumbar vertebrae are much thicker than vertebrae located higher on the spine. Schultz sees this as a human "adaptation" to uprightness and compares the spinal column to the trunk of a tree, which is thickest at the bottom where it supports the greatest weight. From the functional perspective, the greater thickness of the lower vertebrae in humans is a logical consequence of uprightness, so it is strange to find the same trait in both anthropoid apes and lower primates. As the figures below indicate, the massiveness of the vertebrae increases from lower primates to gibbons to large anthropoid apes to humans.

	Thoracic vertebra	*Lumbar vertebra*
proboscis monkey (female)	3.1	5.0
proboscis monkey (male)	3.1	5.7
gibbon (female)	3.9	6.4
gibbon (male)	4.0	6.6
orangutan (female)	4.9	6.5
orangutan (male)	5.2	6.9
chimpanzee (female)	4.4	6.4
chimpanzee (male)	4.5	6.9
human (female)	5.2	7.8
human (male)	5.5	8.2

(Average thickness of the middle thoracic and lumbar vertebrae of adult primates, expressed as percentages of torso length.)

Human proportions are prefigured, as it were, throughout the phylogenetic primate series but become meaningful only in connection with uprightness, a typically human circumstance. In this respect, the massive appearance of the human thoracic and lumbar vertebrae must be considered a Type II characteristic.

The Sacral Vertebrae

The sacral vertebrae, which fuse to form the sacrum, become increasingly massive in the ascending phylogenetic series of primates, a trend that climaxes in human beings. In lower catarrhines, the sacral vertebrae seldom constitute more than 10 percent of the total weight of the spine. This number rises to 14 percent in gibbons, 14.4 percent in orangutans, 17.0 and 17.4 percent in chimpanzees and gorillas, respectively, and as much as 20.3 percent in humans (average values from Schultz, 1961). This trend persists even after correcting for the absence of the tail in higher primates (Schultz, 1962). The purpose of this increasing massiveness is revealed in human beings (see above, page 165).

In humans the cervical vertebrae, although long (which preserves fetal proportions), are surprisingly light in comparison to the other vertebrae, while the sacrum is very heavy. In the tree shrew (*Tupaia*), which some scientists consider the most primitive primate, the exact opposite is true; its sacrum is light and its cervical vertebrae heavy in comparison to the other vertebrae. The proportional weights of human vertebrae in different parts of the spine seem to be determined primarily by retardation (prolongation) of the second, cephalocaudal developmental tendency, which allows the vertebrae farther from the head to become heavier (an advantage for upright posture). Many phenomena—beginning with the fact that the growth of the lower torso catches up with that of the upper torso (see p. 67)—confirm that the spine's growth pattern is largely cephalocaudal. In the course of growth, the relative length of the cervical region in humans (and other primates) actually decreases, while the lumbar region becomes disproportionally longer (O'Rahilly et al., 1980). Admittedly, this general phenomenon may be eclipsed by various specializations. For example, although the highly retarded anthropoid apes have heavy sacrums, their cervical vertebrae are also heavy because of the extensive processes they develop.

The Foot

The final shape of the human foot develops very quickly during the embryonic period. The most conspicuous feature of the foot in its earliest stages is the primitive, splayed position of the toes and the still relatively undeveloped heel (see *Figures 91* and *92*). The heel's initial underdevelopment is related to the limbs' distoproximal growth pattern (see Chapter 3). Tax (1980) writes, "The anterior part of the foot in fetal life is proportionally much wider than in the newborn, and even more so than in the adult. The foot thus becomes longer and narrower from birth and during the earlier postnatal period, on through the adult stage."

As overall growth continues, the heel constitutes an increasingly large percentage of total foot length—22 to 23 percent at the age of one year, 27 percent at age six, and 29 to 30 percent at age eighteen (Anderson et al., 1956). Meanwhile, the growth of the foot as a whole lags behind the increasing length of the lower leg. The foot is often longer than the tibia at age one, between 82 and 84 percent as long as the tibia at age six, and only approximately 72 percent as long in adults (Anderson et al., 1956; see *Figure 33*, p. 73). This is another example of the extent to which early fetal growth patterns are preserved in later developmental phases.

If we measure the width of the embryonic human foot at the base of the toes and compare it to the length of the toes (specifically, the middle toe), we find that the proportions are already essentially the same as those of the adult foot. As *Figures 91* and *92* show, the feet of ape and human fetuses are very similar. Both have short, narrow heels, the bases of the toes form a similar curve, and the ratio of foot width to toe length is the same. But although they begin with very similar initial foot forms, humans and apes undergo very different developments. In both human and ape ontogeny, the heel becomes more massive and the foot's initial triangular shape becomes longer and more rectangular; that is, the heel makes up for its initial developmental delay. In humans, however, pronounced retardation (prolongation) of this catch-up growth results in a relatively more massive heel. Furthermore, the human big toe, which plays a major role in supporting body weight, becomes relatively more massive than the other toes, as we will discuss below, while in all other respects the relative proportions of the phalanges and metatarsal bones remain roughly the same. Specifically, the phalanges remain much shorter than the metatarsal bones. Bolk (1922) also mentions that the arch of the human foot (which plays a significant part in walking) is a fetal characteristic that persists in humans, but he seems not to have pursued this point in detail.

In apes, the foot undergoes more comprehensive alteration. Two significant changes transform it into a hand of sorts: the phalanges begin to lengthen faster than the rest of the foot—that is, the toes become fingers—and the position of the big toe begins to shift. In all primate embryos, the big toe is located near the base of the second toe when the toes first emerge from the foot plate. In all monkeys and anthropoid apes, the big toe's point of attachment then begins to shift in the direction of the ankle joint. Only in humans does the big toe remain in its original fetal position next to the other toes. In apes, this shift leads to the development of an opposable "thumb-toe." As a result, feet as well as hands can be used for grasping. In this respect, the human foot can be seen as a "retarded" hand that retains a number of traits from the earliest fetal phases and therefore fails to develop the grasping function. Short toes and the big toe's location at the end of the foot next to the other toes are Type I human characteristics.

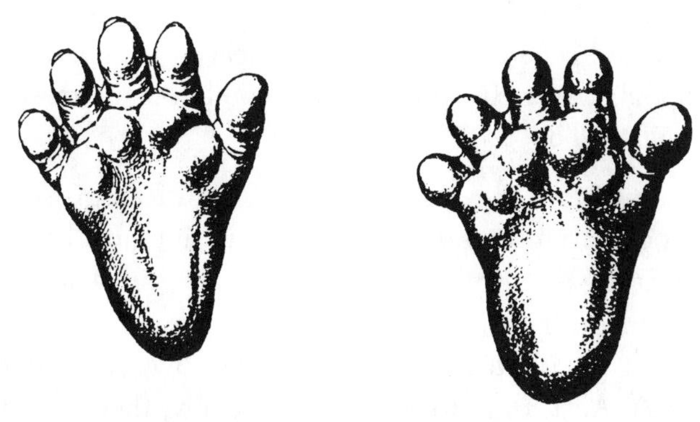

Fig. 91: Exact drawings of the right foot of a macaque fetus (left) and of a human embryo (right). The torso length of both the fetus and the embryo was 24 mm. The two feet are quite similar. In both, the heel is less developed than the toes and the five toes splay out in a fanlike shape (Schultz, 1968).

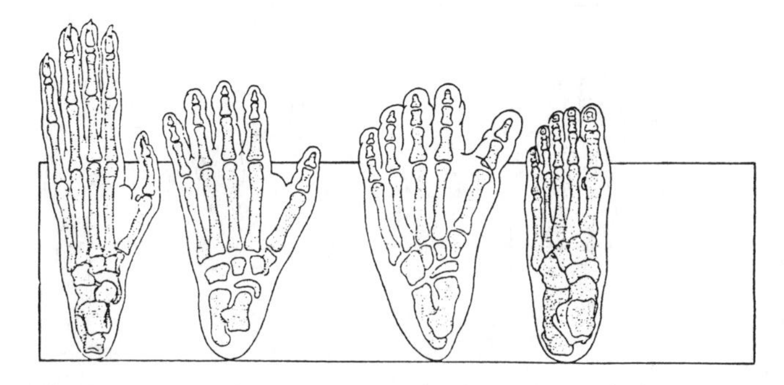

*Fig. 92: Embryonic foot skeletons (two center drawings) and adult foot skeletons of a howler monkey (*Alouatta*, a platyrrhine, left) and of a human (right). The torso lengths of the two fetuses were 30 and 36 mm, respectively. The relative proportions of the digital phalanges common to both fetuses are preserved in the adult human foot, but in the monkey the toes grow into "fingers" (Schultz, 1926a).*

In this context, the foot of the mountain gorilla deserves special attention. The shape of this foot varies considerably among individuals. The big toe is located far forward on the foot, and in many cases the proportions are surprisingly humanlike (see *Figure 93*). Fetal foot characteristics—short toes and the location of the big toe—appear to be better preserved in this ape than in most other anthropoid apes. Straus comments:

> If the foot musculature as a whole is considered, the gorilla resembles man more closely than does any other primate. In certain important muscular details, however, the human foot is much more primitive and generalized than are the feet of the gorillas and other anthropoid apes and approaches the conditions in low catarrhines (in Schultz, 1936).

But in actual fact, the development of adult foot proportions is a very different process in mountain gorillas than in humans. In humans, the big toe remains at the front of the foot, but in this gorilla subspecies, it shifts toward the heel during the fetal phase, producing a typical ape foot. After birth, however, the big toe slowly shifts back toward the front of the foot (see *Figure 94*). According to the biogenetic law, therefore, it is clear that mountain gorillas (unlike humans) are descended from ancestral species that had ape feet with opposable big toes.

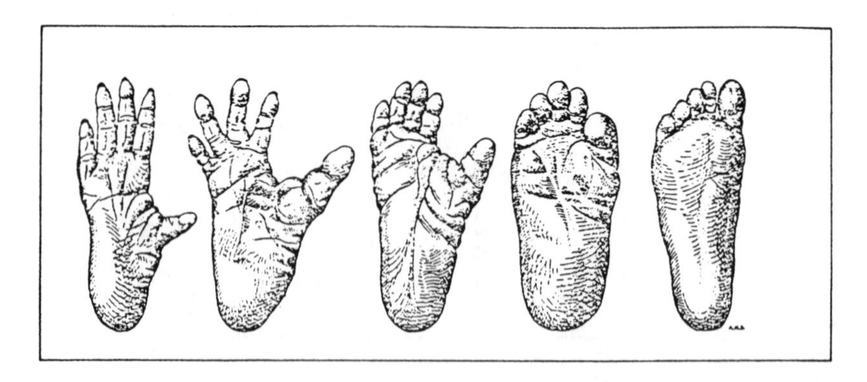

Fig. 93: Adult feet of (left to right): orangutan, chimpanzee, lowland gorilla, mountain gorilla, human. In the apes, the toes have grown long, the big toe is offset (and, in the orangutan, greatly reduced in size, as is its thumb). Although the base of its big toe is closer to the heel, the foot of the mountain gorilla looks surprisingly similar to the human foot (Schultz, 1926a).

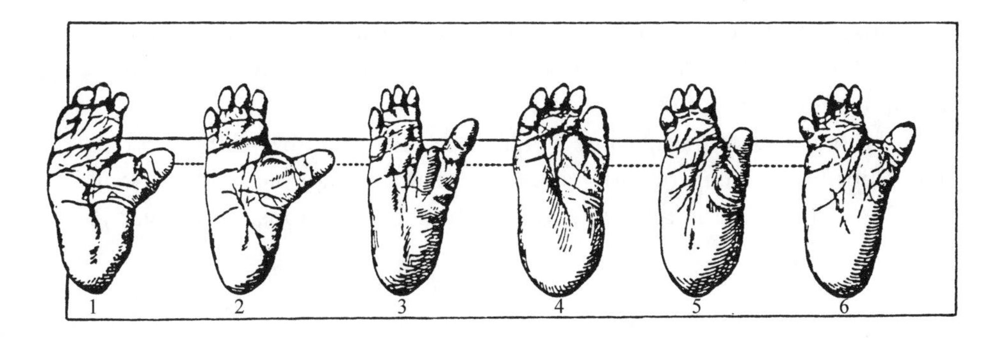

Fig. 94: Metamorphosis of the foot of the mountain gorilla. Right feet of (from left to right): (1) fetus, (2) juvenile, (3) adolescent, (4) adult female, (5, 6) adult male, two variations. After birth, the big toe shifts back toward the front of the foot, achieving humanlike proportions again by way of a detour (Schultz, 1934).

The splayed position of human toes during the fetal stage (*Figure 91*) is sometimes used to support the argument that the human foot evolved from an apelike foot-hand. In all probability, however, this trait is simply related to the relative underdevelopment of the heel at this stage, which gives the foot its initial V-shape and tends to increase the divergence of the toes. As *Figure 91* shows, at a certain stage the little toe angles outward even more sharply than the big toe, although it is aligned with the other toes in adult primates. Incidentally, Mijsberg (1922) points out that the outward angle of the human big toe is not to be confused with the opposability that is observed in adult apes. No trace of an opposable big toe is found in human ontogeny. "The voluminous big toe splays outward but remains in the same plane as the other toes" (Hinrichsen, 1990). We will return to this issue later.

A detailed comparison of foot proportions in humans and apes yields interesting data. Schultz (1936) points out that human toes are short in comparison to ape toes because the second phalanx is relatively short in humans. In the third toe of human adults, the middle bone makes up 24 percent of the total length of the three toe phalanges. In orangutans, gibbons, and lower catarrhines, this number hovers around 33 percent. If we look at fetal proportions, we find that the middle phalanx accounts for not quite 25 percent of the total length in both humans and apes.

Humans, therefore, retain fetal primate proportions while apes abandon them. Once again we see a general and conspicuous retardation tendency in the development of the human foot.

Humans also share some details of foot structure with lower primates but not with anthropoid apes. One simple example is the so-called cuboid bone, which is relatively long in humans, gibbons, and lower primates (see *Figure 95*).

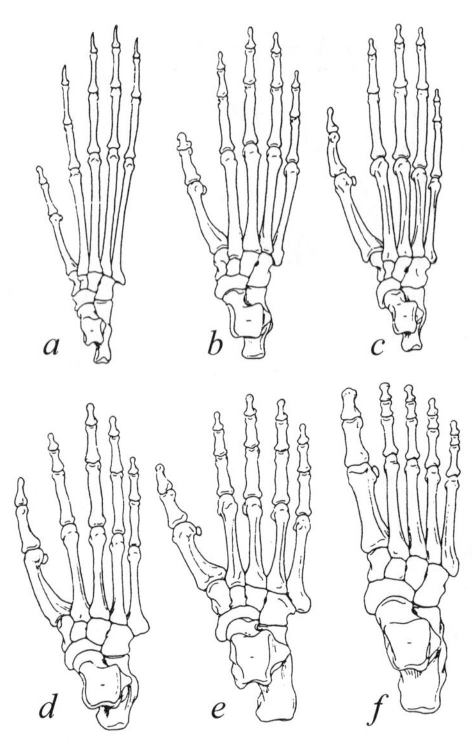

*Fig. 95: Shape of the skeletal foot of: (a) a platyrrhine (*Callithrix*), (b) macaque, (c) gibbon, (d) chimpanzee, (e) gorilla, and (f) human. The human foot differs in that the second through fifth toes remain short while the big toe is relatively well developed and has retained its position next to the other toes. The metatarsal bone of the fourth and fifth toes rests on the so-called cuboid, which is shorter in the large anthropoid ape species and elongated in humans and monkeys. The human big toe becomes massive in comparison to the other toes—a phenomenon related to human upright walking. The same tendency, however, is apparent in African anthropoid apes (Schultz, 1963).*

Straus (1949) gives several examples of the primitive character of human foot muscles in comparison to those of the specialized anthropoid apes. His comparative studies of primate feet, among other subjects, led him to the conclusion quoted earlier in this book (page 26):

> In a considerable number of important characters the human being can only be regarded as essentially generalized or unspecialized . . . In these characters humans find their counterparts not in the anthropoid apes but in animals that are clearly to be regarded on both paleontological and comparative-anatomical grounds as more primitive, namely, such primates as the monkeys and prosimians, and even mammals of other orders. In other words, in many characters, and particularly in those that define an anthropoid ape, the anthropoids (and the great apes especially) can only be considered as far more specialized than the human being (1949).

Straus concludes that the common assumption that human beings are descended from ancestors similar to anthropoid apes cannot possibly be true, and Mijsberg, on the basis of his comparative study of humans and apes, goes so far as to formulate the opposite assumption, namely, that anthropoid apes are descended from a more humanlike species. Many other scientists, including Schultz (1951), have expressed the same idea. Kortlandt and Kooij (1963) also apply this dehumanization hypothesis to anthropoid apes, especially chimpanzees. According to these authors, unused and degenerated remnants of higher capabilities are evident in the behavior of these animals. This claim has suggested to some researchers that African anthropoid apes may be descended from australopithecines—extinct, upright African primates that are generally considered quite closely related to modern humans (Kleindienst, 1975). A few authors continue to entertain this notion, although Kortlandt himself rejected it (1975).

1. Lessertisseur and Jouffroy (1975), in a continuation of Schultz's work, distinguish four typical characteristics of the human foot: of all primate feet, the human foot is the shortest in proportion to the length of the leg. Foot length as a percentage of leg length is 22 to 23 percent in humans, 25 to 28 percent in

anthropoid apes, and 32 percent in macaques. Thus the more humanlike an animal's form, the shorter its feet tend to be in comparison to its legs. According to scientific studies, foot length and the length of the femur are inversely proportional in primates.

2. In comparison to the feet of other primates, the human foot has a very large heel, very short toes, and an average metatarsus. For example, foot proportions in humans and macaques are:

Percentage of foot length

	Humans	*Macaques*
toes	20	36
metatarsus	30	32
heel	50	33

3. The human big toe is exceptionally large in comparison to the other toes.

4. In the human foot, the morphological axis, which is defined by the position of the longest digital ray, falls between the first and second toes. (A digital ray is a toe plus the corresponding metatarsal bone.) The toes are numbered from the big toe (I) to the little toe (V). In prosimians, the morphological axis passes through the fourth toe. In platyrrhines and lower catarrhines, rays III and IV tend to be the same length. In anthropoid apes, the morphological axis passes through toe III. Finally, in humans, rays I and II tend to be the same length; looking only at the toe phalanges, the length sequence is: I > II > III > IV > V (Martin and Saller, 1959). *Figure 95* also shows that ray II is much shorter than ray III in monkeys but somewhat longer than ray III in humans. The more humanlike primate bodies become, the more the morphological axis of the foot shifts toward the big toe. In higher primates, the first few toes tend to become larger than the others, and this trend climaxes in humans.

Clearly, traits 1 and 2 are direct consequences of retardation of the caudocephalic growth pattern in the limbs and must therefore be considered hypermorphic effects. But what about traits 3 and 4? The long, heavy, human big toe is directly related to upright walking. With each step, the body sways from side to side like a pendulum, and its center of gravity moves back and forth (see *Figure 96b* and *Figure 80*, p. 155). This motion is due to the fact that the weight of the body is supported first by one foot and then by the other. From the mechanical perspective, this swaying is a necessary evil in that the energy it consumes does not contribute to forward motion. Hence, to minimize swaying, as much weight as possible is placed on the inner edge of the foot, where the big toe is located. Thus, the stronger bones of the human big toe are logically consistent with the human mode of locomotion (Napier, 1967).

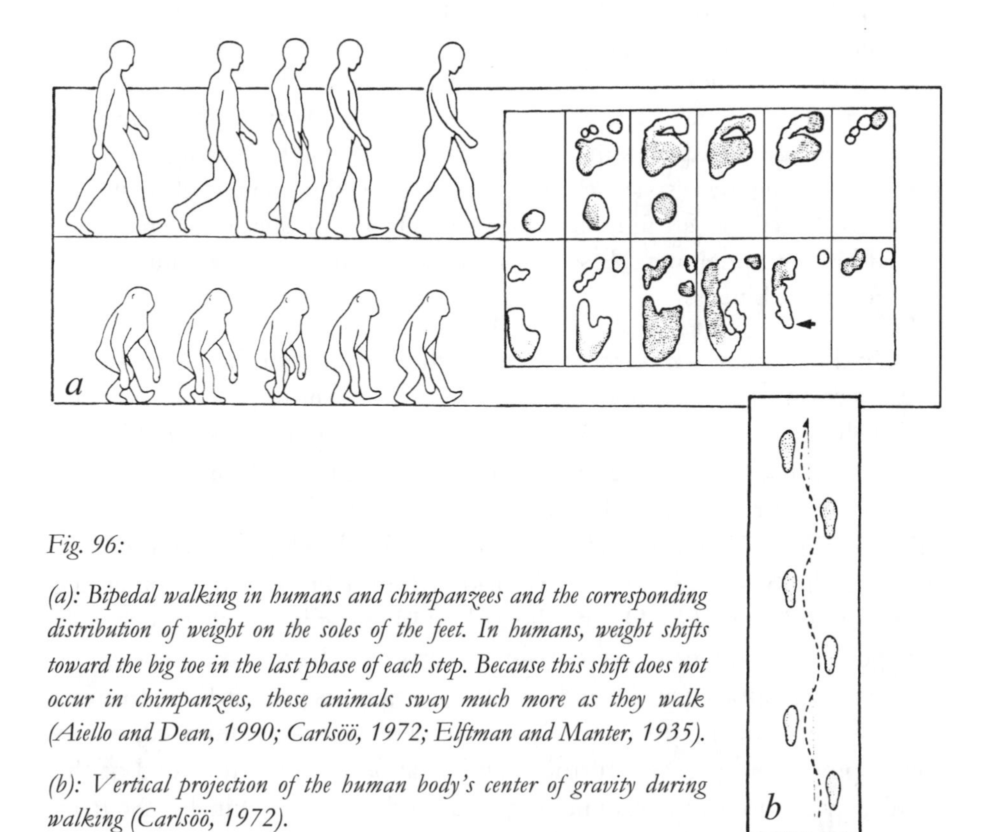

Fig. 96:

(a): Bipedal walking in humans and chimpanzees and the corresponding distribution of weight on the soles of the feet. In humans, weight shifts toward the big toe in the last phase of each step. Because this shift does not occur in chimpanzees, these animals sway much more as they walk (Aiello and Dean, 1990; Carlsöö, 1972; Elftman and Manter, 1935).

(b): Vertical projection of the human body's center of gravity during walking (Carlsöö, 1972).

It would be obvious to conclude that the extensive development of the human big toe is a classic example of specialization, an adaptation to bipedal walking, and therefore not fundamentally different from any specialization in animals. *Figure 95* gives a first indication that this is not the case. Chimpanzees—and, to a still greater extent, gorillas—also have large, strong big toes, even though their body weight rests primarily on the outer edge of the foot (see *Figure 96a*). In view of this fact, the increased size and strength of the big toe can be considered a Type II characteristic in that it becomes meaningful only through human uprightness, although it is clearly prefigured in higher primates. For this reason, Mijsberg (1922) saw the well-developed big toe of the gorilla as an argument in favor of the dehumanization hypothesis mentioned above:

> Anthropoid apes initially follow the human path of development but later abandon it . . . It is a justified question whether these apes may also have undergone an initial upright stage. In this context, the gorilla's thick, well-developed toe bones become especially significant. This trait clearly distinguishes the gorilla from the other higher apes and places this species closer to the human being. The gorilla's big toe is so strong and its upright walking so ungainly that it is difficult to attribute the toe's strength to the species' current degree of uprightness. It seems more likely that the gorilla's ancestors were bipeds.

In the primate series, the systematic shift of the morphological axis in the direction of the big toe suggests that the human big toe develops its relatively massive proportions as a result of hypermorphosis. Shubin and Alberch (1986) investigated the development of the cartilaginous skeleton of terrestrial vertebrate limbs and discovered that the big toe is indeed a latecomer in the universal pattern governing the development of the fore- and hind limbs (see *Figure 97*). Consider the skeleton of the human leg (see *Figures 97b* and *98*). During the earliest phase of its development, only one portion of the leg is present—a precursor of the femur, as it were. A division then appears at the end of this cartilaginous primordium, forming the rudiments of the tibia and fibula. The tibia remains undivided, but as development progresses a further division appears at the foot end of the fibula, giving rise to the rudimentary talus and calcaneus. Finally, the talus divides, forming the primordium of the navicular bone, while the calcaneus develops into a complete arch. Through a series of bifurcations and separations, the cuboid bone and the three cuneiform bones emerge; at each of these stages the primordium of the metatarsal bone develops simultaneously. The fourth toe ray develops first. Both it and the little toe, which appears slightly later, develop out of the cuboid bone. Later still, the middle and second toes and, finally, the first (big) toe develop out of the three cuneiform bones (see *Figure 98*).

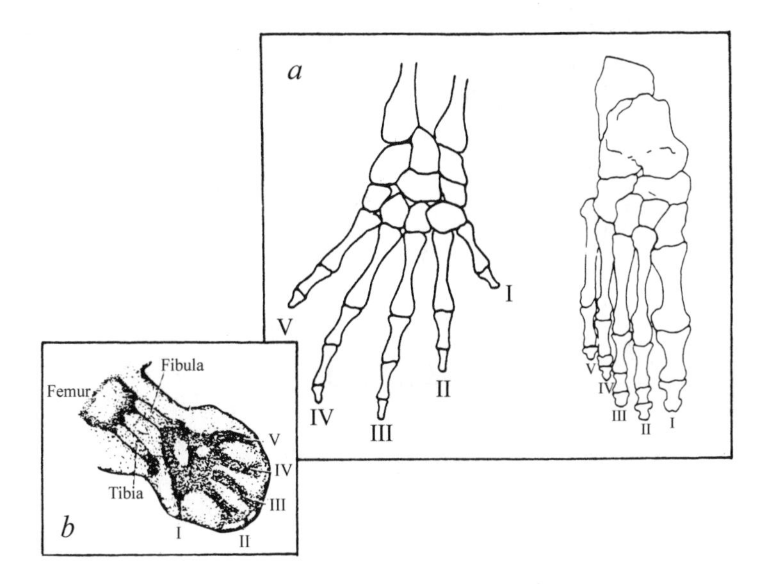

Fig. 97:

a: Foot of a salamander (left) and of a human (right). The toes are numbered from the big toe (I) to the little toe (V). The salamander foot is relatively unspecialized but differs markedly from the human foot in certain respects. The toes are long in comparison to the heel and the big toe is poorly developed, while the third and fourth toes are very large. These differences can be explained by the retarded (protracted) growth of the human foot (Holder, 1983).

b: Rudimentary foot of a 15.5-mm-long human embryo. The development of the toes is quite advanced in comparison to the narrow heel; only the big toe (I) lags behind. The femur, tibia, and fibula are still very small in comparison to the foot, and the foot itself is very broad. Through retardation, body parts that initially lag behind—the big toe as compared to the other toes or the leg as compared to the foot—ultimately become larger. Note that the foot skeleton develops out of the fibula (Jarvik, 1980, adapted from Schmidt-Ehrenberg).

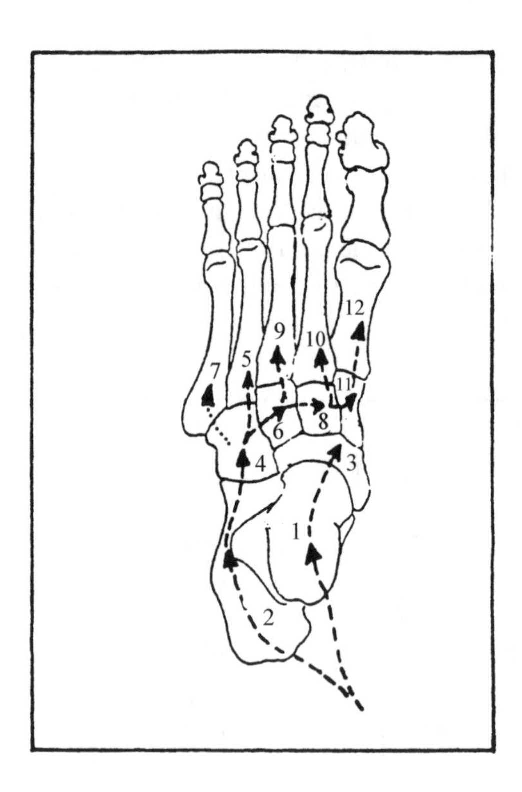

Fig. 98: Developmental sequence of the bones of the foot (Shubin and Alberch, 1986). The talus (1) and calcaneus (2) result from bifurcation of the fibula. The navicular bone (3) develops out of the talus through segmentation. Next to it, also through segmentation, the cuboid bone (4) develops out of the calcaneus. Bifurcation of the cuboid bone leads to the development of the fourth toe (5) and the lateral cuneiform bone (6). Somewhat later, the fifth toe (7) also develops out of the cuboid bone. Bifurcation of the lateral cuneiform bone produces the intermediate cuneiform bone (8) and the third toe (9). In turn, bifurcation of the intermediate cuneiform bone produces the second toe (10) and the medial cuneiform bone (11), out of which the last foot digit (the big toe, 12) develops.

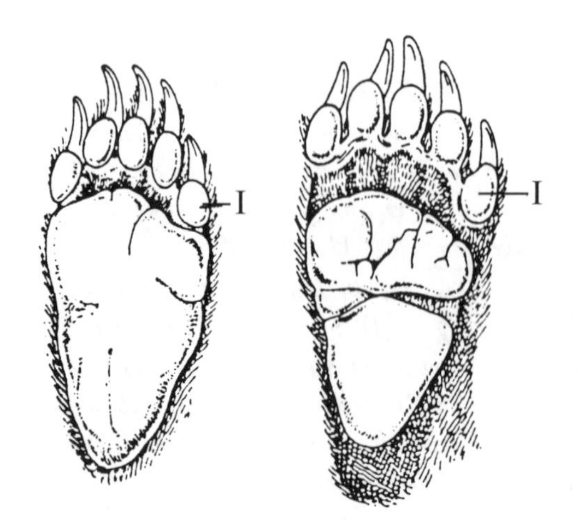

Fig. 99: Right feet of two bear species (Böker, 1935). In certain significant respects, the relatively unspecialized feet of these species are more similar to human feet than to those of apes. The big toe is not opposable and the toes are short. In humans, the heel and the big toe (I) are more developed than they are in bears, differences that can be explained by retardation of the universal growth pattern in humans. It is worth noting that bears are better at walking upright than other carnivores.

In this universal developmental pattern, the big toe appears last (*Figures 97* and *98*). Retardation of this pattern produces the massive proportions of the human big toe and the shift of the morphological axis toward the inner toes (see also *Figure 99*). The long, massive human big toe is a Type II characteristic that develops through hypermorphosis. Schultz (1925) presents data showing how the inner toes catch up with the outer ones in size as embryonic development progresses. During weeks 8 and 9 of gestation, the relative lengths of the toes are: III > IV > II > I > V (I is the big toe; V is the little toe); that is, toes III and IV, which begin to develop first, are the longest. In week 10, the sequence is: II > III > IV = I > V; the second toe, which begins to develop later, and (to a lesser extent) the big toe, which develops last, have already partially caught up. Around week 12, the sequence is: II > III = I > IV > V; around week 18, it is: II = I > III > IV > V. The big toe continues to catch up as long as growth continues. The length of the big toe phalanges in the ten-week-old fetus is approximately 80 percent of that of the second toe. This percentage increases to 87.7 in the four-month-old fetus, to 97.4 in the newborn, and to 105.7 in the adult. These growth patterns, seen as a whole, can be explained by hypermorphosis (retardation or protraction of the universal growth pattern illustrated in *Figure 98*). In apes, the relative lengths of the toes reverse after a certain point in development. In a newborn chimpanzee, the first digital ray is still 80.9 percent as long as the second; in the adult animal, this percentage falls to 66.3, primarily as a result of increased growth of the phalanges of the second ray. In contrast, the human big toe seems to continue to catch up in size even into adulthood (Hawkes, 1914; Huizinga and De Vetten, 1967).

The big toe continues to grow faster than the other toes throughout fetal development and finally catches up with the second toe around the time of birth (Bernhardt, 1988). The marginally greater postnatal growth of the first and second toes in comparison to the others is probably associated with the longitudinal arch of the foot, another typically human characteristic. The diagonal orientation assumed by the metatarsal bones of the first two rays produces a longitudinal concavity in the sole of the adult human foot on the side of the big toe. This diagonal orientation means that the inner rays must become relatively longer than the outer rays (*Figure 100*). A longitudinally arched foot, which is already evident in newborns but becomes more pronounced in the course of childhood, offers various advantages for bipedal walking (Aiello and Dean, 1990). To the extent that it is related to the retarded and therefore increased growth of the first several toes, this arch must be considered a retardation effect. Divergence of the big toe in the early fetal stage (see above, p. 179) may be related to tension on the inner side of the foot, which is caused by the delayed growth of the big toe. Thus both the initial angle of the big toe and the later arching of the foot may be due to the same phenomenon, namely, unequal growth rates of the individual digital rays. The outer rays have an initial head start but are gradually overtaken by the inner rays.

In summary, we can say that hypermorphosis accounts for all of the human foot "specializations" mentioned by Lessertisseur and Jouffroy (1975), namely, a short foot relative to the length of the leg, a very pronounced heel, and enhanced development of the big toe. This entire complex of retardation effects is also clearly advantageous for upright posture, which from this perspective is a potential that emerges spontaneously through retardation of the universal mammalian pattern.

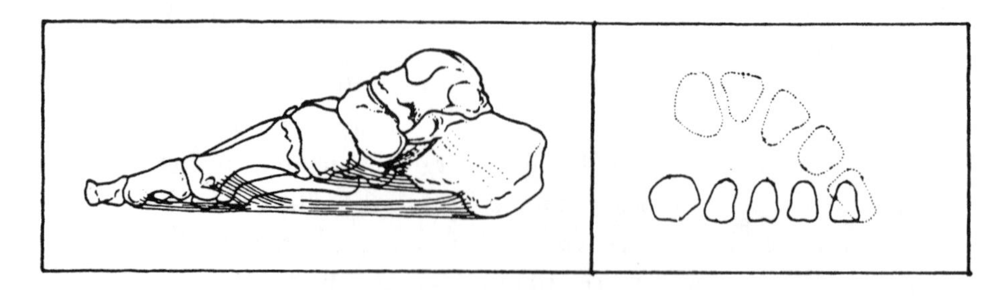

Fig. 100: The arch of the human foot. Left: The bones of the human foot as seen from the side of the big toe. Right: Cross-sections through the human foot at the base of the toes (solid line) and the base of the metatarsus (dotted lines). The fifth metatarsal bone is almost horizontal, while the rays on the other side of the foot are oriented diagonally (Aiello and Dean, 1990).

Pentadactyly as a Retardation Phenomenon

Pentadactyly (having five fingers and five toes) is an ancient and universal trait among terrestrial tetrapods. Of course there are many vertebrates with fewer fingers, and some (such as whales and snakes) have even lost entire limbs in the course of evolution. Embryological and paleontological data, however, clearly indicate that the ancestors of these species also had five digital rays.

We have already discussed the universal sequence of digit development (see *Figure 98*). The basal bones (in the foot, the cuboid bone and the three cuneiform bones) develop in an orderly progression from the outside to the inside, and the fingers or toes (from IV to I) always develop out of these bones in the same sequence. Thus ray I, which corresponds to the thumb or big toe, develops last. Ray IV (in the hand, the ring finger) is the oldest. Ray V (the little finger or toe) falls out of the sequence and begins to develop shortly after ray IV.

Some tetrapods (such as certain frogs, moles, and the giant panda) appear to have more than five fingers or toes. In these species, the extra "finger"— or pseudofinger—is a specialization and develops in a different way. Extending the universal development pattern does not add a finger on the thumb side in any of these species (Gould, 1977b). The earliest tetrapods, however, also had more than five fingers or toes, a phenomenon known as polydactyly. The three tetrapods known to have existed at the end of the Devonian period (*Ichthyostega*, *Acanthostega*, and *Tulerpeton*) had fingers and toes that were very long in proportion to their short, compact legs. These species had six to eight digits (see *Figure 101*). It is interesting to note that the structure of these polydactyl hands and feet nonetheless suggests the number five (see caption, *Figure 101*).

According to Coates and Clack (1990), the decrease in the number of digits, which occurred in the earliest phase of tetrapod evolution, is easy to explain in terms of the mechanism that shapes successive digits—the process simply ceases earlier. In Gould's words, "If digits form . . . in temporal order, then reduction can be readily achieved by an earlier shutdown. The principle is obvious and pervasive: stop sooner." (1977b).

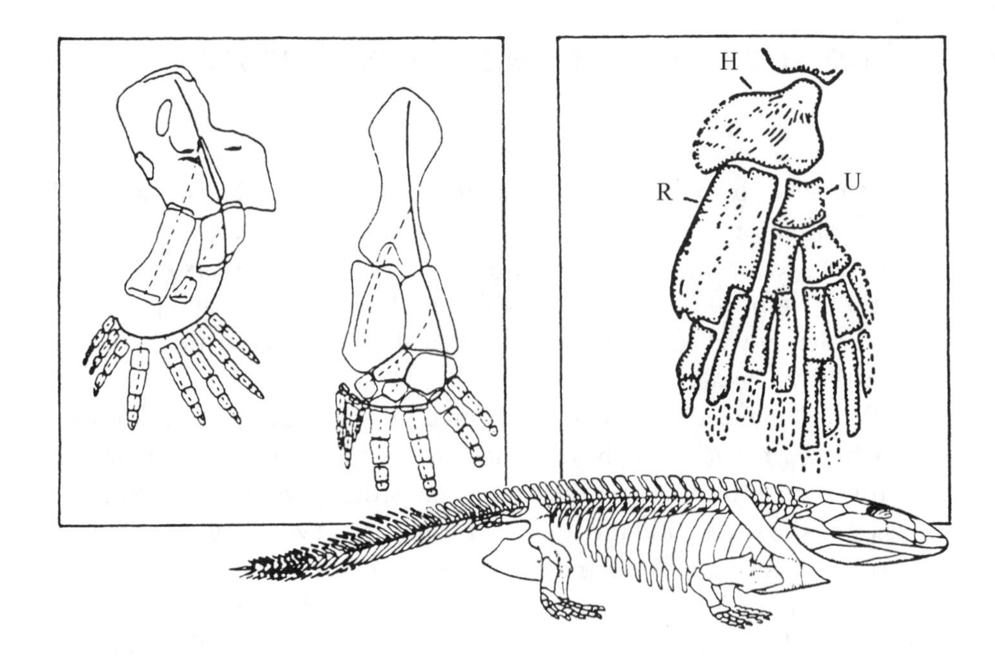

Fig. 101: Left: The forefoot of Acanthostega *and the hind foot of* Ichthyostega *(Coates and Clack, 1990). Both species date from the late Devonion period and are among the oldest known tetrapods. The solid line indicates the primary axis of development of successive portions of the leg and fingers; the broken lines indicate secondary axes (cf.* Figure 98*). Because extra fingers are eliminated if development along the primary axis ceases earlier, the development of the five-digit hand from older polydactyl forms can be considered a fetalization phenomenon. Note that in* Acanthostega *the three extra fingers are relatively small and have fewer phalanges. In* Ichthyostega, *the fifth, sixth, and seventh toes are very close together, as if they formed a single big toe. Thus both examples appear to prefigure pentadactyly, although in different ways.*

Right: The pectoral fin of Sauripterus, *a Devonian fish (Vandebroek, 1961, adapted from Broom). The tendencies we observed in the limbs of the earliest amphibians are even more pronounced here. The number of "fingers" is greater, and although the long bones are extremely compact, elements corresponding to the humerus (H), radius (R), and ulna (U) are already present. The earliest amphibians are presumed to be descended from fish species like* Sauripterus.

Below: Reconstruction of Acanthostega *(Jarvik, 1980). The species had a fin-shaped tail, and its legs were very short in proportion to its torso. When this illustration was first published, polydactyly in the earliest amphibians had not yet been confirmed, so the limbs are incorrectly depicted as pentadactyl.*

This means that the development of pentadactyly can be considered a fetalization phenomenon; that is, the process of development simply stops short at an earlier stage. The extra fingers and toes never appear, just as in humans a full coat of fur never develops. This view, however, gives rise to another question, which according to Gould (1977b) still remains unanswered:

> The greatest puzzle of all [is] the recalcitrant stability of five once it evolves . . . Why should five, once attained by whatever route and for whatever reason, be so stubbornly intractable as an upper limit thereafter—so that any lineage again evolving six or more must do so by a different path? The inquiry could not be more important, for this issue of digits is a microcosm for the grandest question of all about the history of animal life: Why, following a burst of anatomical exploration in the Cambrian explosion some 550 million years ago, have anatomies so stabilized that not a single new phylum (major new body plan) has evolved since?

Why is pentadactyly so stable? To answer Gould's question, we must turn to the Bolkian "intrinsic factor" that guides evolution as a whole. Evolution works toward an ultimate form, the human gestalt. Through successive steps of retardation and fetalization, the outline of this form becomes increasingly explicit. The legs of the earliest tetrapods were very short in comparison to their extensive extremities ("hands" and "feet"), which had more than five digits each. At a very early stage, fetalization reduced the number of digits to five. Once the human number was achieved, it was persistently preserved because it coincided with the ultimate human form toward which evolution was working. Consequently, the additional fingers of the oldest tetrapods must be considered specializations.

Retardation of the course of growth in vertebrates causes progressive increases in the length of the limbs relative to the trunk. This hypermorphic tendency culminates in the lower limbs of humans. In addition, retardation of distoproximal limb growth changes their proportions. The distal extremities ("hands" and "feet"), which were originally very large, become smaller in proportion to the proximal portions of the limbs. The first terrestrial vertebrates represent very early stages in this retardation process, through which the human gestalt ultimately materialized.

Amphibians probably evolved from certain fossil fish species with fins in which the structure of the future limbs and digits is already apparent. The digits were very numerous, however, and the limbs extremely short (*Figure 101*). Even the very first amphibians, with fewer toes and longer limbs, can be considered retarded in comparison to these fish species. Incidentally, there is some indication that a latent potential for pentadactyly is already present in some fish species. In an experiment conducted by Harms, *Periophthalmus variabilis* (a fish with highly developed pectoral fins that allow it to move through mud) developed five-rayed limbs when treated with thyroxine, the primary component of thyroid hormone (Vandebroek, 1961).

From a holistic perspective, the relationship of humans to apes is characterized by fetalization of the skull and hypermorphosis in the proportions of the limbs; see Chapters 2 and 3 and also Chaline et al. (1986) or McKinney and McNamara (1991). It is very interesting to note that a similar phenomenon characterized the transition from fish to amphibians. As Long (1990) demonstrates, the skulls of the oldest known fossil amphibians appear fetalized in comparison to those of closely related fish species. Amphibian legs, however, cannot be understood in the same way. In comparison to fish, they represent a further step in the emergence of the human gestalt. In fact, limbs are later evolutionary phenomena, and retarded ontogenesis permits more extensive limb development—which, however, is under constant pressure to specialize. Pentadactyly must also be considered a fetalization phenomenon or Type I characteristic because it appeared very early in the evolution of terrestrial vertebrates and has been preserved unchanged throughout the human line of evolution.

The number of phalanges in reptiles and mammals

Another step in fetalization is apparent in the transition from reptiles to mammals. Humans, like all mammals (which are unspecialized in this respect), have three phalanges in each digit, with the exception of the thumb and the big toe, which have only two. This universal mammalian pattern is represented schematically by the phalanx formula 2-3-3-3-3.

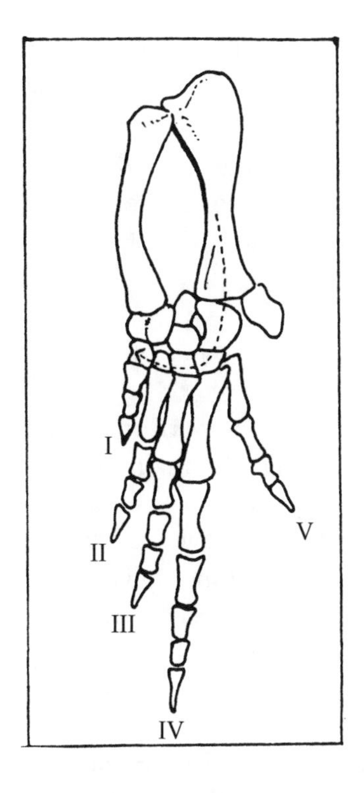

Fig. 102: Bones of the hand and lower arm of Varanosaurus, *a primitive reptile. The broken line indicates the main axis along which the hand bones develop in succession. In this primitive hand, fingers IV and III, which develop earliest, are the longest and have more phalanges (2-3-4-5-3). Note also the length of the hand in comparison to the lower arm.*

In the transition from reptiles to mammals to humans, finger I (the thumb or big toe), which develops last, becomes disproportionally larger. This is exactly the sequence that we would expect from increasing retardation. The additional phalanges of fingers IV and III in the primitive reptile are absent in mammals, resulting in a phalanx formula of 2-3-3-3-3. The disappearance of these additional phalanges can be considered a fetalization phenomenon that is in complete harmony with the fact that early-developing fingers are shortened relatively more than later-developing digits.

A universal, primitive phalanx formula also exists among reptiles, but in this case the pattern is 2-3-4-5-3; that is, the fourth and third digits have more phalanges (*Figure 102*). These specific fingers or toes, however, are the earliest to begin to develop. Whenever limb development is retarded, these digits are disadvantaged in comparison to the ones that begin to develop later. The absence of extra phalanges is more or less what we would expect of retarded growth and coincides well with the shift of the foot's morphological axis in the direction of the big toe, a phenomenon that is apparent throughout the primate series and culminates in humans.

On the smaller scale of the phalanges, the transition from reptiles to mammals repeats the evolution that affected entire digits in the transition from fish to amphibians. In the latter instance, the last digits to develop were eliminated; in the former, the last phalanges of specific digits failed to develop. In several independent lines of evolution within the reptile branch (for example, in turtles and various extinct mammal-like reptiles), a tendency toward the human phalanx formula 2-3-3-3-3 can be observed.

The breakthrough of the human phalanx formula is a good illustration of how fetalization and the increasing emergence of the human gestalt are two complementary aspects of one and the same process. Through fetalization, certain final and purely animal stages of development are eliminated, leaving time and energy available for more explicit emergence of the human gestalt, that is, for the continued and more complete unfolding of early, humanlike developmental steps.

Bokian Retardation

The human foot is highly fetalized, not only in comparison to ape feet but also in comparison to the human hand. Toes remain shorter than fingers, and the big toe, unlike the thumb, does not move downward (or backward) and cannot rotate. Furthermore fewer sesamoid bones (nodular masses of cartilage or bone in the tendons) develop in the foot (Mottershed, 1988). Nonetheless, hands and feet obviously develop according to the same basic pattern. Like fore- and hind limbs, for example, or successive fingers in a series, they can be considered equivalent organs.

S. T. Bok (not to be confused with Bolk) discovered a principle of fetalization that he related directly to Bolk's fetalization theory: *Within a series of equivalent organs, the organ that develops latest will also be the most fetalized.* This new fetalization principle describes the lawful connection between different organs within an organism rather than the relationship between different species. Bok first discovered this principle as it applies to the brain (Verhulst, 1999b); see Chapter 10 below.

We can distinguish two aspects of growth. First, in the process known as *qualitative growth*, a plethora of new traits and structures appear during the complex transitions leading from the fertilized ovum to the adult organism. The organs and body parts of the adult organism are not present in the fertilized ovum, but appear only gradually. Second, simple growth of already existing structures occurs; that is, these structures become larger. Growth unaccompanied by the appearance of new traits or structures is called *quantitative growth*.

The essential point that needs to be made here is that *qualitative growth precedes quantitative growth*. An organ must first appear before it can grow quantitatively. In animal and human ontogeny (the development of an individual), new traits and structures obviously appear primarily during early (i.e., embryonic) phases of development, while quantitative growth occurs primarily postnatally, during childhood. This means that in the course of ontogeny, quantitative growth becomes increasingly strong in comparison to qualitative growth. *In a later phase of ontogeny, an equivalent quantum of qualitative growth will be accompanied by more extensive quantitative growth.*

Let's consider a series of equivalent organs in a complete idealized or generalized ontogeny. These organs display the same qualitative growth because they develop according to the same structural pattern. Within this series, the latecomers will display relatively more quantitative growth because the same quantum of qualitative growth in later developmental phases is accompanied by more quantitative growth (see *Figure 103*).

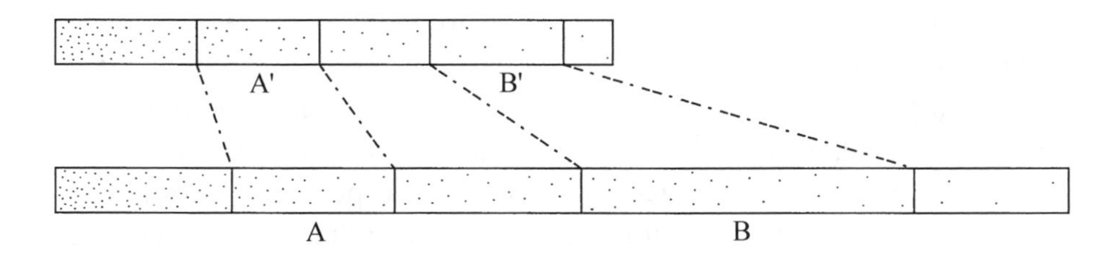

Fig. 103: The lower bar represents the time line of an idealized, completely retarded ontogeny. The length of each segment is a measure of quantitative growth, while the density or number of dots represents qualitative growth during the same time period. As we see, less and less qualitative growth occurs in later developmental phases. Consider two equivalent organs, (A) (a forerunner) and (B) (a latecomer). Because they are equivalent, both organs in this completed ontogeny have the same qualitative growth (i.e., the same number of dots in segments (A) and (B)). Latecomer (B), however, displays more quantitative growth (due to hypermorphosis, indicated by the greater length of segment (B)).

The upper bar illustrates the same concept with regard to a moderately compressed, humanlike ontogeny, in which later developmental phases of the universal ontogeny are suppressed to a greater extent than earlier ones. The growth of organ (B') relative to that of the forerunner (A') has decreased. Although the quantitative growth of (B') remains greater (that is, (B') is still hypermorphic in comparison to (A')), (B') is qualitatively fetalized in comparison to (A') (Bokian retardation). Thus (B') has fewer dots than (A').

Compressing some aspects of this generalized pattern produces a humanlike pattern of development (see *Figure 23c*, p. 55). The development of a latecomer (such as the foot in comparison to the hand or the thumb in comparison to the other fingers) will be suppressed to a greater extent than that of a forerunner (see *Figures 25* and *26*, p. 62). On the one hand, the latecomer's qualitative growth will be relatively smaller; that is, the latecomer will be paedomorphic or fetalized in comparison with the forerunner (for example, the foot in contrast to the hand). This is precisely the effect that Bok discovered. On the other hand, the latecomer's quantitative growth will also be suppressed, but since this growth was very extensive in the generalized developmental pattern, the latecomer can still become relatively large in a moderately compressed ontogeny such as that of humans. *Figure 103* summarizes these relationships.

When the generalized pattern of development is accelerated, an increasing amount of species-linked specialization occurs (see *Figure 23b*, p. 55). For example, consider a typical quadruped. On the one hand, the qualitative growth of the hands and feet is suppressed; for example, latecomers in the series of fingers and toes (specifically, the thumb and big toe) may be eliminated. Both sets of incompletely developed distal extremities are then specialized to serve a single purpose (locomotion) and therefore develop approximately similar shapes. The strange fact that human hands are very different from human feet is generally explained as a specialization, an adaptation to upright posture, but this view does not account for why the ontogenetic latecomer (the foot) appears fetalized in comparison to the forerunner (the hand). Bok's principle, however, successfully incorporates the strange relationship between human hands and feet into the overall pattern of retardation. Here again we recognize that pure retardation leads to a body adapted to upright posture.

Bokian retardation in the fingers and toes

Bok's principle predicts that within the series of fingers and toes, the latecomers (thumb, forefinger, big toe) will be relatively fetalized in comparison to the other digits. Several examples demonstrate that this effect actually does occur:

- *Hair on fingers and toes:* As we already know, fetalization is accompanied by regression of the hair covering the body (see pages 37 and 38). This phenomenon can also be observed on the fingers and toes. Danforth (1921) discovered that within the primate series, hair on the backs of the hands tends to decrease; this process culminates in humans. This, of course, is another example of Bolkian retardation. Danforth also observed, however, that the amount of hair is not the same on all fingers and all phalanges. On each finger, the proximal phalanges have the most hair. Since finger development follows a proximodistal pattern, this is an example of Bokian retardation; the phalanges that appear first in individual development also develop the most hair. Furthermore, we know that the fingers appear in a specific sequence; the ring finger is first, while the forefinger is a relative latecomer (see *Figures 98* and *101*, pages 186 and 190). Thus Bok's principle predicts that there will be less hair on the forefinger. This decrease in the amount of hair on successive fingers is especially evident on their middle phalanges (see *Figure 104*). Of course it is difficult to include the thumb and big toe in this sequence, since they have only two phalanges each.

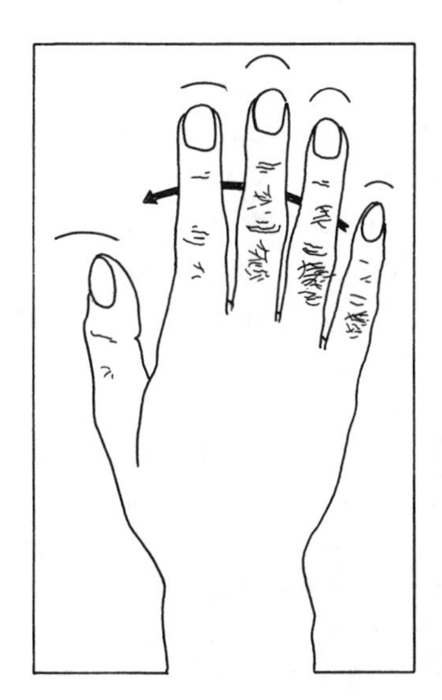

Fig. 104: Degrees of fetalization in the human hand. The ring finger has the most hair, while the forefinger is relatively bare. This difference is most pronounced on the second phalanges. The same degree of fetalization is evident in the lateral curvature of the nails, which is most pronounced in the ring finger (see p. 199).

- *Lateral nail curvature:* Transverse or lateral curvature is much greater in claws than in nails. The pronounced curvature of claws, however, develops only gradually; fetal claws are nail-like (see *Figures 71* and *72*, p. 141). The relative flatness of nails in comparison to claws can be considered a Type I characteristic. Bok's principle predicts that the ring finger's development will be relatively advanced compared to that of the forefinger (because the ring finger is an ontogenetic forerunner in comparison to the forefinger). Because lateral curvature is greater in earlier-developing and therefore more clawlike nails, we would expect to observe more curvature in the nail of the ring finger and a flatter nail on the forefinger, and in fact that is exactly what we find on most human hands. We have already mentioned the strange fact that in primates the nail of the big toe is always flat, never curved like a claw. Bok's law explains why this is so. In comparison to the hand, the foot is an ontogenetic latecomer, and the big toe is also the latecomer among the toes. This means that the fetalizing tendency (i.e., the tendency to remain flat) is most pronounced in the nail of the big toe. According to Bok's principle, nail curvature in humans—but not in most apes—can be understood as a retardation effect. Not only is lateral nail curvature in apes more pronounced than in humans but the sequence of the digits with regard to nail curvature, as shown in *Figure 104*, is usually interrupted. Vigner (1896) and Bruhns (1910), who conducted extensive studies of nail curvature in primates, discovered very pronounced transverse curvature in the second toe of many species (*Figure 105*). Nail proportions in animals, unlike those of humans, are specializations and therefore cannot be explained by Bok's law (i.e., simple retardation).

- *Bending of the fingers in the resting position:* When the fingers first appear, they are straight. Their tendency to bend develops only later, although it is already evident in a 5 cm-long fetus. Bok's principle predicts that the youngest finger, which is the forefinger, will bend the least, as we can see from any hand at rest (*Figure 106*).

• *Zygodactyly*: Zygodactyly (see page 107) occurs when the membrane that connects the digits during the early embryonic period becomes permanent. Strangely enough, the membrane tends to persist between the second and third toes, which are the two ontogenetic latecomers (if we disregard the thumb and big toe, which have fewer phalanges and therefore constitute an exception in certain respects). Zygodactyly can also be considered a Bokian phenomenon.

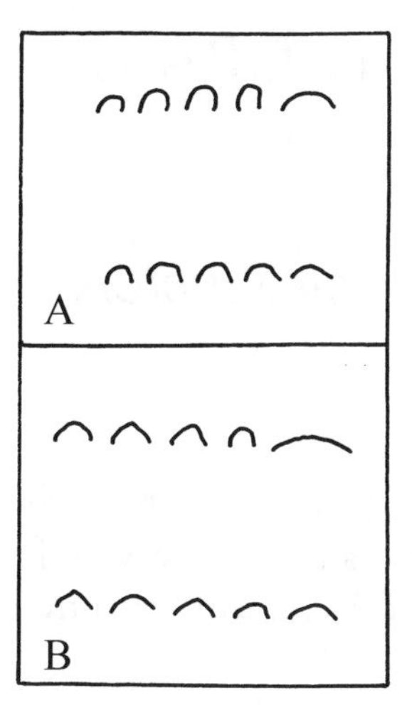

Fig. 105: Cross-sections of fingernails and toenails in two primates. Within each box, the upper row shows the fingernails and the lower row the toenails; nails of the thumb and big toe are on the right. (A): lemur; (B): macaque (Bruhns, 1910). Note the flat nail of the big toe and the clawlike curvature of the nail of the second toe. The latter characteristic cannot be explained by Bokian retardation.

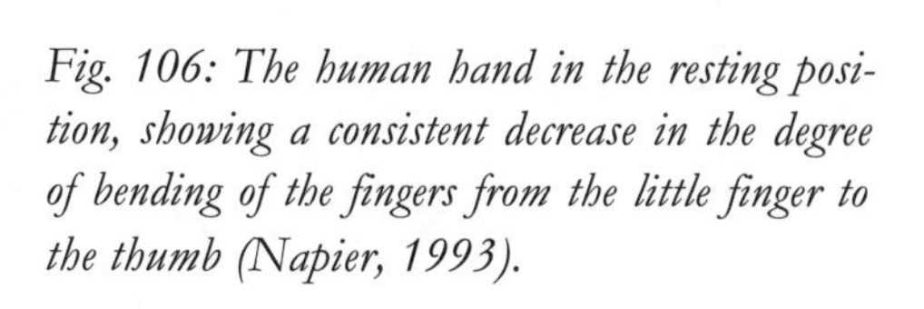

Fig. 106: The human hand in the resting position, showing a consistent decrease in the degree of bending of the fingers from the little finger to the thumb (Napier, 1993).

7

THE ORGANS OF THE THORAX

The Cardiovascular System: Introductory Remarks

The human circulatory system appears retarded in certain respects in comparison to that of mammals. Humans, for example, have very large red blood cells (erythrocytes). Volumes of these cells (in μm^3) in different species are:

Species	*Volume (μm^3) of a red blood cell*
goat	19
sheep	31
mouse	49
cat	57
rat	61
pig	61
dog	66
hamster	70
guinea pig	77
chimpanzee	81
human	87

Large red blood cells are a fetal characteristic. All of the mammals studied to date have large red blood cells during the fetal stage. The following sizes have been reported at different developmental stages in humans (Altman and Dittmer, 1974):

Stage of human development	*Volume (μm^3) of a red blood cell*
fetus (10 weeks)	180
fetus (22 weeks)	140
fetus (36 weeks)	120
newborn	113
adult	87

Betticher and Geiser (1989) discovered a close relationship between the size of red blood cells and their resistance to hemoglobin loss, which occurs in a hypertonic milieu. Kidney damage occurs when free hemoglobin circulates in the blood plasma. When red blood cells are exposed to excessive salt concentrations, they begin to break apart and release hemoglobin. Large erythrocytes disintegrate more readily than small ones, which are resistant to high concentrations of salt. The highest concentrations are found in the kidneys, where urine is formed. Betticher and Geiser discovered a systematic correlation between red blood cell size and urine concentration in mammals. (The lower the concentration of salt in the urine, the larger the red blood cells.) Remarkably, these authors also discovered the same relationship within individual species, between newborns and adults. Young animals have both larger red blood cells and less concentrated urine.

Human urine is less concentrated than that of most terrestrial mammals; with a maximum concentration of 1.43 OSM U, it falls between the urine of freshwater mammals and that of marine mammals. Freshwater dwellers have diluted urine (beaver urine, for example, has a maximum concentration of 0.52 OSM U) because fresh water is never a limiting factor. The maximum concentration is higher in marine mammals (1.70 OSM U in dolphins, for example) because only saltwater is available to them. The maximum concentration is higher still in typical terrestrial mammals (3.20 OSM U in cats and 4.60 OSM U in the kangaroo rat, *Dipodomys phillipsii*, a desert animal). The conspicuously low maximum concentration in humans is cited as a rationale for the "aquatic ape" theory, which claims that humans are descended from primates displaying extensive adaptation to life in the water (Verhaegen, 1985 and 1991). Maximum urine concentrations in monkeys, however, are generally roughly the same as in humans (see, for example, J. Goodman et al., 1977).

In humans, low maximum urine concentrations, like the related size of the red blood cells, appear to be a Type I characteristic or fetalization phenomenon. The concentrations typically found in humans—and, to a large extent, in all higher primates—more closely approximate the lower levels found in all juvenile mammals. Because juvenile terrestrial mammals drink milk, they run a lower risk of fluid shortage than adults do. Thus the highly concentrated urine of adult terrestrial mammals is clearly a specialization, an adaptation to water shortage. In this respect, humans appear to have remained unadapted.

The structure of the human heart and blood vessels also suggests that human development is retarded in comparison to that of other mammals (see below). Furthermore, mammals are already retarded in certain respects in comparison to other terrestrial vertebrates.

Protein and DNA Structure

Hemoglobin plays an important role in comparative biochemistry. A comparison of amino acid sequences of hemoglobins in different animal species reveals more similar sequences in more closely related species. A certain average rate of mutation seems to have prevailed throughout the evolution of mammals; as a rule, the longer two lines of evolution have been separated, the more significant the biochemical differences between them. This phenomenon, first discovered in hemoglobins, has since been found in many other types of molecules.

More precise observation, however, reveals that mutation rates are not the same in all species. There are fewer biochemical differences among higher primates than there are among other mammals. This means that biomolecular evolution appears to have been delayed in primates (especially higher primates) in comparison to other mammals. For example, J. R. and M. Goodman and their colleagues discovered that alterations in hemoglobin occur nearly three times faster in horse species than in higher primates (J. Goodman et al., 1977; M. Goodman et al., 1983). This means that hemoglobin has remained more primitive in primates, while in other mammals its amino acid sequence has undergone more alteration from the original common structure. A number of studies confirm that the mutation rate of many types of molecules is even slower in humans than in anthropoid apes (Hasegawa et al., 1987, 1989; Koop et al., 1986; Li and Tanimura, 1987; Verhulst and Jaspers, 1997).

From a Bolkian perspective, therefore, we can say that the biochemistry of higher primates is markedly retarded. Researchers have related this phenomenon to the primitive morphology of humans. For example, M. Goodman (1977) writes, "The genealogical reconstructions [indicate] that humans have diverged less at the molecular level than the extant African apes . . . Even at the morphological level, the common hominine ancestor may have had a stronger tendency towards certain human-like features such as bipedalism than is generally supposed." Among others, Hasegawa et al. (1985) have expressed similar thoughts.

There is no consensus among scientists on the reality and extent of biochemical retardation (Verhulst and Jaspers, 1997). Sarich and Wilson, for example, maintain that the separation between humans and anthropoid apes occurred very late and that the subsequent brief period of separate evolution definitively accounts for the slight biomolecular differences between them (M. Goodman et al., 1990). Conversely, Goodman (1996) maintains that even if the separation occurred as late as possible, the biomolecular data can be explained only by a very significant slowing of the rate of molecular evolution in the human evolutionary line.

Although research scientists do not agree on this issue, we can nonetheless assume that on the whole, human evolution on the molecular level is retarded in comparison to that of the average mammal. Consequently, human biochemistry tends to be more similar to the biochemistry of other primate species than is the case, for example, among anthropoid apes. On the biochemical level, we rediscover a phenomenon that is already familiar from the morphological perspective. The human species, because it has remained so primitive, shares a surprising number of traits with species that are not closely related (cf. *Figures 13, 14*, and *15*).

These relationships are evident, for example, in the degree of divergence among the DNA of different primate species (expressed as percentages). The following figures are taken from a paper by Barriel and Darlu (1990).

	Human	Chimpanzee	Gorilla	Orangutan	Macaque
Chimpanzee	1.46	—	—	—	—
Gorilla	1.45	1.82	—	—	—
Orangutan	2.96	3.37	3.32	—	—
Macaque	6.94	7.10	7.10	7.23	—
Spider monkey	10.12	10.29	10.29	10.45	11.73

(Divergence in the nucleotide sequences of ε-globin in higher primates, expressed in percentages. The human DNA sequence is always the most primitive in the sense of being most similar to the sequences of the other species.)

Systemic and Pulmonary Circulation

Like all warm-blooded species, humans have a dual circulatory system (*Figure 107*). The heart itself also consists of the two separate systems of its left and right sides. Each half of the heart has an atrium, through which blood enters the heart, and an atrioventricular valve connecting the atrium to the corresponding ventricle. Other valves allow the blood to pass from the ventricles into the pulmonary artery and the aorta. Oxygen-poor, carbon dioxide-rich blood coming from the organs enters the right atrium through two veins and then flows on into the right ventricle and from there through the pulmonary artery to the lungs. In the lungs, the blood picks up oxygen and releases carbon dioxide gas. This oxygen-rich blood then flows through the pulmonary veins to the left atrium. The circulation of blood between the right ventricle and the left atrium is called *pulmonary circulation.*

The blood then flows through the left atrium to the left ventricle, leaves the heart through the main artery known as the aorta, and is then distributed among the organs of the body, releasing oxygen and picking up carbon dioxide gas. This blood, which has become poor in oxygen and rich in CO_2, is then collected again in the right atrium. Circulation between the left ventricle and the right atrium is called *systemic circulation.*

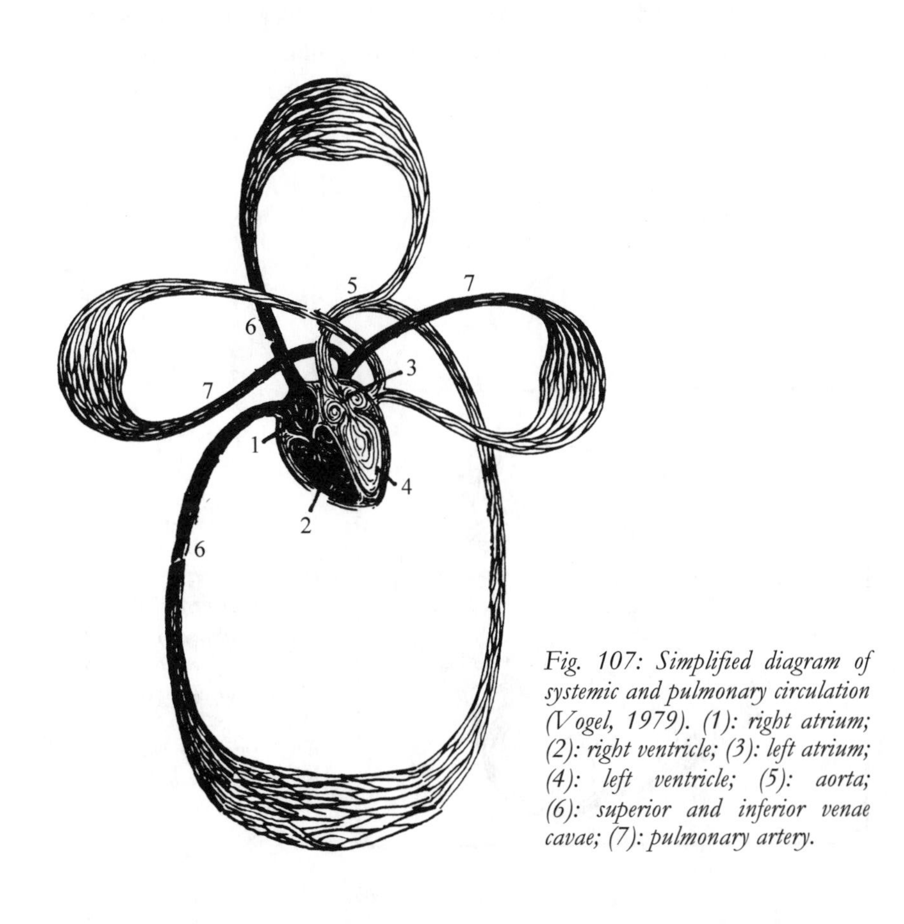

Fig. 107: Simplified diagram of systemic and pulmonary circulation (Vogel, 1979). (1): right atrium; (2): right ventricle; (3): left atrium; (4): left ventricle; (5): aorta; (6): superior and inferior venae cavae; (7): pulmonary artery.

The Cardiac Skeleton

The cardiac skeleton, a more or less flat plate of connective tissue penetrated by four valves, lies between the atria and the ventricles (*Figure 108*). The left and right atrioventricular valves (the bicuspid and tricuspid valves) have two and three cusps, respectively. On the ventricle side, musculotendinous structures called the chordae tendineae hold each of these cusps in place, permitting blood to flow only from the atrium to the ventricle. Any pressure differential in the opposite direction closes the valve. The other two valves that breach the wall of the cardiac skeleton belong to the aorta and the pulmonary artery, which also breach the wall. These valves have three cusps each.

The connective tissue of the cardiac skeleton is very dense, especially in its center (between the atrioventricular valves and the aortic valve), where it is breached only by the atrioventricular node of the heart's conduction system. The muscles of the ventricle walls attach to the cardiac skeleton, which almost completely separates the atrial walls from those of the ventricles.

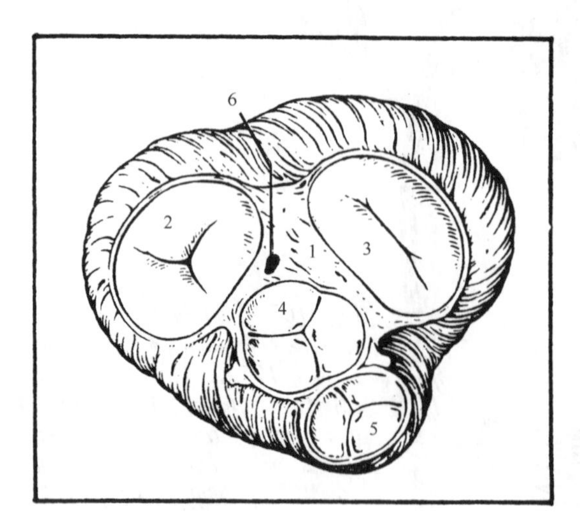

Fig. 108: The cardiac skeleton, seen from the atrial side. (1): connective tissue of the cardiac skeleton; (2): the tricuspid valve (right atrioventricular valve); (3): the bicuspid valve (left atrioventricular valve); (4): the three-cusped aortic semilunar valve; (5): the three-cusped pulmonary semilunar valve; (6): the atrioventricular node of the conduction system; (7): ventricle wall (Kahle et al., 1975).

In many mammals, the proverbial "heart of stone" becomes a material reality in that an ossification tendency is apparent in the center of the cardiac skeleton. The so-called os cordis has not yet been researched in all mammals but has been studied extensively in cattle. In the beef heart, one or two irregularly shaped bits of bone, complete with marrow and several centimeters in diameter, are found near the aortic valve. Their function, however, remains unclear (James, 1968).

The os cordis of older animals often develops out of cartilage formations in juveniles. The hearts of young porcupines, for example, contain cartilage that may become ossified in older specimens. Older pigs may have one or two small bones in the heart, whereas juveniles have only cartilage. At the transition to the aorta, deer species, chamois, and sheep have two bits of cartilage that ossify in older specimens. However, the hearts of many mammals—horses, dogs, cats, bears, rabbits, and some marine mammals—contain only cartilage. The loris (a prosimian) has one small medullated bone; other prosimians have only cartilage. Frick (1960) cites a remarkable study of two chimpanzee hearts conducted by Mourgues; one of the hearts contained a small medullated bone.

Although the research is still incomplete, cartilage formation and ossification in the heart are obviously common in mammals and possible even in chimpanzees, the closest animal relatives of human beings. Moreover, the absence of ossification in the heart is a juvenile trait; the os cordis develops only in older mammals. The consistent absence of bone and cartilage in human hearts is therefore a Type I characteristic. It is also possible that the absence of certain ossified structures (such as the os cordis or the penis bone) in humans is related to the typically low calcium phosphate levels in human blood (see *Figure 38* and pages 82-83).

The Embryological Development of the Heart

When the human heart first develops, at the beginning of the third week of gestation, it is tube-shaped and joined in front to two aortic arches that connect to two dorsal aortas. All vertebrates pass through this stage.

The heart—which has already begun to beat—grows faster than its surroundings and therefore begins to assume a characteristic bent shape (see *Figure 109*). The upper part of the cardiac tube, the bulbus cordis, moves to a lower spot along the ventral side. The tube's mid-section, the ventricular space, rapidly increases in size, while its lower section, the atrial space, moves higher in back. The atrium, although initially smaller than the ventricle, eventually catches up and surpasses the ventricle's growth, which accelerates again, however, once the atrium has risen above the level of the ventricle.

Around week 4, the heart's four chambers are already visible. The atrium develops two outgrowths, one on either side of the ventrally located bulbus cordis. The lower portion of the bulbus later develops into the right ventricle; the middle portion or conus cordis is also incorporated into the ventricle at a later stage. The upper portion of the bulbus is transformed into the arterial trunk.

Although the rudiments of all four chambers are visible in week 4, the heart is still a continuous bent tube. Several transformations occur before its final configuration is achieved. The arterial trunk divides into the aorta and the pulmonary artery (see *Figure 110b*). Next, a septum separating the two ventricles develops, so that each half of the heart has its own outlet. This separation process will be discussed in greater detail below (p. 215). A septum also begins to develop between the atria, but remains incomplete until birth due to idiosyncrasies of fetal blood circulation. The arterial system that supplies both atria develops out of the original end of the cardiac tube, the double sinus venosus.

The Aortic Arch and the Development of the Major Arteries

In a three-week-old embryo, the newly formed cardiac tube is connected at the head end to two dorsal aortas whose growth catches up at a somewhat later stage. The quickly growing head bends sharply forward, resulting in six so-called branchial arches that are all fed by a pair of newly developed aortic arches (*Figure 109*). The function of some of these aortic arches is temporary and is lost once the branchial arches develop into important parts of the jaws, ears, and vocal apparatus.

In this same phase, other aortic arches develop into major arteries (*Figure 110*). In most humans, three major arterial branches emerge from the upper portion of the aortic arch—the left subclavian artery, the left carotid artery, and the common trunk of the right carotid and subclavian arteries. We will later demonstrate that this branching pattern constitutes a Type I characteristic with regard to comparable structures in most mammals (page 220).

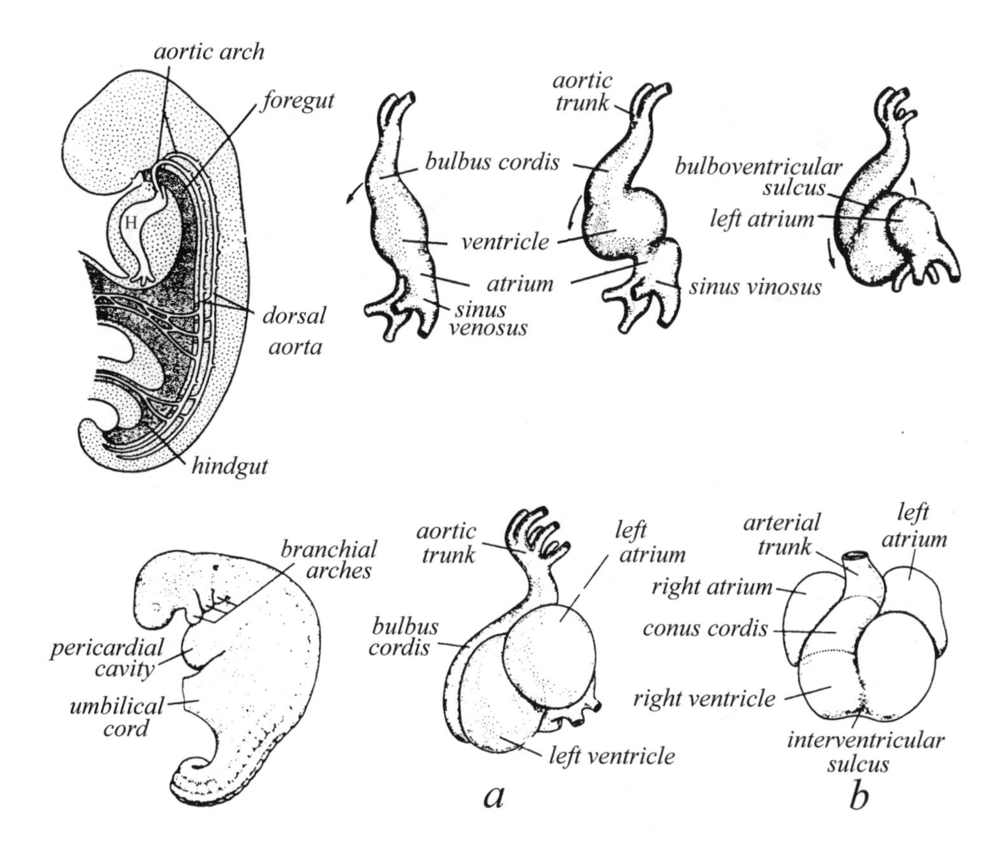

Fig. 109:

Above left: The location of the heart (H), the first pair of aortic arches, and the dorsal aortas of a three-week-old human embryo. Above right: Three stages in the development of the heart in the first few days of week 3 of gestation (side view; the ventral side is on the left). A second pair of aortic arches appears during the last of these stages.

Below left: A four-week-old embryo with three pairs of branchial arches. Below middle (a): the heart with three pairs of aortic arches (left side). Below right (b): The atrium has already risen above the ventricles; outgrowths (left and right atrium) have developed on either side of the bulbus (front view). The lower portion of the bulbus becomes the right ventricle and the upper portion the arterial trunk, which later divides into the aorta and the pulmonary artery (see Figure 110*). The mid-portion of the bulbus, the conus, is absorbed into the right ventricle at a later stage (Langman, 1966).*

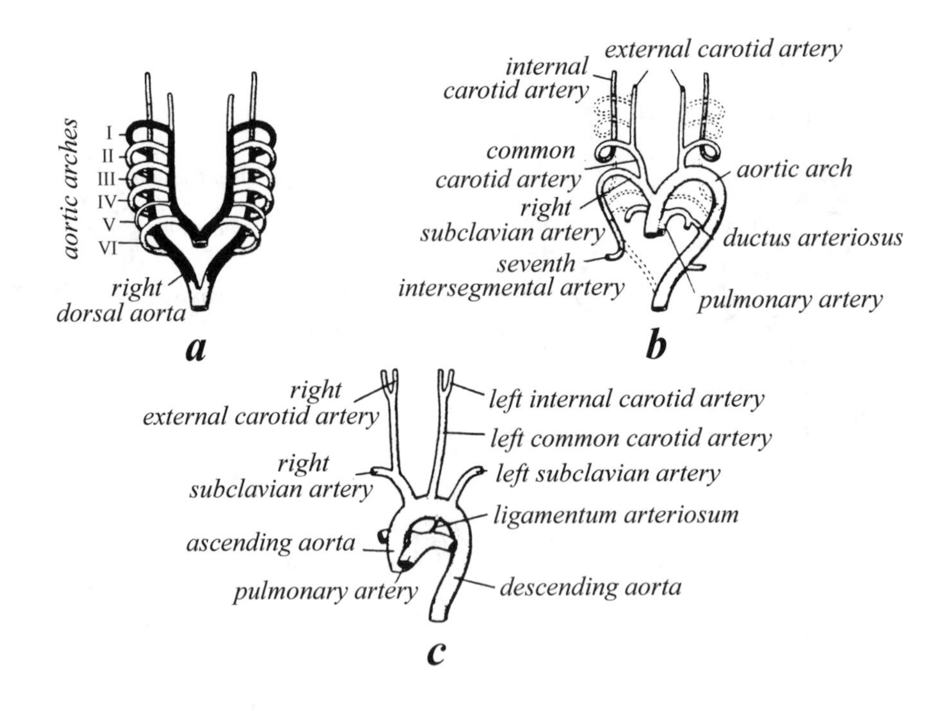

Fig. 110:

(a): Theoretical diagram of the original arterial pattern (ventral view, head at top). In the oldest of these structures (see Figure 109*), shown here in black, the arterial trunk divides into the first pair of aortic arches (I), which empty into the dorsal aortas (which are joined below). As the branchial arches develop, the five subsequent pairs of aortic arches also develop, and the ascending arches and dorsal aortas develop branches that ascend toward the head. This representation is purely theoretical; in human development the six aortic arches are never all present at the same time.*

(b): The solid lines indicate the parts of the original system that are transformed into major arteries (the first, second, and fifth pairs of aortic arches are not). The carotid arteries develop out of the third pair of aortic arches. In adults, the common carotid artery divides near the throat into the external and internal carotid arteries, which supply the face and the brain, respectively. The fourth pair of aortic arches develops into the major aortic arch on the left and the brachiocephalic trunk and right subclavian artery on the right. The sixth arch develops into the pulmonary arteries, whose common trunk branches off from the aortic trunk. Pages 216-217 compare this branching process in different terrestrial vertebrates.

(c): Normal locations (in 77% of cases) of the major arteries in adults (Langman, 1966).

The Mammalian Heart as a Fetalized Organ

In humans, the ventricles of the heart are completely separated. We find this trait in three groups of animals, namely, mammals, birds, and crocodilians. Other reptiles, amphibians, and fish species do not have separated chambers.

The degree of relationship among mammals, birds, and crocodiles has been a matter of debate since the nineteenth century. According to the orthodox view, birds and crocodile species are relatively closely related, while the connection between these two groups and the mammals is much more tenuous. Birds are supposedly descended from (or at least very closely related to) dinosaurs, which in turn are said to be descended from the group of primitive reptiles known as thecodonts (order Thecodontia of the subclass Archosauria), a group of primitive reptiles. Cotylosaurs are said to be the common ancestors of thecodonts and mammals. It is interesting to note that according to this current view, birds and crocodiles are more closely related than, for example, crocodiles and lizards (*Figure 111*).

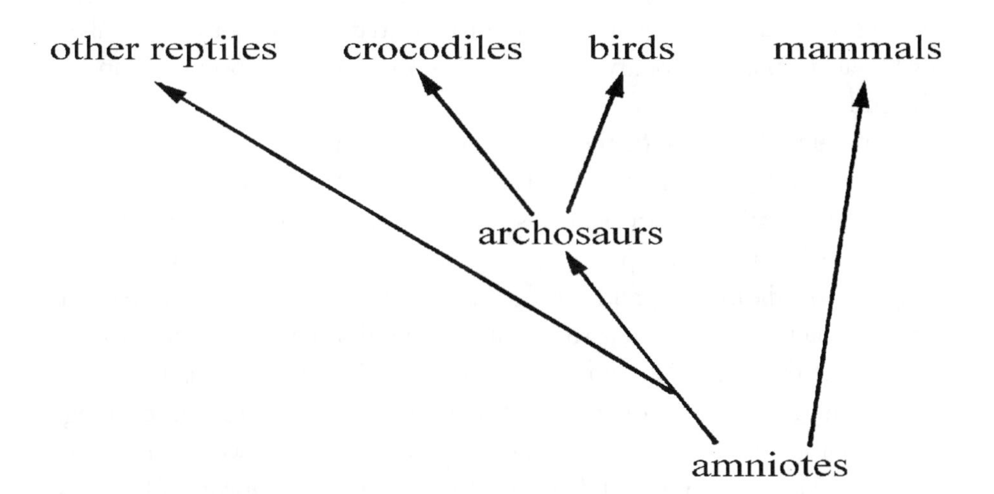

Fig. 111: The evolutionary tree of reptiles, birds, and mammals, according to Benton (1990).

This view, however, has been contested in at least two respects. First, certain researchers deny that mammals evolved from primitive reptiles. As early as the nineteenth century, Thomas Huxley, Darwin's great champion, traced the descent of mammals directly to amphibians. As we will see later, mammalian morphology and ontogeny are often surprisingly simple and straightforward in comparison to reptiles. Second, birds and mammals share a whole series of traits (many of them related to the structure of the circulatory system) that are absent in all reptiles. Hence, a group of unorthodox researchers concludes that the generally accepted view is incorrect and that birds and mammals are more closely related (see Bishop and Friday [1987] and the references they list). A comparison of the heart and circulatory system in mammals, birds, and crocodiles strongly supports these opponents of the orthodox view. The next three sections of this chapter will explore several points in detail.

Development of the Arterial Valves

The bulbus, which forms during the heart's early embryonic development, is attached directly to the aortic trunk (see *Figure 109*). Four areas of thickened tissue, which will develop into the triangular cusps of the valves of the aorta and pulmonary artery (see *Figure 108*), develop inside and at the end of the bulbus.

Initially, the shape of these four bulges is very similar in amphibians and mammals (see *Figure 112*). In both cases, the pattern is bilaterally symmetrical with two larger bulges and two smaller ones. In amphibians, the original shape is not preserved as development continues; one of the bulges soon disappears, and the final shape of the valve is asymmetrical. In mammals, the elongated shape of the opening paves the way for the division of the original arterial trunk (see *Figure 113*) into the aorta and the pulmonary artery. As soon as this separation reaches the four thickenings in the bulbus, each of the two larger ones are split into two smaller ones (*Figure 112*), producing the two times three cusps of the aortic and pulmonary artery valves.

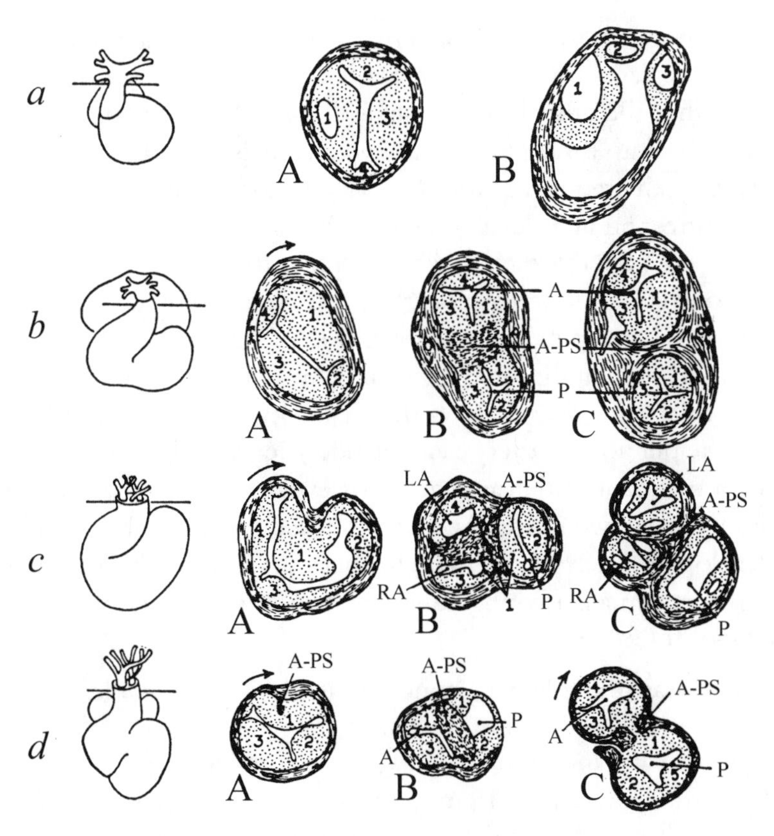

Fig. 112: Location of the endocardial thickenings (labeled 1 to 4) from which the valves of the aorta and pulmonary artery develop. On the left is an outline of the respective heart; the line indicates the level of the cross-sections. (A): aorta, (P): pulmonary artery, (A-PS): aortopulmonary septum, (RA): right aorta, (LA): left aorta.

(a) Tadpole at 19 mm (A) and 30 mm (B). The symmetrical pattern of the four thickenings in stage A is lost in stage B.

(b) Pig, three embryonic stages. The initial form (A) is the same as in amphibians. In subsequent stages, the septum between the aorta and the pulmonary artery develops perpendicular to the two large thickenings. In this instance, the symmetry and different sizes of the two pairs of thickenings in stage A are functional and predictive of subsequent stages.

(c) Alligator, three embryonic stages. The symmetry initially present in amphibians and mammals is absent here. In later phases (B and C) the aorta is not only separated from the pulmonary artery but is also subdivided within itself, a separation that does not occur in mammals.

(d) Chicken, three embryonic stages. Beginning with a reptile-like initial form, the bird achieves a mammal-like final form (separation of the aorta and the pulmonary artery, valves with three cusps each) (Shaner, 1962).

In alligators, the original pattern as it appears in amphibians and mammals is replaced by four asymmetrical thickenings, one of which is very large, while the other three are small (*Figure 112*). The original arterial trunk separates into three rather than two channels; in addition to the pulmonary artery, the reptile has not one but two aortas (see below).

The corresponding development in birds is quite remarkable. The original shape of the thickenings is very similar to the arrangement in reptiles, with one large and three small thickenings, one of which soon disappears. (Birds, like mammals, have only one aorta.) A single wall divides the arterial trunk, splitting the large thickening into two parts on one side and separating thickenings 2 and 3 on the other. In this stage, therefore, the aorta and the pulmonary artery each include one small thickening (either 2 or 3) and a fragment of thickening 1. However, the valves that develop in birds do not have two cusps. At a later stage, an additional thickening (4 or 5) develops in each of the two channels. As a result, birds ultimately achieve the mammalian configuration of two arteries, each with a valve consisting of three cusps.

We can draw the following conclusions from this example:

- Mammalian (and human) ontogeny is consistent and straightforward. The initial pattern is the same as in amphibians, but mammalian development (unlike that of amphibians) continues in a straight line. Shaner (1962) writes, "The mammalian bulbus is a good example of fetalization: it retains the embryonic amphibian distal cushions and adapts them to mammal needs. The reptilian bulbus is an example of progressive deviation: it develops a new, large bulbar cushion and adds a Y-shaped aortico-pulmonary septum found only in reptiles." The original pattern (two large and two small thickenings, arranged in bilateral symmetry) makes sense only in view of the subsequent separation of the aorta and pulmonary artery, which does not occur in amphibians.

- The specialized pattern of thickenings in reptiles serves a complicated division of the arterial trunk into three channels.

- At the beginning of their embryonic development, birds reveal their relationship to reptiles. Subsequently, however, birds follow a complicated route (dissolution of one of the original thickenings and the formation of two additional ones) to arrive at exactly the same final form as mammals. This phenomenon seems to point to the existence of a "guiding factor" that aspires to a specific form, which is achieved by direct means in humans and mammals and indirectly, by way of a unique detour, in birds.

Development of the Interventricular Septum (Shaner, 1962)

After undergoing metamorphosis, amphibians have separate systemic and pulmonary circulatory systems, although the right and left atria of the amphibian heart are separated while the left and right ventricles are not. In mammals, birds, and crocodilians, however, the ventricles are separated; hence, the aorta is supplied with blood by the left ventricle and the pulmonary artery by the right.

In mammals, this separation develops very simply. Beginning at thickenings 1 and 3 (which are cut in half when the arterial trunk divides; see *Figure 112*), two ridges of tissue grow from the interior of the bulbus in the direction of the ventricles. After the aorta and the pulmonary artery separate, the ventricles also separate as the two bulbar ridges grow together. It is interesting to note that the ridge that begins at thickening 1 is visible even in amphibians, where no ventricular separation occurs.

In crocodilians, the separation is a much more complicated process. These species have two aortas, each arising from one of the ventricles. In crocodiles, the two bulbar ridges (which begin at thickenings 1 and 4), together with a tripartite septum that grows from the arterial trunk into the bulbus, accomplish the necessary separations.

To a great extent, the separation process runs parallel in birds and mammals. In birds, however, as we have already seen, the original shape of the thickenings is clearly reptilian. Birds have only a single aorta, and the two bulbar ridges emerge from thickenings 1 and 3. As in mammals, ventricular separation occurs when these two ridges grow together.

Here again, we see that the mammalian pattern is a simple and consistent continuation of the primitive stage that appears in amphibians, while reptiles undergo a much more complicated development. Birds overcome their original reptilian potentials to follow a mammal-like course of development.

Division of the Arterial Trunk (E. Holmes, 1975)

Reptiles are sometimes excluded as possible ancestors of mammals on the basis of the location of their major arteries. *Figure 113* gives a schematic representation of the location of the aortas and pulmonary and carotid arteries in mammals, birds, and reptiles. Reptiles have two aortas, and their carotid arteries branch off from the *right* aorta. The situation is similar in birds; the only difference is the disappearance of the *left* aorta. Mammals, in contrast, have no *right* aorta, and their carotid arteries branch off from the left aorta. In addition, mammals have subclavian arteries.

Any trace of a second aorta is conspicuously absent in the embryonic development of birds and mammals. Holmes (1975) is unable to give a simple explanation for this absence but points out that in reptiles the division of the arterial trunk takes place in two steps. (See *Figure 113*.) First a separation develops between the future pulmonary artery and the aortic channel, which in turn divides into two aortas as a second wall develops. In mammals, the second step does not occur. Holmes writes, "The only modification necessary is elimination of the subdivision of the ventral aorta into right and left aortas. This transformation could be accomplished readily by a minor neotenic change." (In this case, "neotenic change" means fetalization in the Bolkian sense.)

We saw earlier that, with regard to many features, human beings are both paedomorphic and unspecialized in comparison to anthropoid apes. The final stage in the development of a full coat of fur, for example, does not occur in humans. Similarly, mammals are neotenic in comparison to reptiles in that the final stage of arterial separation does not occur. According to Holmes, the same retardation process occurs—completely independently—in birds. Here again, birds develop in the direction of mammals but from a reptilelike point of departure. As we said earlier, phenomena of this sort suggest the presence of a "guiding factor" whose effect, although evident primarily in the mammalian line of evolution, also has an impact on the separate evolutionary path of birds.

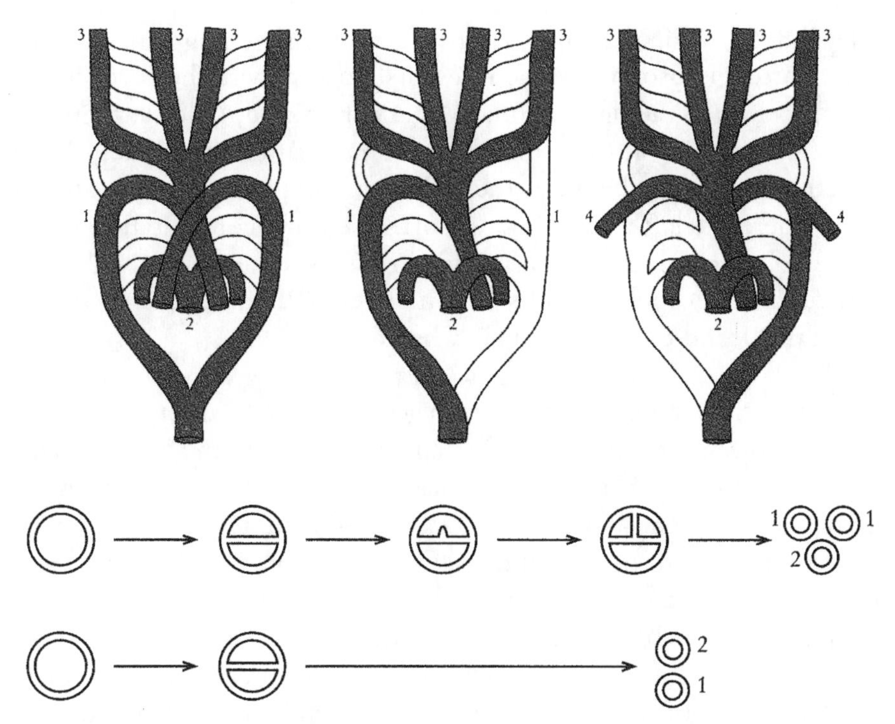

Fig. 113: Above: Schematic representation (ventral view) of the location of the major arteries in (left to right): reptiles, birds, and mammals. (1): aortas, (2): pulmonary arteries, (3): carotid arteries, (4): subclavian arteries. The aortic arches that do not develop into major arteries are indicated in white (cf. Fig 109*) (Ihle et al., 1924).*

Below: Diagram of division of the arterial trunk in reptiles (upper row) and mammals or birds (lower row). (1): aorta, (2): pulmonary artery.

The Aortic Arch in Humans and Animals

Bolk (1920) and Ariëns-Kappers (1931) both cite the configuration of the major arteries in humans as an example of retardation. As early as 1895, Keith conducted an extensive comparative study of the arterial system in humans and monkeys and discovered that in lower primates the four major arteries (the two carotid and two subclavian arteries) generally arise from a common trunk, which in turn branches off from the aortic arch. In baboons and gibbons, the left subclavian artery is already independent, arising directly from the aorta. In orangutans, the left carotid artery is also independent, but its juncture with the aortic arch is still very close to the right common arterial trunk. In African anthropoid apes, this trend in the direction of the human pattern continues; that is, there is more space between the right common arterial trunk and the point where the left arteries branch off. The same configuration is also found in humans and— remarkably—in the most highly evolved platyrrhines, namely, capuchin and spider monkeys (Parsons, 1902). (See *Figure 13*, p. 29, and the comments in Chapter 1.)

Keith interpreted his findings to mean that a progressive line of evolution ultimately culminates in the human gestalt. From the embryological perspective, however, the human configuration is essentially the most primitive.

Lewis (1923) published an especially fine example of the sequence of arterial development in sheep (see *Figure 114*). In a 14-mm-long sheep embryo, the pattern of arterial division is still qualitatively the same as in humans; that is, the left carotid and left subclavian arteries branch off from the aortic arch at separate points, while on the right side both arteries are connected to the aortic arch by a short common trunk. In an 18-mm-long embryo, the left carotid artery already shares a long common trunk with the two right arteries, while the left subclavian artery is still directly connected to the aorta. And finally, in the 47.5-mm-long embryo, the adult configuration is already achieved; that is, both carotid arteries emerge from the top of a common trunk, while the subclavian arteries branch off lower on the same trunk. (See Seidel and Steding [1981] for information on how such a shift occurs.) From the qualitative perspective, therefore, humans remain at the stage of a 14-mm-long sheep embryo.

Studies by Parsons (1902) and Meinertz (1975) demonstrate that the primitive, minimally branched structure of the human arteries is rare among mammals. In most animal species, either the four major arteries share a common trunk (as they do in sheep) or only the left subclavian artery is independent. Most authors associate the tendency toward increased branching with the narrower, wedge-shaped ribcage of quadrupeds. The fact that the forelimbs attach close to the neck results in a relatively narrower front portion of the ribcage (where the four major arteries that supply the head and forelimbs diverge), so it makes sense that the arteries should remain united longer in a common trunk. This effect is intensified by the fact that narrowing of the upper thorax almost automatically entails a downward (i.e., caudal) shift in the location of the heart. If the thorax is wide, however, and the arteries must move out in very different directions, the less branched early embryonic configuration is more suitable.

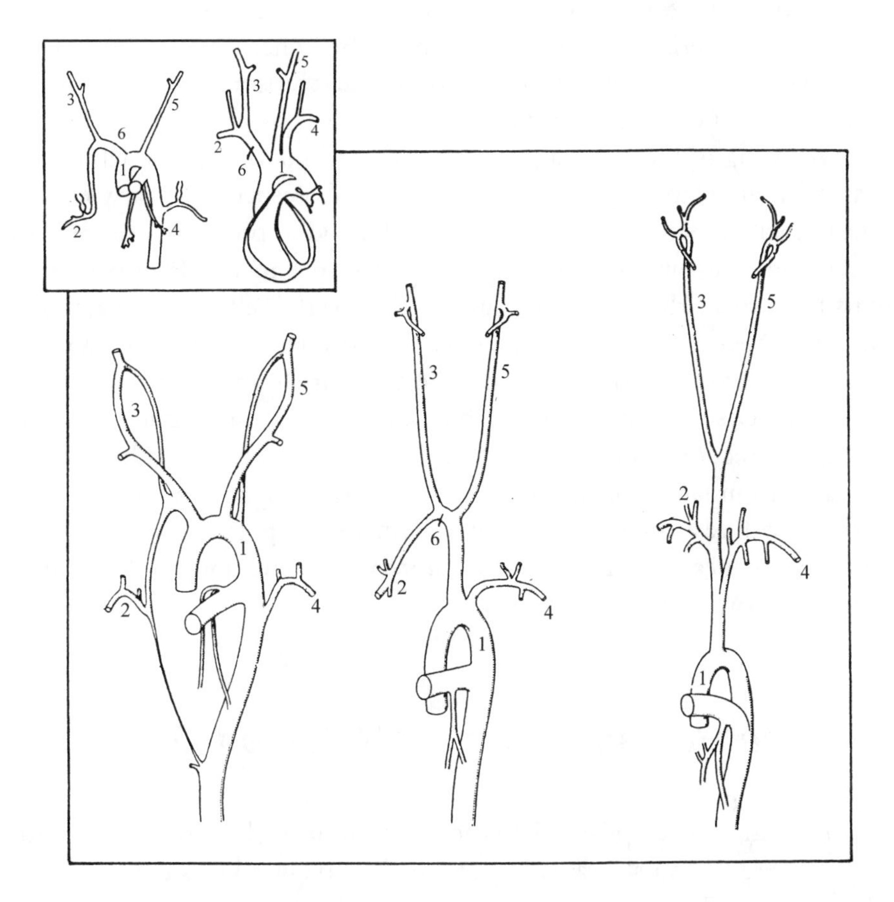

Fig. 114: Above: Aortic arch of an 8-mm-long human embryo (left) and in a premature fetus (right) (Congdon, 1902). The branching pattern in a full-term embryo is the same as in the adult (see Figure 110a*).*

Below: The aortic arch in the sheep embryo at lengths of (left to right) 14, 18, and 47.5 mm (Lewis, 1923). The last branching pattern is retained in the adult sheep. (1): aortic arch, (2): right subclavian artery, (3): right carotid artery, (4): left subclavian artery, (5): left carotid artery, (6): common trunk of the right carotid and subclavian arteries.

It is worth noting that the most primitive mammals, the monotremes (cloacal mammals), have wide ribcages and the same arterial configuration as humans. The human pattern is also evident in some marsupial and rodent species (e.g., beaver), in the Edentata (anteaters, armadillos, and sloths), and in seals and manatees. Most of these animals are either swimmers or burrowers, so a fairly flat ribcage is advantageous.

Human arterial structure, in which the branches do not unite to any great extent, can be considered a Type I characteristic, while the greater arterial unification typical of many mammals is a specialization related to the narrowing of the ribcage (see *Figures 65* and *66*, p. 132), which in turn is an adaptation to walking on all fours. In this respect, the early embryonic branching pattern of the arteries can be said to prefigure uprightness.

It becomes apparent that quadrupedalism entails additional deviations from the early embryonic pattern. Golub (1931), for example, points out that the bovine heart rotates as development progresses. When a bovine embryo is 5- to 6-mm long, the tip of its heart points in a caudal direction. Subsequently, however, the heart begins to turn until it points in a ventral direction. This rotation does not occur in humans. The heart of an upright human being and the heart of a standing cow are in the same position with respect to gravity, but in humans this position is achieved by retaining the heart's universal early embryonic orientation, while in quadrupeds this orientation shifts.

The Human Lung as a Retarded Organ

The lungs develop later than the heart. Initially, the lower zone of the lungs, where the most branching occurs, lags behind in its development. During the earliest phase of growth, therefore, the lungs tend to grow broad rather than long, and this pattern is preserved throughout the growth period. At age six, the heart has achieved 78 percent of its adult diameter in girls and 81 percent in boys. In comparison, the lungs have achieved only 66 to 67 percent of their adult width (in line with the relative size of the body as a whole) and only 62 to 63 percent of their adult length. That is, the heart preserves its head start over the lungs, and lung width remains ahead of lung length. In puberty, all three dimensions are subject to a growth spurt, which affects lung length roughly six months later than heart diameter and lung width (Simon et al., 1972).

General Comments on the Lungs' Location Within the Thorax

Pulmonary tissue is very light and compressible. Logically enough, it is also located above the other organs. In the upright human being, this location is achieved simply by retaining fetal circumstances, with the centrally located diaphragm almost perpendicular to the spine and the heart directly above the diaphragm.

In marine mammals, the lungs are dorsally located and the sternum is very short. This shift in the location of the lungs is related to stability while swimming. It is possible to trace the gradual reorientation of the diaphragm in the ontogeny of aquatic mammals. Even in these animals, the diaphragm is initially nearly perpendicular to the spine. Later, its orientation becomes more diagonal so that the lungs can elongate along the spine. Incidentally, the lungs of quadrupeds also tend to elongate along the backbone (see *Figure 115*).

The location of the lungs in the upper part of the body favors uprightness. Since the lungs are light, situating them higher lowers the center of gravity, which increases stability. In this respect, too, the prototypic mammalian structural plan, in which the centrally located diaphragm develops at right angles to the spine, prefigures uprightness.

When the lungs shift to the dorsal side, the airways must become longer. In the human body, where this shift does not occur, several other factors also contribute to reducing the average length of the airways:

- Because the face is parallel to the spine in humans and muzzle development is suppressed, the length of the air passage through the skull is reduced.
- Upright body posture allows the upper part of the human thorax to increase in volume (Tredgold, 1897). In quadrupeds, the ribcage is very narrow as a result of the location of the forelimbs, and the lungs are generally located lower in the chest.

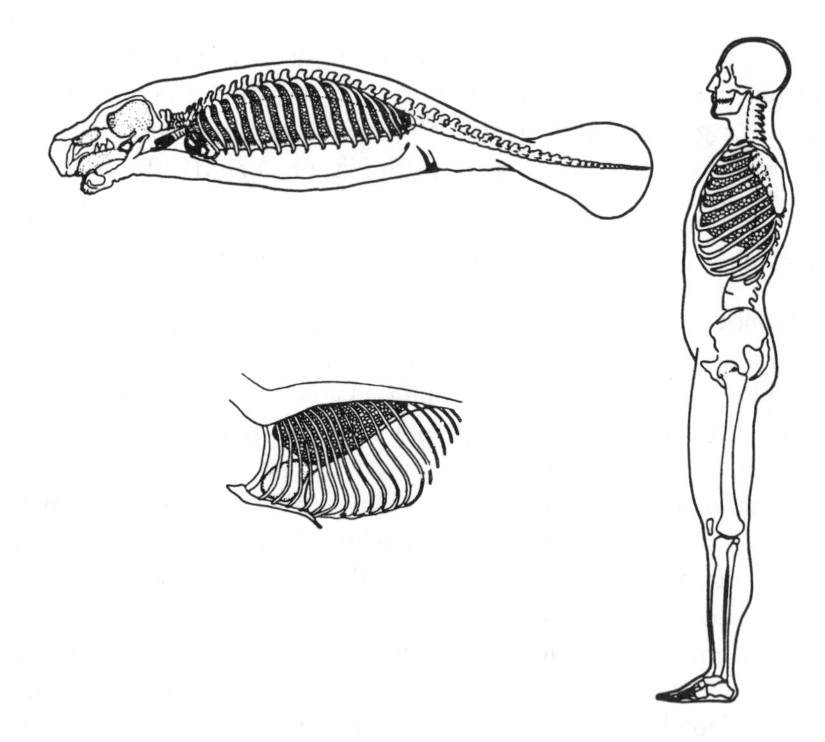

Fig. 115: Location of the lungs in manatees, elephants, and humans (von Hayek, 1956). In large mammals, the lungs tend to shift to a location above the heavier organs. This dorsal location, which guarantees stability while swimming, makes good sense in aquatic mammals. In humans, the lungs are located closer to the head, in the broader part of the thorax.

Retardation Phenomena in the Human Lungs

Location of the Human Heart; the Infracardiac Lobe

The heart begins to develop earlier than the lungs. In the fourth week of gestation, when the heart has begun to beat and its general form is already visible, the lungs are only a bud with two caudal outgrowths. The more the lungs catch up, the more they tend to grow around the heart and alter its position. The size of the heart relative to the ribcage decreases consistently as growth continues. In the embryo, the heart occupies 61 percent of the width of the ribcage; this figure drops to 57 percent in the newborn and 47 percent in the adult (Walmsley and Monkhouse, 1988).

In early stages of development, the heart of any mammal touches both the sternum and the diaphragm. This contact may be lost in later stages due to expansion of the lungs. In certain mammals such as guinea pigs, the adult heart is surrounded on all sides by lung tissue. In humans, however, the heart's original contact with sternum and diaphragm is maintained.

Loss of contact between the heart and the diaphragm, which represents a qualitative break with the fetal situation, occurs in almost all mammals except humans and higher primates. Von Hayek (1960) repeatedly points out that during the early stages of development, the pericardium and the diaphragm constitute a single layer of tissue in all mammals and that this situation is permanent only in higher primates. In most mammal species, one pulmonary lobe grows between the heart and the diaphragm, forming an indentation known as the infracardiac recess or sinus.

> In all vertebrates, the base of the pericardium develops out of the transverse septum, which also forms part of the rudimentary diaphragm. The pericardium remains connected to the diaphragm in humans and in a few ape species, but in most monkeys (as in most other mammals) this contact dissolves in the course of development as the infracardiac recess inserts itself between pericardium and diaphragm . . . Thus in the ontogeny of humans and anthropoid apes, in whom this recess is absent, there can be no talk of 'regression' of the infracardiac recess or of the pericardium 'growing together' with the diaphragm. Rather, the ontogeny of humans, chimpanzees, orangutans, and gorillas remains primitive with regard to this particular developmental process, while the other primate species transcend this primitive stage by developing an infracardiac recess.

Figure 116 shows this relationship in primates. In adult prosimians, pericardium and diaphragm are connected only by thin ligaments (dense, tendon-like bands of connective tissue). In macaques and *Hylobates* species (gibbons and siamangs), the connection is stronger. Development of the infracardiac recess is minimal in young *Hylobates* but quite advanced in adults. And finally, in anthropoid apes the lungs protrude only slightly beneath the heart. Thus we see increasing fetalization as we move through the primate series in the direction of the human being.

In the gorilla, this fetalization process is most advanced; the lungs do not penetrate beneath the heart at all. Ruge (1893) comments, however, that the infracardiac lobe is more visible in the gorilla than in any other anthropoid ape species (see *Figure 117*). Thus the lobe that is located below the heart in other mammals is probably specialized in gorillas. It seems that in this respect, too, the gorilla's species-specific ontogeny initially deviates from the human gestalt but then tends to revert to it, sometimes quite extensively. The corresponding pulmonary lobe is much less developed in chimpanzees and is either totally absent or barely visible in humans.

The absence of a pulmonary lobe below the heart is directly related to human uprightness, since in the upright position the heart would exert pressure on lung tissue located below it, hampering it in its function. This problem does not exist in quadrupeds. The original contiguity of the pericardium and diaphragm can therefore be considered a Type I characteristic that anticipates human uprightness.

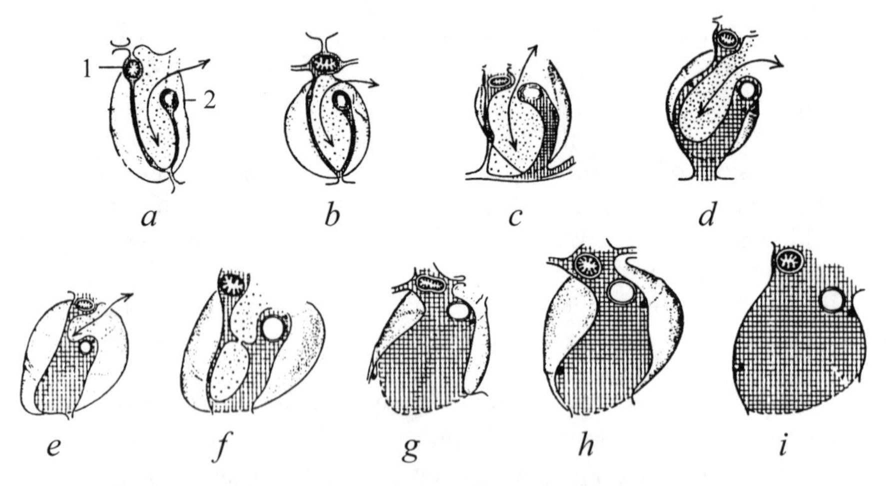

Fig. 116: Surface contact between the diaphragm on the one hand and the heart and lungs (infracardiac recess) on the other. The infracardiac recess (double arrow) and the area of contact between diaphragm and pericardium (cross-hatching) in: two lower primates (a, b); two lower catarrhines (c, d); a young siamang (e); an adult siamang (f); an orangutan (g); a chimpanzee (h); and a gorilla (i). (1): esophagus, (2): inferior vena cava (Ruge, 1893). The infracardiac recess is either mostly or wholly absent in the large anthropoid apes. It is less developed in the juvenile siamang than in the adult.

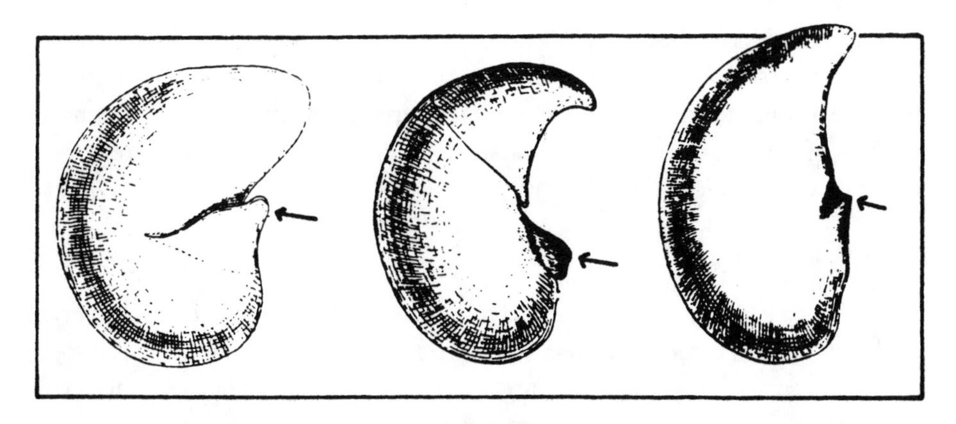

Fig. 117: Right pulmonary lobes (seen from below) of (left to right): gorilla, chimpanzee, orangutan. A rudimentary infracardiac lobe (arrow) is still visible in these three species. The development of this lobe is greatest in the gorilla, although the infracardiac recess is completely absent in this species. In humans, the infracardiac lobe is either very small or nonexistent (Ruge, 1893).

Lobulization of the Lungs

The lobes of the lungs are divided into little sections that Cuvier called the "lobuli pulmonis." These lobules are visible on the surface of the lungs as little polygons whose edges range in length from 0.5 to 3 cm. Lobules can shift position relative to each other. Von Hayek (1939, 1956) makes it clear that these lobules, although often numerous and clearly visible in fetuses and juveniles, often disappear completely in mature animals. One exception is the domestic pig, whose lungs still contain clearly visible lobules. (Domestication of a species is often associated with the retention of fetal or juvenile traits.) In general, pulmonary lobules disappear almost completely in adult primates, including chimpanzees, orangutans, and gibbons (von Hayek, 1960). Human beings are the only exception. Lobulization, although more apparent in the human fetus, also persists throughout the life of the adult. Engel (1959) notes that in the early stage of its development, lung tissue is still finely subdivided: "We know that in fetal development the bronchial tree appears before respiratory tissue is formed . . . Fusion of the small units is obviously a secondary process which may be arrested when some isolated lobes have been formed, or fusion may finally lead to a coherent organ." Quite possibly, the permanent lobulization of the human pulmonary lobes might therefore be understood as a Type I characteristic. Additional comparative studies of fetal, juvenile, and adult specimens of a variety of different primate species would clarify this point.

The Size of the Alveoli

The size of the pulmonary alveoli is only loosely related to overall body size but is directly related to the ratio of total alveolar gas exchange surface area to lung volume. Larger alveoli reduce the size of the total alveolar surface area. Hence we might expect less active mammal species to have larger alveoli, since they consume less energy per unit of time and need less oxygen. And in fact, of all mammals, the sloth has the largest alveoli, with diameters of nearly 400 μm. Surprisingly, humans have the next largest alveoli, followed by anthropoid apes and several very large whale species (von Hayek, 1956).

Von Hayek (1956) describes a radical transformation in rat lungs during the first few months after birth. The older alveoli, with diameters of nearly 100 μm, are replaced by newly developing ones with diameters of approximately 50 μm. As a result, the respiratory surface area increases considerably. This phenomenon is observed in several animal species (Burry, 1982; Dunnill, 1982). In humans, the size of the alveoli remains almost unchanged during the first three years of life; subsequent increases in size are in proportion to the overall size of the lungs—a phenomenon that is also observed in mammals. Postnatal transformation of lung tissue of the type that is observed during the first few weeks of a rat's life is not found in humans (Zeltner and Burri, 1987).

On the basis of exercise experiments conducted on rats, von Hayek (1956) concludes that postnatal transformation of rat lung tissue is caused by increased oxygen demand due to greater activity after birth. If this conclusion is correct, however, it means that earlier lung development in rats anticipates low oxygen demand. In this connection, Marcus (1928) notes, "The low index [respiratory surface per gram of body weight] in humans is striking because it is not much greater than that of reptiles and similar to that of sloths. This phenomenon is probably related to upright walking, which conserves energy." And in fact, as we saw in Chapter 6 (page 152ff.), upright walking is much more energy-efficient in humans than it is in quadrupeds, who must constantly exert energy in the muscles of their legs and neck to remain upright. From the Bolkian perspective, we can consider the transformation of lung tissue in rats a specialization that becomes necessary because the prototypic upright posture has been abandoned in favor of the more energy-intensive, specialized posture of quadrupeds. Von Hayek (1956) sees a connection between the size of the alveoli and that of the red blood cells: large alveoli have larger capillaries and correspondingly large red blood cells. The large diameter of human erythrocytes can also be considered a fetalization phenomenon (see page 200 above). Hence different retardation effects appear to be harmoniously linked.

The Bronchial Tree

The lungs develop through a process of successive branchings (see *Figure 118*) in which we can identify successive "generations." If we call the trachea itself generation 0 and the two main bronchi generation 1, and so on, we can count roughly twenty-two generations in the human lungs. From each generation to the next, the airways double in number but become increasingly narrow. The bronchial tree that develops through these systematic doublings can be measured very precisely if castings (of silicon, for example) are made. For example, it is possible to determine how the lengths or cross-sections of the air passages change over generations. The many interesting data compiled using this technique reveal a hidden lawfulness in the seemingly chaotic structure of the pulmonary tree. With regard to the length of their branches, pulmonary ramifications in the lungs are not at all symmetrical; as a rule, one branch in each pair is significantly longer than the other. Goldberger et al. (1985) discovered that in humans and at least some other mammals, the ratio of the lengths of the two branches of a ramification is always the golden section. In terms of the relative lengths of each pair of branches, the bronchial tree is constructed on the pattern of a Fibonacci series. The authors make a significant comment: "From an evolutionary point of view such scaling mechanisms are of interest because they lead to complex anatomic structures without apparent recourse to natural selection" (Goldberger et al., 1985).

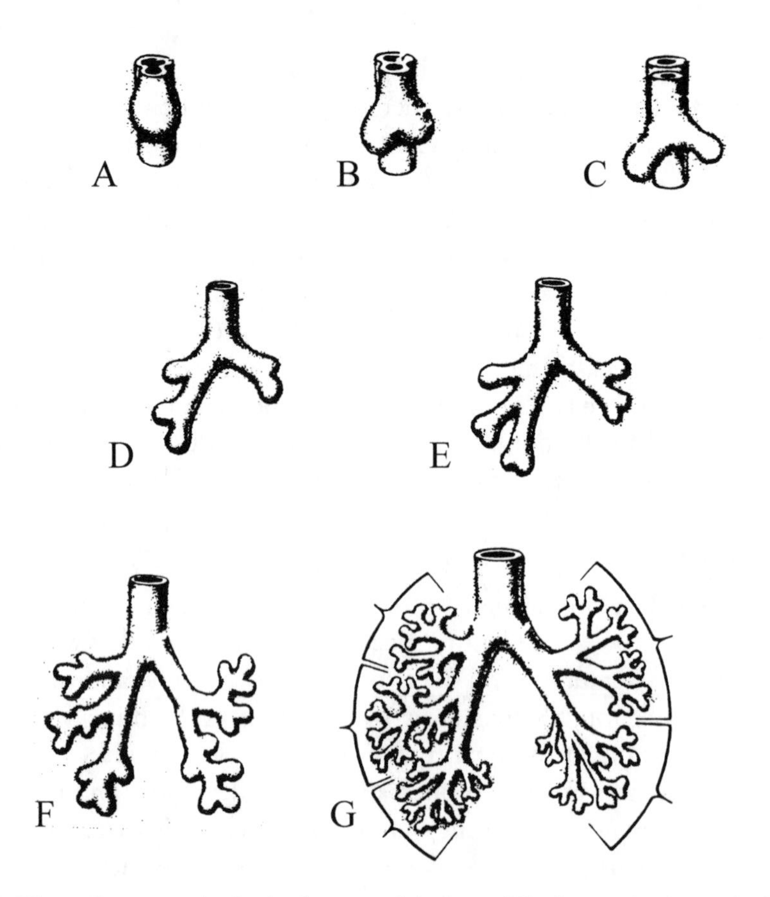

Fig. 118: The earliest stages in the development of the lungs. The lungs arise from a bud on the front side of the esophagus. Successive ramifications produce an increasingly finely branched pattern. (A-C): development in the course of the fourth week of gestation; (D-E): development during the fifth week; (F): sixth week; (G): eighth week. The lungs have three lobes on the right and two on the left; clearly, this distribution is prefigured from the earliest stages of development (Moore, 1982).

Phalen and Oldham (1983), who made informative discoveries about the characteristics of the human bronchial tree in comparison to the airways of specific mammals, state that "of all the studied mammalian lungs, those of humans appear to be the least heterogeneous." The most important distinguishing characteristics are:

- The overall shape of the human lungs is quite spherical, while typical mammalian lungs tend to be more elongated.

- The ramification pattern in human lungs is fairly symmetrical with regard to the diameter of the branches; that is, the diameters of the two branches in each pair are comparable. In contrast, animal lungs tend to develop a single main branch with many smaller, "bushy" laterals.

- The length-to-diameter ratio of the trachea varies widely from species to species. Phalen and Oldham (1983) set this figure at 20 for goats and ferrets, 14 for guinea pigs, 12 for rabbits, and 8.6-10.6 for dogs, monkeys, rats, and hamsters. "The human is again extreme at 6."

The very uniform shape of the human lungs is related to fairly simple scaling mechanisms that have nothing to do with natural selection. Now that fundamental scaling relationships have been discovered with regard to the length ratios of branching pairs of airways, similar relationships between their branching angles or relative diameters may also be discovered. The more heterogeneous tendency of animal lungs can be understood as the result of the pressure of natural selection following the transition to the quadrupedal stance. As we discussed earlier (see pages 130-133), using the upper limbs for locomotion narrows the front of the ribcage. On the other hand, this posture encourages a more dorsal lung orientation, above the heavy organs such as the heart and liver. Both factors cause the lungs to lengthen, resulting in more asymmetrical branching in which one branch emerges as the main channel and continues relatively unchanged in its original caudal direction. It seems probable, therefore, that the branching pattern of the human lungs is governed primarily by inherent developmental principles, while nonhuman lungs are also influenced by the external pressure of natural selection.

When West et al. studied the scaling principle that governs the average diameter of the airways throughout the entire series of bronchial generations, they discovered that the branching pattern is influenced by not one but an infinite series of regularly distributed factors (West and

Goldberger, 1987; Nelson, West and Goldberger, 1990). The result of the generalized scaling principle is an exponential reduction in average bronchial diameter from one generation to the next, modulated by a periodic variation (see *Figure 119*). West and Goldberger make the important comment that "physiological structures like the bronchial tree are static only in that they represent the 'fossil remnants,' or casts, of an active developmental process." It should be possible, therefore, to determine compression or retardation relationships among the ontogenies of different species by tracing the variation in one parameter, such as the average diameter of the airways, through successive generations of the bronchial tree.

And in fact, the periodic variations in airway diameter as compared to the best linear fit, which describes a purely exponential relationship, reveal a clear retardation (*Figure 119*). In humans and animals, the overall pattern is the same. The first generations of airways correspond to the values of the best linear fit, but after that, diameters increase until they reach a maximum that is greater than the best linear fit. This happens in the third or fourth bronchial generation in animal lungs but only in the seventh generation in humans. Then the diameters begin to narrow again (that is, in relationship to the best linear fit; of course the absolute value always decreases with each successive branching). The minimum is achieved in the ninth to twelfth generations in animals and beginning with the fourteenth in humans. At the minimum, the human curve is initially very flat, and the airways remain close to the best linear fit until the final generation of branching. The animal pattern, however, continues to change after the minimum has been reached. The airways become wider again, and the average values for the last generations again rise above the best linear fit. This means that the human lung, which also has fewer bronchial generations, omits the final steps that occur in animals. Hence the branching pattern in animals is clearly compressed in comparison to the human pattern; maximums and minimums appear earlier in animals, and the later stages in the development of animal lungs are absent in humans. The human branching pattern can therefore be considered a Type I characteristic. The later developmental phases of the animal lungs, however, are a specialization related to the lungs' elongation due to the shift in position necessitated by quadrupedalism.

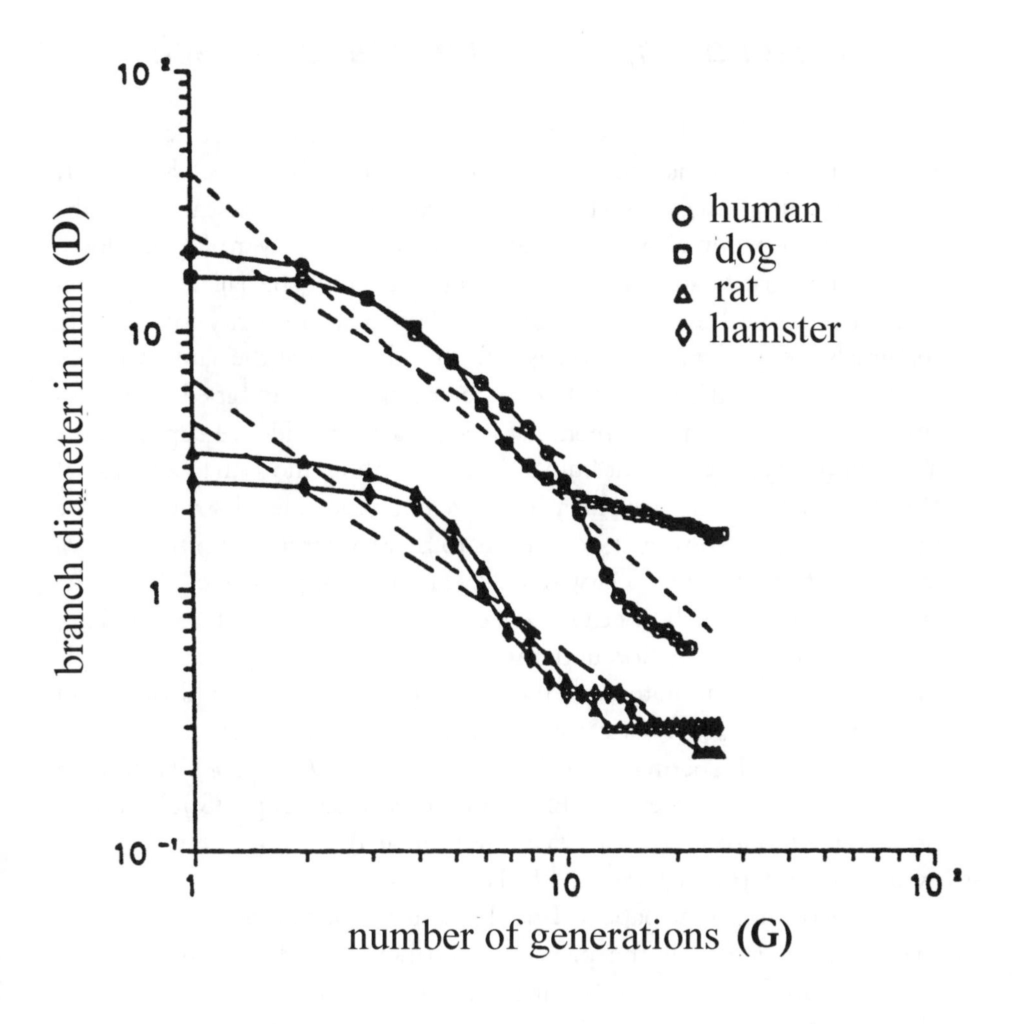

Fig. 119: Average diameter D of a bronchial branch (vertical axis, in mm) as a function of the number of generations G (horizontal axis) of the same branch, for humans, dogs, rats, and hamsters; the units on both axes are logarithmic. The broken line indicates the best linear fit, which is obtained by applying the formula ($D_G = f \times G^{-a}$) to generations 4 through 16. For humans, dogs, rats, and hamsters, constant a is 1.26, 0.86, 1.05, and 0.9, respectively, and constant f is 40.5 mm, 23.8 mm, 6.5 mm, and 4.5 mm (that is, f is the diameter of the first bronchial branch). All four data sets show harmonic modulation of the actual diameter of the bronchial tubes around the best linear fit (from Nelson, West and Goldberger, 1990).

Respiration, Uprightness, and Speech

Some of the above-mentioned retardation characteristics of the human lungs (such as the absence of the infracardiac lobe) are clearly related to uprightness. Such characteristics seem to indicate that the unspecialized mammalian structural plan is oriented toward bipedal locomotion and that quadrupedalism is a derivative and specialized trait. Initially, this hypothesis seems absurd. The great majority of modern mammals are quadrupeds, and fossil finds indicate that the same was true in the past. From the paleontological perspective, human bipedal locomotion is a very recent phenomenon. A great deal of evidence supports the assumption that the upright human body is nothing more than a marginal variant on an ancient and still very prevalent quadrupedal structure. The history of science, however, teaches us to be very careful with this type of evidence. For example, Darwin believed that lungs evolved from the swim bladders of fish species—a view that seems almost self-evident, since lungs cannot function under water, fish species are paleontologically older than land vertebrates, and most fish species have swim bladders but very few have lungs. Darwin's view is still propounded in standard recent works such as Lieberman's *Biology and Evolution of Language* (1984). In actual fact, however, the swim bladder evolved from lungs (Gould, 1989). Some unorthodox scientists even felt obliged to conclude that fish evolved from terrestrial animals (Jaekel, 1927).

In two important papers, Bramble and Carrier discuss several facts that speak in favor of the prototypic character of the upright bipedal mode of locomotion (Bramble and Carrier, 1983; Carrier, 1984). Carrier (1984) first determined that "for any given level of aerobic exertion, the tissues' need for oxygen will be met by a specific lung minute volume, which is the product of the breathing frequency and the tidal (lung) volume. However, as a consequence of the elastic resistance of the thorax, airway resistance, and dead-space volume, there is a specific rate and depth of breathing, for any minute volume, that maximizes ventilatory efficiency." To a great extent, upright walking frees human respiration from the constraints of locomotion. Human beings have no difficulty adapting their respiration rate to their oxygen needs by taking two, three, or four steps per breath or three steps for every two breaths, and so on. The transition to a different respiratory rhythm, with or without a change in walking speed, is effortless. Humans resort to the rhythm of one breath per step, which is so typical of animals, only during very heavy exertion, such as walking up a steep hill (Bramble and Carrier, 1983). In contrast, the respiratory rhythm of quadrupeds is strictly bound to the motions of walking or running. "When a quadruped trots or gallops, the muscles and bones of the thorax must absorb the shock of impacting on the front limbs. In addition, during a gallop all mammals exhibit some degree of axial bending in the dorso-ventral direction. This motion compresses and then expands the thoracic space with each stride. These aspects of gait restrict the ventilation of quadrupeds to one breath per locomotor cycle . . . In contrast, the upright bipedal gait in humans makes possible a large number of breathing patterns" (Carrier, 1984).

This biomechanical merging of respiratory and locomotor functions forces quadrupeds to specialize in a narrow range of energy-efficient speeds, while humans remain unspecialized in this respect and have no optimum running speeds. In humans and in animals, different speeds require different specific amounts of oxygen per minute. Each amount is best absorbed at a specific rate of respiration, which depends on the resistance of the airways, the elasticity of the ribcage, and so on. The ideal respiratory rhythm is generally not the same as the associated rhythm of walking or running. Humans can adjust each of these rhythms independently, and are therefore not bound to specific, optimally energy-efficient running speeds. For example, the amount of energy a human being needs to run one kilometer is the same at a speed of 8 km/h as it is at 15 km/h. In other mammals, these two rhythms are strictly linked. Unlike humans, quadrupeds have certain preferred, specialized speeds that represent the best possible compromises between the differing demands of locomotion and respiration. *The constraints imposed on respiration by the movements of running are a direct consequence of imposing quadrupedal locomotion on the universal structural plan of mammals.*

Carrier also points out that the need for thermoregulation while running imposes an additional limitation on quadruped respiration. Large mammals eliminate excess heat primarily through respiration (think of the panting of a dog, for example). Human beings, however, have exceptionally well-developed eccrine sweat glands and can produce more sweat per unit of surface area than any other mammal. The absence of a full coat of fur means that a walking human can eliminate excess heat very efficiently by perspiring. As Carrier mentions, "The human mechanisms of thermoregulation [sweating, as opposed to panting, and hairlessness] are probably more effective at dissipating metabolic heat produced in running than are the regulatory mechanisms used by other mammalian cursors..." "Since evaporative cooling in humans is not limited, as it is in panting mammals, by the rate and volume of lung ventilation, it represents a means of thermoregulation that is compatible with running."

As a result of minimal hair (which Bolk considered an example of human fetalization) and the exceptionally well-developed system of sweat glands in humans, the human lungs remain unconstrained by both thermoregulation on the one hand and locomotion on the other. The skin glands of most mammals are primarily so-called apocrine glands, whose secretions include not only fluid but also cytoplasm from the secreting cells. (Odor production is also a very evident part of the function of apocrine glands.) In contrast, eccrine sweat glands (whose primary function is to produce moisture as such) are most prevalent in humans. This phenomenon is often considered an obvious example of human specialization, but there is good reason to question this perspective. In an overview of the presence of eccrine sweat glands in mammals, Sokolov (1982) writes, "Eccrine glands are most numerous and best developed in higher primates. In other mammals they are found only on a thickened epidermis in areas subject to wear . . . In man, the eccrine sudoriferous glands achieve the highest development. They differ sufficiently from such glands in other mammals to constitute a human characteristic." The extraordinarily well-developed eccrine system in humans is not a species-linked specialization but rather the culmination of a development that becomes ever more apparent in higher primates. The Lemuridae, Lorisidae, and Tarsiidae, as well as the primitive platyrrhines, have eccrine sweat glands only on the soles of their feet and the pads of their fingers. In more highly evolved platyrrhines, such as howler monkeys (*Alouatta* spp.) or spider monkeys (*Ateles* spp.), eccrine glands are found on other parts of the body, although apocrine glands still predominate. In catarrhines, finally, eccrine glands are found on all parts of the body where hair also occurs. In macaques, apocrine glands are still much more common than eccrine glands, but the proportion is reversed in African anthropoid apes. This development reaches its culmination in human beings (for details, see Sokolov, 1982).

From the evolutionary perspective, however, body-wide distribution of eccrine sweat glands is not a very recent phenomenon. Strangely enough, such glands are found all over the body in certain primitive mammals such as the platypus or the tree shrew. The disappearance of this trait in most mammals can be attributed to compression of animal ontogenies, since the eccrine glands—with the exceptions of those on the soles of the feet and the palms of the hands—are the skin structures that appear latest in human development (Montagna, 1965). When the universal developmental pattern is compressed, these structures can easily disappear, as has indeed happened in most mammals. Only certain primitive mammals whose evolution has led in other directions retain the original trait, which otherwise appears only in more retarded primates. The human system of eccrine sweat glands, therefore, is a Type II characteristic. It appears as the culmination of a development that becomes increasingly evident in higher primates but makes sense only in humans because, in combination with sparsely haired human skin, it allows human hunters to exhaust animals like deer, kangaroos, or zebras in uninterrupted chase (Carrier, 1984). In a commentary included in Carrier's article, Scott wonders what nonhuman primates do with their efficient system of warmth distribution. Quite possibly, the answer is, "little or nothing." In higher primates, the system of eccrine sweat glands remains relatively nonfunctional. It appears in these species simply because their development is significantly retarded; as a result, the human gestalt is more explicitly prefigured in them.

Thus the human lung is not merely a retarded organ. From both a morphological and a physiological perspective, it is also a *liberated* organ. On the one hand, both the human lungs' retarded morphology (which is evident in the absence of the infracardiac lobe, for example) and the fact that human respiration is not restricted by locomotion are directly related to human uprightness. On the other hand, the absence of respiratory restrictions due to warmth regulation depends on retardation characteristics of both Type I (reduced hairiness) and Type II (a well-developed system of eccrine sweat glands). In turn, by liberating the lungs and their functions, this complex of retardation effects creates the physiological framework for the emergence of speech.

8

HUMAN REPRODUCTION

Rhythms in Human Reproduction

Reproductive capacity is one area in which the fetalization tendency in humans is highly advanced. The infertility of the fetus and child cannot, of course, persist in the adult.

Let's look at a few striking phenomena that contribute to a low reproductive rate in humans:

- The growth of the human sexual organs is significantly delayed in comparison to the growth of the body as a whole (see *Figure 36*, page 79).
- Human sperm has a notoriously high percentage of incompletely developed spermatozoids (often more than 40 percent), more than in any other primates except gorillas.
- The rate of early spontaneous abortion (miscarriage) is also extremely high in humans.
- In cattle, any single attempt at artificial insemination has a 75 percent chance of success, but in humans the chance of successful conception in any one cycle is often less than 25 percent (Voit, 1921).

- Women achieve optimal fertility very late (see *Figures 120* and *121*). No ovulation occurs in 60 percent of all menstrual cycles between the ages of 12 and 14, and this figure is still as high as 25 percent between the ages of 18 and 29. Maximum fertility is achieved only at age 26 to 30 (see *Figure 121*) (Döring, 1969).

It is interesting to note that a significant phase of adolescent infertility has also been documented in anthropoid apes. In chimpanzees, there is an interim period of one to two-and-a-half years when females menstruate and engage in regular sexual activity but are unlikely to conceive. The comparable period of infertility in humans appears to be extremely protracted. Short (1976) therefore concludes that humans are an extremely infertile species.

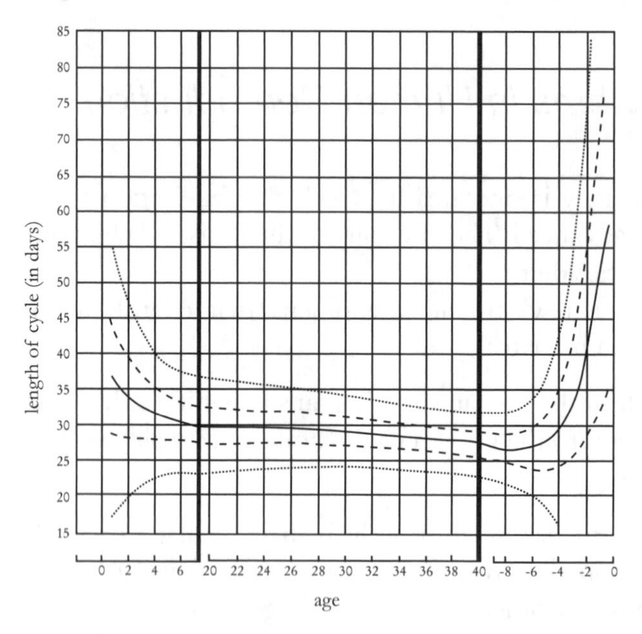

Fig. 120: Length of the human menstrual cycle as a function of age. The human reproductive span is divided into three blocks. Left, the first seven years after menarche; center, ages 20 to 40; right, the last eight years before the beginning of menopause. The solid line indicates average cycle length at each age. 50 percent of cycles fall within the range defined by the broken lines; 90 percent fall within the dotted lines. The first and last seven years are characterized by longer, less regular cycles, while great regularity is typical of the block from 20 to 40 years of age. Average cycle length during the most fertile span is approximately 29 days (Treolar et al., 1967).

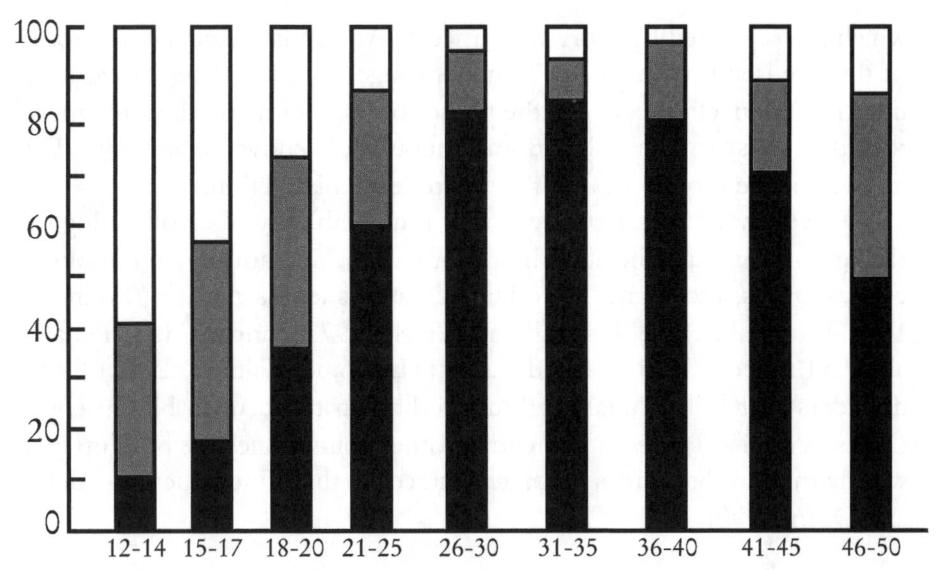

Fig. 121: Fertility in women as a function of age. For each age group (12-14 years, 15-17 years, etc.), the diagram shows the percentage of normal menstrual cycles (black), menstrual cycles without ovulation (white) and presumed sterile menstrual cycles (gray). Body temperature was used in delineating the three categories. (For details, see Döring, 1969.)

Reproduction and the Lunar Cycle

In his book *The Descent of Man*, Darwin pointed out that certain processes in human beings are linked to the lunar cycle, a fact that he saw as an argument in support of his hypothesis that humans are descended from aquatic, fishlike ancestors (1874).

As reflected in the tides, the monthly lunar rhythm is further subdivided, with strong (spring) and weak (neap) tides falling in alternate weeks. The tides and this clear weekly rhythm are caused by the combined effect of lunar and solar gravitational fields on the earth's surface (Jelley, 1986). The moon's effect on the tides is stronger than the sun's by a ratio of 1 to 0.46. Spring tide occurs shortly after full moon and new moon, when the effects of the sun and moon supplement each other. When the moon is in the first or last quarter, however, sun and moon counteract each other; neap tides occur, and high and low tides are less extreme. There is a considerable difference between spring tide and neap tide, which are separated by more than a week. At a rough estimate, the force of the tide is approximately 2.7 times stronger at spring tide than at neap tide. The connection between the phases of the moon and the strength of the tides is easy to establish and was undoubtedly known in antiquity. The division of the month into four weeks reflects this rhythm.

A weekly pattern is evident in a remarkable number of biological rhythms—in wound healing, blood pressure fluctuations, immunological responses, etc.—and in certain diseases such as malaria (Pöllman, 1984; Wu et al., 1990; Romball and Weigle, 1973; Sanchez de la Pena et al., 1984). (See *Figures 122* and *123*; cf. Levi and Halberg, 1982.) Sleep disorders that follow tidal rhythms are known to occur in blind people (Miles et al., 1977). Teeth and certain other solid tissues are built up in a weekly rhythm that leaves permanent traces in their fine structure (Aiello and Dean, 1990).

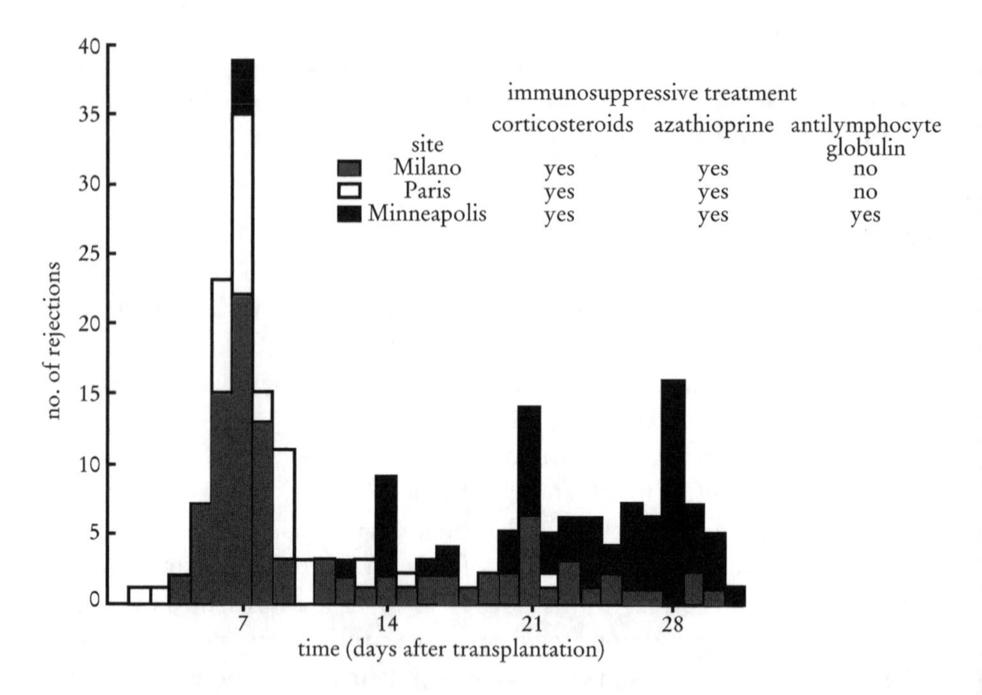

Fig. 122: Example of a weekly rhythm in humans: rejection of kidney transplants as a function of number of days after surgery; a summary of results from three transplant centers (Levi and Halberg, 1982).

Fig. 123: Two examples of weekly rhythms in animals.

Opposite:

Increase and decrease in the number of antibody-producing cells (PFC) in the spleen of rabbits after injecting a foreign protein on Day 0 (Romball and Weigle, 1973).

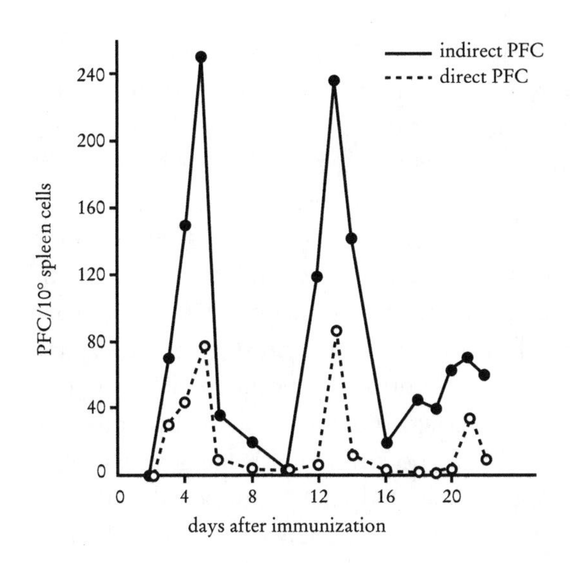

Opposite:

The consequences of infecting mice with malaria. Horizonal axis: days since infection. Vertical axis: number of mice who died on the corresponding day. The experiment was conducted with two strains of mice with different degrees of resistance to infection. The first strain (upper graph) generally died after an average of one week. In the more resistant strain (center graph) many mice survived the first week but died approximately 14 or 21 days later. The lower graph combines both results. There is a clear weekly rhythm in resistance to the disease (Sanchez de la Pena et al., 1984).

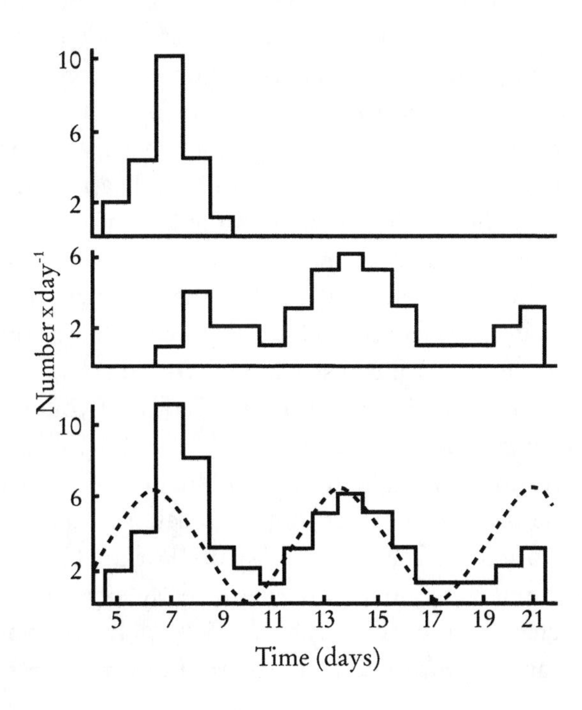

Lunar rhythms also play a very large role in the life of marine animals. The reproductive activity of many different fish and invertebrate species is intimately related to the phases of the moon, although the advantage of such behavior is not always obvious (Carter, 1989).

One extremely interesting example is the coho salmon (*Oncorhynchus kisutch*), whose growth follows a pattern of alternating weeks of faster and slower growth (Farbridge and Leatherland, 1987). This pattern is linked to the phases of the moon, with the slowest growth occurring at new moon and full moon and the fastest at the first and last quarters. The vernal equinox is an important point in time for this species, and the greatest change in growth rate occurs shortly before March 21. Remarkably, the investigators who studied this species observed this phenomenon even in fish raised in an artificially lighted laboratory and unable to directly perceive the moon or experience the tides. The mechanism that allows fish to respond to solar and lunar influences remains unexplained.

Evidently, other primitive animal species also possess a mysterious ability to regulate their behavior according to the phases of the moon. Brown and Park (1975), for example, discovered a lunar rhythm in the activity of flatworms raised in the laboratory. Brown (1954) also observed the same phenomenon in oysters. Admittedly, these reports have been strongly contested, but clear morphological traces of lunar rhythms have also been discovered in mussels, corals, and various fish organs (Pannella, 1971).

The moon's influence is also clearly apparent in the migratory behavior of salmon. Salmon hatch in the headwaters of rivers and undergo fundamental behavioral and metabolic changes shortly before returning to the ocean—an essential process that enables the young salmon to live in salt water. Concentrations of thyroxine, the hormone that promotes this metamorphosis, increase sharply around the new moon closest to the vernal equinox, and the seaward migration of young salmon peaks at the following full moon (Grau et al., 1981). (In this connection it is also interesting to recall that the combination of the vernal equinox and phases of the moon also determines the date of Easter and that the first Christians often used the fish as a symbol of the Christ.)

Lunar rhythms, however, clearly persist in terrestrial species not affected by the tides. Richter (1968) demonstrated that lunar rhythms appear not only in rats with thyroid disorders or brain damage but also in healthy specimens. Lunar cycles are also evident in humans (Verhulst, 1999a). The symptoms of many illnesses are influenced by lunar rhythms. Richter lists schizophrenia, epilepsy, and manic depression as examples, although these findings were heavily contested. Duration of menstruation and the length of the estrous cycle are not linked to the lunar cycle in all species. Richter discovered clear lunar cycles in male and female squirrel monkeys (*Saimiri*), a platyrrhine species (1968). The *Saimiri* estrous cycle, however, is only ten to twelve days long (Martin, 1990). Cycles of approximately thirty days are very common in both humans and animals. Hence Richter, like Darwin, concluded that the lunar rhythm must be an ancient phenomenon in natural history. Brown (1988) came to the same conclusion. He compared the gestation periods of 213 species of terrestrial mammals and discovered that the duration of pregnancy tends to approximate multiples of thirty days. This is also the case with human beings. The synodic month (the interval between two full moons or new moons) averages 29.53 days, and the average duration of pregnancy (from conception to birth) in humans is 266 days, or exactly nine synodic months (9 x 29.53 days = 265.77 days) (Menaker and Menaker, 1959). Medical literature, however, often lists the duration of human pregnancy as 270 days.

The length of the human menstrual cycle varies with age and among individuals, but its average duration is almost 29 days (see *Figure 120*) (Menaker and Menaker, 1959; Treolar et al., 1967). The results of several different studies, summarized below, confirm the relationship between the course of the menstrual cycle and the phases of the moon (Cutler, 1980; Cutler et al., 1987; Friedman, 1981):

- Approximately 30 percent of all women have regular cycles of 29.5 days (±1 day).
- Within this group, 65 to 70 percent of all menstrual periods occur in the "light" half of the synodic month; that is, in the two weeks around full moon. Most menstrual periods occur during or shortly after full moon.
- A similar but weaker tendency is evident in women with *irregular* cycles. Of course no relationship is evident in women with *regular* cycles of more or less than 29.5 days.

Some evidence also suggests a connection between the male urogenital system and phases of the moon, and there is a significant connection between lunar phases and hospitalizations for ureter blockage due to prostate swelling (Payne et al., 1989).

Seasonal and location-dependent rhythms (rutting season and the like), which are common among mammals, must be considered species-linked specializations. In contrast, the prototypic and seemingly very widespread lunar rhythm of reproduction is preserved to an exceptional extent in humans and most primates (*Saimiri* spp. are an exception). With the exception of most primate species, no mammals have estrous cycles that approximate the length of the synodic month. Cycles average approximately 11 days in sheep, 21 days in cows, five to six days in mice, etc. The lack of connection between most mammalian estrous cycles and the synodic month is often cited as an argument against the existence of such a connection in humans, but this objection is based on the unproven assumption that humans do not constitute a special case among mammals. The human being, however, is an exceptionally retarded being in whom a prototypic and universally distributed trait, such as the physiological lunar rhythm, is better preserved than it is in other mammals. The relationship of the human menstrual cycle (and the estrous cycles of many primates) to the synodic month can be considered a Type I characteristic.

Seven-Year Periodicity

Figure 121 shows that female cycles, which as such are governed by monthly and weekly rhythms, follow a seven-year pattern as they change character in the course of life. Girls typically undergo menarche around age twelve or thirteen (Amundsen and Diers, 1969 and 1973; Ellison, 1981). A study of girls in the former German Democratic Republic disclosed an additional seasonal rhythm. First menstrual periods occur primarily during winter or summer and much less frequently in spring or fall (Richter and Kern, 1987). This phenomenon remains unexplained. Menopause occurs around age fifty, that is, after the end of the seventh seven-year period. A remarkable symmetry is also evident in the time between menarche and menopause. During the seven years after menarche (the third seven-year period in a woman's lifetime), cycles are irregular and the pelvis and birth canal continue to undergo significant changes. Similarly, menopause is preceded by a seven-year period in which cycles again become irregular. Between these times of transition lies the period of greatest cyclical regularity, between the ages of 21 and 42. Furthermore, the evidence suggests that within this span of time, the central seven-year period (ages 28 to 35) is the most fertile and includes the greatest number of regular cycles. It is also the time when the average duration of a cycle most closely approximates the length of the synodic month.

Although humans share the lunar rhythm with animals, the seven-year rhythm seems to be unique to human beings. The idea of a seven-year periodicity in human development was already familiar to the ancient Greeks and Romans. According to Seneca, for example, "every seventh year imprints its characteristic attribute on human life" (Webster, 1951). The observers of antiquity also associated an astronomical rhythm with this seven-year periodicity. Of the seven traditional planets, the Moon is the fastest. Saturn is the slowest, returning to its original position in relationship to the fixed stars only after 29.5 years (= 29.5 x 365 days). This means that 365 synodic months elapse in the course of one Saturn revolution; that is, the ratio of Saturn's period of revolution to the synodic month is the same as that of one year to one day. This remarkable state of affairs undoubtedly made a great impression on the first people to precisely observe the mysterious movements of the planets against the background of the fixed stars.

Furthermore, Saturn's degree of brightness varies in a unique rhythm of seven-plus years, which attentive observers may have noted. This rhythm is caused by the planet's system of rings (see *Figure 124*). Because these extremely thin rings lie in a plane that intersects the plane of the Earth's orbit at an angle of 28°, we see the ring system from different angles over the course of a "Saturn year." When Saturn is between the zodiac constellations of Leo and Virgo, the plane of the rings coincides almost exactly with an observer's line of sight, and the rings are almost invisible even with a strong telescope (see *Figure 124*). Roughly seven years later, the plane of the rings lies at its greatest possible angle to our line of view, and although the rings are still not visible to the naked eye, the planet is noticeably brighter. Over the course of the next seven years, Saturn's brightness again declines to its minimum level, and so on.

Although the interval between minimum and maximum brightness varies from slightly less than seven years to nearly eight, it averages 29.5 years divided by four, or 7.375 years. Thus a full cycle of changing degrees of brightness takes 2 times 7.375 years or 14.75 years.

It is remarkable that the brightness cycles of the very fastest and the very slowest of the seven planets of classical astronomy (those visible with the naked eye) relate to each other as a day relates to a year, that their rhythms include oscillations of somewhat more than seven and fourteen days or seven and fourteen years, respectively, and that corresponding rhythms manifest in the human life span.

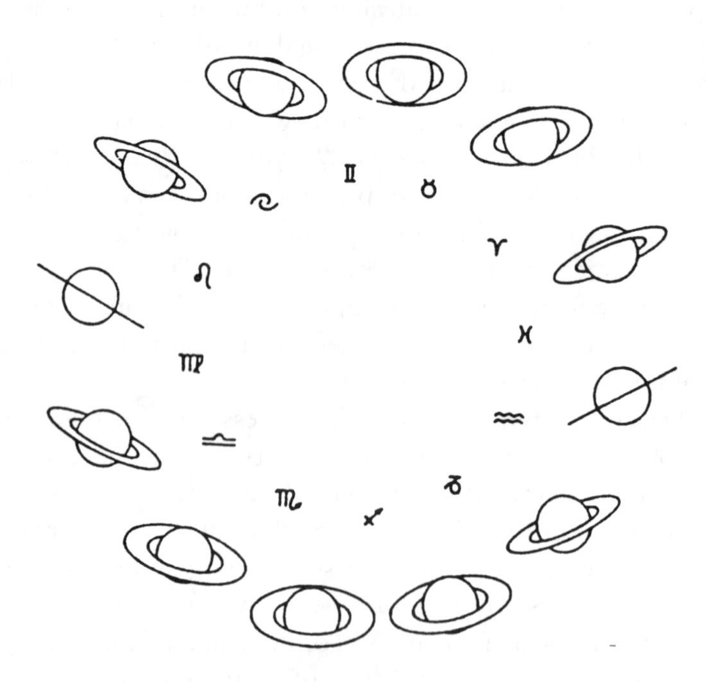

Fig. 124: The angle at which the planet Saturn and its ring system are viewed from the Earth. Saturn passes through the entire zodiac in 29.5 years. When the angle between the plane of the rings and the plane of the Earth's orbit is at its greatest, Saturn is significantly brighter than it is in the other extreme location, when the Earth is aligned with the plane of the ring system and the rings become temporarily invisible (J. Schultz, 1986).

Darwin believed that lunar rhythms in humans were an atavism, proof that our early ancestors were primitive fish species living in tidal zones. But how can we explain the fact that this atavism is so well preserved only in humans and other primates? And what is the Darwinian explanation for the seven-year periods in the human life span? The rhythms of human development and reproduction must have a different origin, and their existence makes it very clear that the human species is not a product of natural selection.

The Male Genitals

In 1926, Bolk published a summary of his theory of retardation in French. A short time later, Neuville responded in a fairly critical essay (1927). Despite his criticism of Bolk's theory, Neuville obviously considered the human penis a very clear example of human retardation. Bolk himself had already formulated similar ideas.

One of the unique features of the human penis is the limited retractility of the foreskin. In newborns, the glans and the foreskin are still firmly attached. Between the ages of one and thirteen—the timing varies widely among individuals—the foreskin gradually separates from the glans. In the final adult stage, the penis is covered by a thin skin that slides over the shaft. This layer of skin is continuous with the foreskin, which forms a double layer over the glans. The foreskin remains firmly attached at the base of the glans; the connecting fold that forms on the underside is called the frenulum (see *Figure 125*).

The final stage in human penile development (the separation of the foreskin from the glans) occurs very early in most mammals, in many instances even before birth. This stage is reached somewhat later in monkeys and later still in humans. In animals (often during the early juvenile period in monkeys) the separation continues, breaking through the frenulum, traces of which remain visible on the underside of the glans. The process of separation continues until most of the penis is free (Neuville, 1927). In adult animals, the foreskin is often much too short to cover the penis, from which it is now fully separated (Izor et al., 1981). In this context, it is interesting to note that the frenulum may persist in animals (such as steers) that are castrated at an early age (Neuville, 1927).

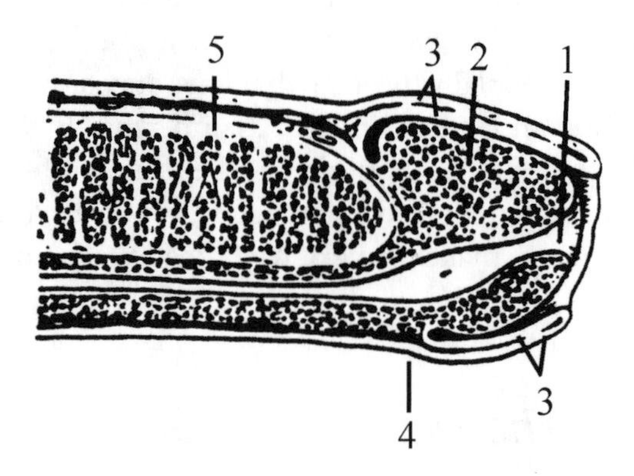

Fig. 125: Longitudinal section of the human penis. 1: orifice of the urethra; 2: glans; 3: double layer of the foreskin; 4: location of the frenulum; 5: erectile tissue or corpus cavernosum (Kahle et al., 1975).

The penises of adult monkeys include structures that are absent during the fetal stage. A frequently cited example is the os penis, one of the so-called sesamoid bones. In adult male primates, there is almost always a bone, nearly 1 cm in length, at the end of the penis. The only exceptions are humans, Eastern tarsiers, and the most highly evolved platyrrhines—that is, the spider monkey, the woolly spider monkey, and the woolly monkey (Hill, 1958). In anthropoid apes, the os penis is no larger than it is in lower primates, which means that it is drastically smaller relative to the size of the body as a whole. Remarkably, although the os penis tends to disappear in both the most highly evolved platyrrhines and the most highly evolved catarrhines, the tendency is more pronounced among platyrrhines. Both Bolk and Neuville comment that the genitals of platyrrhines are in many respects more humanlike than those of catarrhines.

Many animals have cornified structures (in the form of protuberances, spines, or actual barbs) on the glans and shaft of the penis (*Figure 126*). It is interesting to note that the chimpanzee penis is also covered with small, hard spines (Hill, 1946). Such specialized structures are absent in humans.

The development of the gorilla penis is also interesting. Hill and Harrison-Matthews (1949) were astonished by the highly retarded appearance of this penis (*Figure 127*). At all developmental stages, the genitals of male gorillas remain exceptionally small. The foreskin retreats slightly, but much less than in any other nonhuman primate, and the os penis appears only in late adolescence, much later than in other species of anthropoid apes. On the basis of this evidence, the researchers initially concluded that the gorilla penis, like that of the human male, remains unspecialized.

A short time later, however, Hill and Harrison-Matthews had an opportunity to study a one-year-old male gorilla (1950). They discovered that the tip of the animal's penis was covered with a mosaic of slate-gray, smooth, hardened particles of skin (see *Figure 127*). Closer study revealed that these structures were comparable to those found in adult gibbons. In gorillas, however, these structures seem to be a temporary phenomenon that disappears at a later age. Thus the gorilla penis develops by way of a detour; specialization occurs and is later largely abandoned. The hands and feet of gorillas undergo an analogous development (see *Figure 94*, page 179), for example). In humans, no such reversal occurs, and the development of both the hands and the penis takes a direct route from the fetal form to the adult.

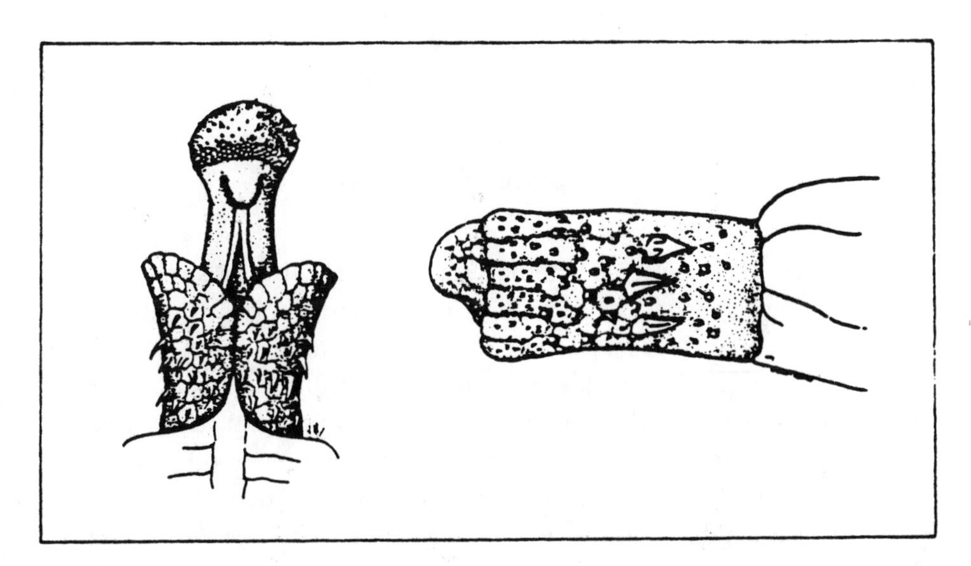

*Fig. 126: Penises of two prosimians. Left, golden potto (*Arctocebus calabarensis*) (view from below); right, brown lemur* (Eulemur fulvus) *(side view). The highly specialized penises of these primates are covered with spiny cornifications (Hill, 1958).*

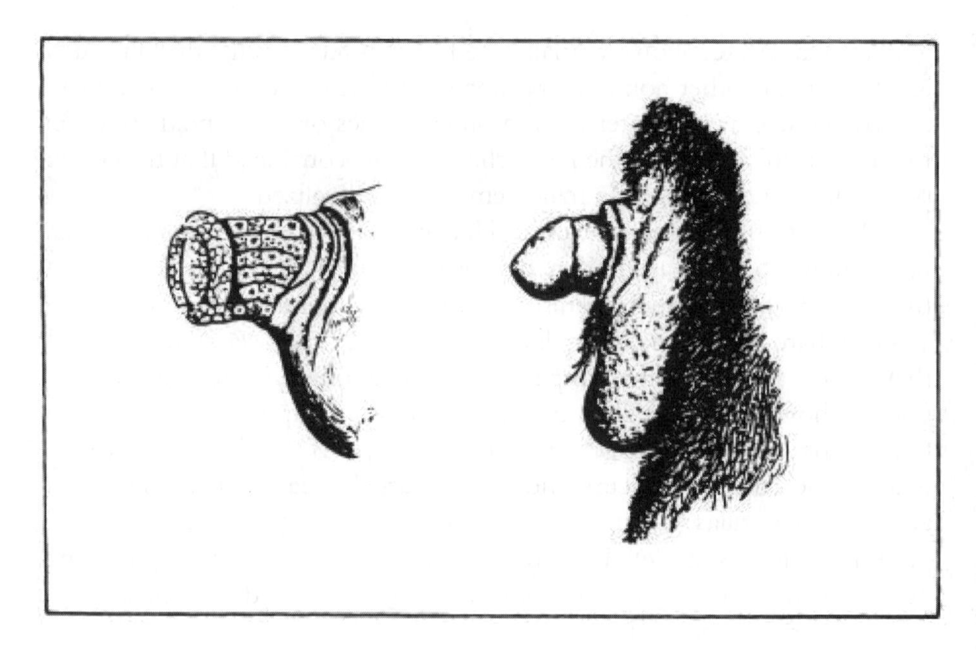

Fig. 127: Penis of a one-year-old gorilla (left) and of an adult specimen (right). The mosaic of hornlike plates on the juvenile penis is a temporary characteristic (Hill and Harrison-Matthews, 1949 and 1950).

The Female Genitals

In 1916, Bolk received a female chimpanzee fetus to study. Shortly thereafter, he published a brief report that included a description of the specimen's genitals (1917). As early as the nineteenth century, it had been noted that the female genitals of humans and anthropoid apes differ in that the latter have no labia majora. Bischoff, an opponent of Darwinism, emphasized this difference in order to contest the relationship between humans and anthropoid apes. Bolk noted, however, that his fetal chimpanzee specimen did indeed have labia majora. In 1917, he did not yet see this fact as an argument in favor of the theory of retardation, which he formulated for the first time only in the following year, but he did make the connection in his book *Hersenen en Cultuur* ("Brain and Culture"):

> Although they are not nearly as highly developed as in humans, rudimentary labia majora are clearly present in this little fetal ape . . . Certain bodily characteristics of human and anthropoid ape fetuses . . . disappear in anthropoid apes but are retained in humans, where their further growth is proportional to the growth of the body as a whole . . . In this respect, too, the human being preserves a fetal character . . . and the anthropoid ape fetus appears more closely related to the human being than the adult ape does (1918).

Bolk goes on to emphasize that the mechanisms of natural selection do not adequately explain fetal similarities between humans and anthropoid apes:

> Clearly, neither the loss of all hair with the exception of that on the skull nor the development of labia majora can be the consequence of adapting a function to changing outer circumstances. It is difficult to ascribe any function at all to the latter characteristic, so it cannot be considered the product of any functional change . . . For myself, I see these two phenomena as expressions of an evolutionary route determined by an intrinsic factor, as symptoms of a predetermined evolutionary direction, and as such therefore already present in principle in the anthropoid ape fetus (1917).

Schultz, confirming Bolk's observation, discovered that the labia majora are well-developed in large anthropoid apes during the late fetal and early juvenile stages but later disappear (1927). Labia majora are also present during the fetal stage in lower apes such as macaques (Hill, 1958). Hill also found that the female genitalia of gibbons are much more humanlike than those of the larger anthropoid apes. Although the labia majora tend to disappear with time in female gibbons, they are still sometimes seen in adult specimens. It is also interesting to note that the labia majora are much better preserved in the pygmy chimpanzee, which is considered a paedomorphic form of the common chimpanzee species (Dahl, 1985).

The hymen is also mentioned as an example of human retardation (Gould, 1983). Monkeys have no hymen. Its existence in humans, long debated, was first confirmed by Aristotle. (Incidentally, the name of the

Greek god of marriage was Hymen.) From the sixteenth to the eighteenth century, an increasing number of authors dismissed the existence of the hymen as nonsense. Many discovered that it was absent even in young girls; others contested its existence because it seemed to serve no purpose (a point that Gould also discusses). Vesalius confirmed the existence of the hymen; Buffon later denied it. (For more historical details, see Gutt, 1975 and Fischer-Homberger, 1977.) Until quite recently, it was questioned whether the hymen actually is a universal phenomenon among human females. A study of 1,131 female newborns, however, found that they all had hymens (Jenny et al., 1987).

In the course of fetal development, the vagina first appears as a solid mass, the so-called vaginal plate. A cavity gradually develops within this organ through vacuolization, a process in which internal cells fill with fluid and lose their walls. The hymen is simply the thin front wall that remains when the process of cavity formation is finished. In monkeys, this wall subsequently disappears, although residual formations such as small raised portions or roughly ring-shaped thickenings often remain visible between the vestibule and the vagina (Hill, 1958). Only in humans, however, does the anterior wall of the fetal vaginal cavity persist as such (see *Figure 128*).

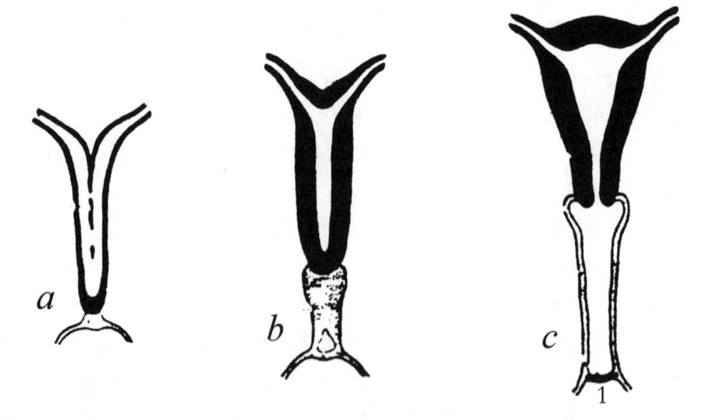

Fig. 128: The vaginal plate (stippling) develops in a caudocephalic direction below the uterus (black). Through vacuolization, the plate gradually becomes hollow, but the anterior wall of the vaginal plate (the hymen, 1) remains intact in newborn human females. (a) fetus at nine weeks, (b) fetus at the end of the third month, (c) newborn.

In addition, monkeys and prosimians undergo various developments that humans do not. For example, in some species, the clitoris grows into a fairly large organ supported by an os clitoris. A unique phenomenon in many catarrhines is the periodic swelling of certain body parts. The female develops a very conspicuous, boldly colored, growth-like swelling that may include the genital area, the buttocks, the base of the tail, or other portions of the torso. The swelling develops slowly after a menstrual bleeding and reaches its maximum size around the time of ovulation, after which it disappears very quickly, often within one day. The female is receptive to mating only when the swelling is present. The phenomenon of genital swelling, which can be observed in many macaque species and all baboons and is especially conspicuous in chimpanzees (see *Figure 129*), is absent in gibbons and present only to a slight extent in gorillas. Remarkably, it occurs in orangutans only during pregnancy (Schultz, 1938a). The phenomenon of cyclical swelling is completely absent in humans, who differ from most primates and mammals in that the moment of ovulation remains totally concealed from external view. Many attempts—often quite far-fetched—have been made to explain this anthropological idiosyncrasy (Daniels, 1983). In fact, however, it can be considered a simple example of retention of juvenile traits.

Fig. 129: Genital swelling in a female chimpanzee.

Bolk found it important that one specific, very common embryonic characteristic is preserved in humans. Human and animal embryos are characterized by ventral curvature in which the head and the pelvic area incline toward each other (see *Figure 130*); humans retain this forward orientation even in adulthood. One consequence, according to Bolk, is the unique location of the vagina in human females, which means that for humans, unlike animals, penetration is easier from the front. In chimpanzees, in contrast, the embryonic orientation undergoes extensive change. The original curvature is abandoned and a straight or even reversed orientation develops.

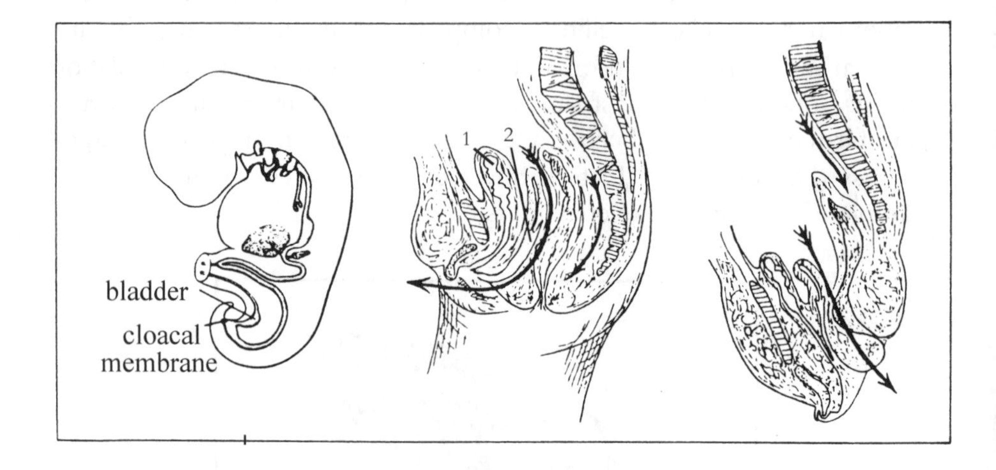

Fig. 130: Pelvic curvature.

Left: Longitudinal section of a five-week-old human embryo (at this stage of development, the situation in the chimpanzee is fully analogous). Note the conspicuous forward orientation of the head. The lower body, however, also curves forward.

Center: Section through the pelvic region of a two-year-old girl. The overall forward curve remains recognizable. (1) bladder and ureter, (2) vagina.

Right: Pelvic section of a half-grown female chimpanzee. The deviation from the embryonic forward curvature is clearly recognizable (Bolk, 1926a).

The Enigma of Human Reproduction

From the zoological perspective, human reproduction has a number of unique features, some of which have not yet been adequately explained, while others have been the subject of boundless speculation. We will look at three examples.

Descent of the Testicles

Descent of the testicles, which are originally located near the kidneys, is characteristic of mammals (including marsupials but not monotremes). In the course of mammalian ontogeny, the testicles migrate in a caudal direction and often come to rest in a scrotum outside the actual torso, as is the case in all primates. In some mammals (whales, dolphins, and armadillos), no scrotum develops, but the internal caudal migration of the testes occurs nonetheless. The scrotum is also absent in many species adapted to aquatic habitats (such as seals, hippopotamuses, and manatees), in some insectivores, and in rodents, marsupials, rhinoceroses, and elephants. Testicular descent is absent in birds.

Among primates, testicular descent is more prominent than it is in other mammals, and in higher primates the process may even be accelerated. It occurs shortly before or during puberty in lower primates, during the juvenile period in anthropoid apes, and as early as shortly after birth in chimpanzees. In humans, finally, the descent occurs before birth (during the eighth month of gestation).

Portmann (1976) takes these observations still further. In higher vertebrates, by and large, there is a heightened tendency toward caudal displacement of the gonads. The more the head is differentiated from the torso, the more the gonads' original location shifts toward the caudal pole. "The strange phenomenon of the testes emerging from the body and entering the scrotum, which we see in higher mammals, appears to be a direct continuation of a process already set in motion by transformations common at lower levels of vertebrate evolution."

With regard to testicular descent, therefore, human beings represent the culmination of a process that can be traced far back into the animal kingdom. The significance of this process, however, remains unknown (Martin, 1990; Schultz, 1952).

Menstruation

The phenomenon of menstruation becomes ever more conspicuous among the higher primates:

> The occurrence of menstruation is a variable feature among primates. It is not found at all in lemurs or lorises . . . and it is only weakly identifiable in tarsiers and New World monkeys, for which microscopic examination is often necessary to confirm the incidence of menstrual bleeding. In Old World monkeys and apes, menstruation is quite easily detected by externally visible signs of bleeding, but in no species is the menstrual flow as copious and as easily recognizable as with the human female. This apparent *Scala naturae* in the occurrence of menstrual bleeding broadly parallels the gradation found in the invasiveness of the processes associated with placentation . . . and this parallelism suggests some functional connection between the two phenomena (Martin, 1990).

And in fact, the more humanlike an animal is, the more maternal and fetal circulations interpenetrate:

> In the Suidae [swine], [there are] six cellular layers . . . between the maternal and the fetal blood (epitheliochorial placenta); in Felidae [cats] and Canidae [dogs] there are four cellular layers (endotheliochorial placenta); in primates, the placenta is hemochorial, that is, the maternal blood flows freely under high pressure around the fetal vessels of the placenta; in primates there are two cellular layers; in humans during the third trimester of pregnancy there is only one cellular layer between maternal and fetal blood and the exchange of oxygen and nutrients is very high (Abitbol, 1990).

According to Abitbol, the greater interpenetration of maternal and fetal circulation in humans is related to the fetal brain's very high requirements for oxygen and glucose. (In a resting human being, the brain uses at least 20 percent of the body's total oxygen requirement and 70 percent of its glucose.) Martin, on the other hand, associates greater circulatory interpenetration with the greater body sizes of higher primates. As the fetus grows, the increase in the area of contact where the substances can be exchanged between mother and fetus is disproportionally smaller than the increase in fetal volume, but the metabolic needs of the fetus are not reduced accordingly. The extremely efficient exchange of substances between a human mother and fetus represents the culmination of a tendency, which can be observed throughout the entire animal kingdom, toward increasingly intensive care of the unborn young. Nonetheless, many questions remain unanswered. Is the structure of the human placenta a primitive or a derivative form? (See Martin, 1990.) Are the structure of the human placenta and the heavy menstrual bleeding found in humans due to a single underlying characteristic, as Martin assumes? The answers remain unknown. We know only that a higher degree of retardation in primates correlates systematically with closer approximation of the human situation.

Hair Patterns and Sweat Glands

Although on the whole humans develop less body hair than other primates, the hair on the human head is exceptionally well developed. During the fetal period, body hair develops in a cephalocaudal direction. In this connection, it is interesting to note that no hair develops on the most distal surfaces—that is, on the palms of the hands and soles of the feet. The starting points of the cephalocaudal and caudocephalic growth tendencies are characterized by maximum and minimum hair development, respectively (cf. *Figure 39*, p. 84). Around puberty, when proximodistal growth tends to predominate in the limbs (see page 82), conspicuous hair development occurs on the proximal parts of the limbs (armpits and pubic area). This phenomenon suggests that hair development is intimately related to the cephalocaudal growth pattern.

Similarly, the caudocephalic growth pattern also seems to be accompanied by a specific type of organ, namely, eccrine sweat glands (glands that produce a watery secretion; see *Figure 131*). In the course of embryonic development, these glands appear first on the palms and soles and only much later on the rest of the body. Just as hair grows best on the head, eccrine sweat glands develop best on the palms and soles, where there are four times as many of them per unit of surface area than anywhere else.

Eccrine sweat glands, found *only* on the palms and soles of lower primates, cover progressively more of the body in higher primates, while hair growth becomes increasingly sparse, especially in the large anthropoid apes. The decrease in hair covering and the increase in the number of eccrine sweat glands in higher primates, especially humans, seem to be symptomatic of the increasingly strong influence of the caudocephalic growth pattern in connection with higher degrees of retardation.

The eccrine glands (which are very important for cooling in humans; see page 234) are not the only type of sweat gland. The so-called apocrine glands, which produce a secretion with higher protein and fat content, are linked to the presence of hair. "In humans dense aggregations of apocrine glands occur in the axillae (armpits), the suprapubic region, the circumanal region and perineum, face, scalp and the umbilical region of the abdomen. It is most significant that these are the parts of the body which have retained substantial growth of hair." (Stoddart, 1990). Apocrine sweat glands empty into hair follicles (see *Figure 131*), while eccrine glands open directly onto the epidermis.

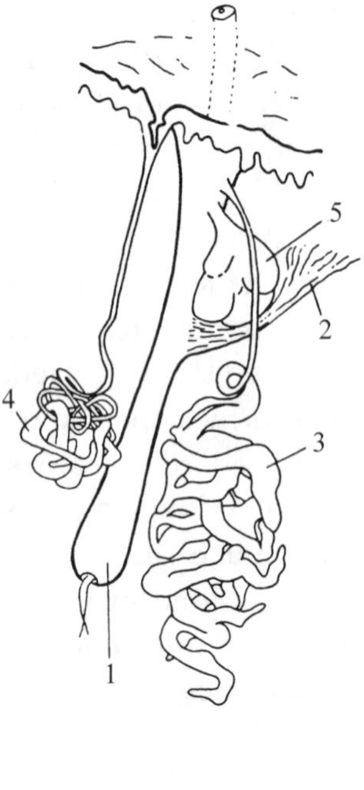

Fig. 131: Structure of the skin and its glands. (1) hair follicle; (2) hair muscle; (3) apocrine gland; (4) eccrine gland; (5) sebaceous gland (McFarland et al., 1985).

The concentration and size of the apocrine glands in the human armpit are extraordinary. This phenomenon is otherwise seen only in African anthropoid apes, where it is less pronounced. The presence of hair and apocrine sweat glands in the human armpit is generally explained as an adaptation to uprightness in that apocrine glands (odor-producing glands with a presumed sexual function) serve their purpose better in the armpit than in the genital area. Hagen (1901) writes:

> The organ of smell is receptive only to objects in its immediate surroundings. Odors that are important in the lives of animals are generally deposited on the ground, where they are readily perceived by other animals that keep their noses close to the ground. Because uprightness raises the level of the human olfactory organ to a height of approximately 1.5 meters, the sense of smell can no longer serve its purpose effectively and therefore gradually regresses. If this interpretation is correct, perhaps I may be permitted the following assumption. In the daily life of civilized human beings, odors that originate in the sexual organs or other parts of the lower body are not perceptible at all. The location of the armpit, however, is much more advantageous and its odors more successful. This is why underarm odor has developed in humans.

This Darwinian explanation, however, seems difficult to reconcile not only with the presence of axillary glands in African apes but also with the abundance of pubic hair in humans.

The apocrine glands become active only around puberty. The close connections between hair and apocrine glands and cephalocaudal growth on the one hand and eccrine glands and caudocephalic growth on the other, along with the incisive change that occurs at puberty, suggest that the entire development of sexual functions is related to the interplay of these two growth patterns.

This picture is further complicated by the presence of a third gland type, the sebaceous glands, whose abundance in humans is equally perplexing. Sebaceous glands often—but not always—appear together with hair. As a rule, large hairs have smaller sebaceous glands and vice versa, and the largest sebaceous glands (such as those on the face) are not associated with hairs at all but empty directly onto the surface of the skin. Sebaceous glands are numerous on the face, neck, and upper torso; for reasons that remain unknown, they can cause acne in these areas during puberty. Remarkably, the sebaceous glands are large and functional in newborns but then enter a dormant phase until they are reactivated in puberty. It is extremely difficult to explain the properties of human sebaceous glands from a Darwinian perspective. In animals, sebaceous glands often serve to keep the hairs oily, but such glands are highly developed in human beings, where this function is irrelevant. Hence, explaining the existence and activity of the three different types of glands in the skin is one of the greatest challenges faced by physical anthropology.

Milk

In several respects, the composition of human milk represents an extreme in comparison to the milk of mammals. Human milk is very low in fat and protein and very high in lactose (milk sugar). Low protein content of milk correlates with slower infant growth (Bounos et al., 1988). Also indicative of a relatively slow growth rate is the fact that a day's milk production in primates contains fewer calories than it does in other mammals (Martin, 1990).

The composition of the protein in human milk is exceptional in and of itself. Milk protein comes in two forms, soluble protein (whey protein) and casein. Casein is insoluble in water. It is present in milk in the form of tiny droplets, which give the milk its white color. The preponderance of whey protein is much greater in human milk than it is in the milk of any animal species. Experiments show that whey protein enhances resistance to disease (Bounos et al., 1988). Hamsters that consume the optimum amount of whey protein live up to 50 percent longer.

Human milk actually does not contain exceptionally large amounts of whey protein; in fact, it contains fairly little, but the ratio of casein to whey protein is greatly reduced. High casein content seems to be associated with prerequisites for rapid growth, while whey proteins, although they enhance resistance and increase life span, permit only slow growth. Bounos et al. (1988) conclude that the composition of human milk reflects retardation in human development.

Human milk also contains much more lactose than animal milk does. Lactose, a specialized sugar found primarily in milk, is a disaccharide, that is, a combination of two monosaccharides or simple sugars, in this case glucose and galactose. The brain consumes a great deal of glucose. Human milk's low protein and fat content is related to the human infant's relatively slow overall growth, while high sugar content is related to the great size of the brain in the small child.

In the first step in the digestion of lactose, the disaccharide is split into its two simple sugars. This transformation requires a specific enzyme called lactase. In mammals, lactase is present only in young, nursing animals. For example, lactase production in rats drops as early as two weeks after birth. Adult mammals, therefore, are no longer able to digest lactose. In many (but not all) mammal species, consuming lactose-containing dairy products causes digestive upsets such as stomach cramps, diarrhea, etc.

Bolk (1929) pointed out that fetalization of certain characteristics in humans is less advanced in some racial groups than in others. He believed that differences in the degree of fetalization of specific characteristics are the primary cause of racial differences. This unfortunately led him to make value judgments that are prejudiced and offensive to the more developed sense of human dignity that is prevalent today. The outstanding characteristic of human beings is that each is a unique individual. The value of an individual is not determined by racial traits but by human qualities, the most important of which are rooted in the human individuality. Racial traits cannot, therefore, be used as criteria for the evaluation of human beings. In our present context, however, it is interesting to note racial differences in the ability to digest lactose. In certain ethnic groups, this ability—normally a juvenile trait—persists into adulthood. This is especially true of Europeans, with the exception of people from Greece and southern Italy (but not Spain). In the countries of eastern Asia, few adults are still able to digest lactose, and many dairy products are less popular. In Bali (Indonesia) milk is even used as a laxative. Lactase deficiency is also very prevalent among certain African ethnic groups as well as in African-Americans (as was discovered in 1963; see Harrison, 1975).

Attempts have been made to interpret lactase persistence in humans as a phenomenon of natural selection. This trait is supposed to have spread rapidly since the advent of animal husbandry, when the ability to digest milk became an advantage (Haegerman and Schenck, 1989; McCracken, 1971). This explanation, however, is untenable in view of the fact that lactase deficiency occurs among ethnic groups (in India, Greece, and eastern Africa, for example) that have practiced animal husbandry for a very long time. Furthermore, as Ferro-Luzzi (1991) points out, better nourished populations do not necessarily reproduce more rapidly; at the moment, the opposite is true on the global level. An enhanced ability to digest lactose is much more compatible with the pattern of progressive retardation in humans.

Why milk contains lactose rather than other sugars such as glucose and sucrose, which are otherwise more common, remains a mystery. It is also not known whether there is any connection between the very high lactose content of human milk and the fact that portions of humanity retain the juvenile ability to digest lactose. Human milk has not yet revealed all of its secrets.

9

THE FACIAL SKULL

Introductory Remarks on the Shape of the Skull

In humans, the oral cavity is exceptionally short because no muzzle develops. A short mouth can be considered a fetalization phenomenon, since the early fetal stages of all mammals have short mouths and compact tongues (see *Figure 132*)—a state of affairs that persists only in humans. The short human oral cavity can also be seen as one aspect of the retarded human digestive system (see Chapter 3, page 88).

The tendency to retain a short facial skull and smaller oral cavity begins far back in the evolutionary series. Martin (1990) compared the ratio of tongue length to skull length (a good reflection of the relative size of the muzzle) in a large number of mammal species. He determined that lower primate palates are almost thirty percent shorter than those of insectivores. This figure increases to forty percent (on average) in monkeys and up to sixty-five percent in humans.

Retaining a shorter facial skull goes hand in hand with a series of other juvenile skull proportions. For example, Flügel et al. (1983), who used computed tomography (CT) scanning and x-rays to study changes in skull proportions in humans and two macaque species, discovered that the viscerocranial ratio (defined in the caption to *Figure 133*) is the same in human and macaque newborns. This juvenile ratio is retained into adulthood in humans but more than doubles in the monkeys. Furthermore, the clivus angle (angle γ in *Figure 133*), which remains relatively constant postnatally in humans, increases by approximately 15° in monkey species. In other words, the base of the skull becomes flatter after birth in monkeys, while its characteristic fetal curve is much better preserved in humans. Between birth and adolescence, the prognathic angle increases by approximately four to eight degrees in humans but by 20 to 30 degrees in macaques (see also *Figure 135*, for example). And finally, the interincisive angle is the same (almost 140°) in the first and second human dentitions, while in macaques the angle of the adult teeth is 30° (or more) smaller than that of the baby teeth; that is, in adult macaques the interincisive angle is less than 90°, so the incisors function more like pincers than scissors. From these observations, the authors conclude:

> Since the newborn skull of the two monkey species studied resemble both in proportion and form much more that of a newborn human, these species might have been developed in the course of evolution from the human line rather than the human from ancestors of these monkeys (Flügel et al., 1983).

Rak and Howell (1978) compared the changes that occur between infancy and adolescence in thirteen different parameters in the skulls of humans, chimpanzees, gorillas, and *Australopithecus boisei.* Post-juvenile change as a percentage of the juvenile value averages 6.77 percent in humans, 21.94 percent in chimpanzees, 33.11 percent in gorillas, and 24.33 percent in *Australopithecus boisei*. Thus the skull of a human infant has already achieved an approximation of its adult form, while the skull of a juvenile African anthropoid ape still undergoes significant change. *Figure 134* illustrates this phenomenon. The original elliptical horizontal cross-section of the braincase is retained into adulthood in humans, but in large anthropoid apes (and other primates), significant post-juvenile deviation from this form occurs. Faint traces of this deviation are evident in adult humans, but it is much more pronounced in apes. Ortmann (1984) comments:

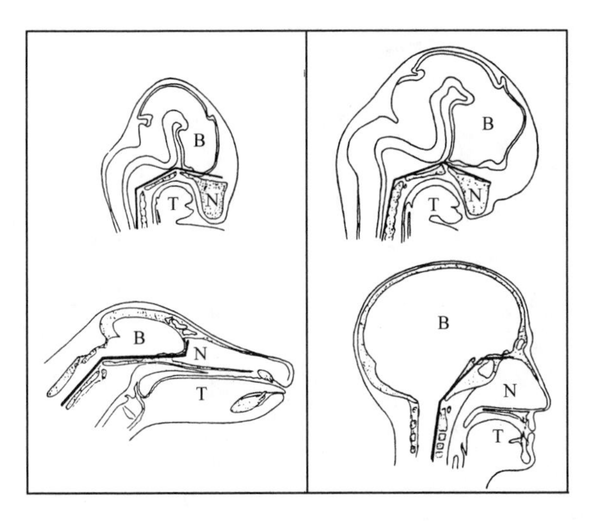

Fig. 132: Cross-section of fetal (upper) and adult (lower) skulls of a dog (left) and a human (right). (T): tongue, (N): nasal cavity, (B): brain cavity. The bold lines indicate the base of the skull. The fetal forms are characterized by short oral cavities and tongues. In the adult dog, the oral cavity has grown into a muzzle, and the tongue has become long and pointed. In the adult human, the oral cavity more closely resembles fetal proportions. In the adult animal, the nasal cavity has rotated and lies in front of the brain cavity, while in the adult human the nasal cavity retains its original position between the oral cavity and the brain cavity (Bolk, 1926a).

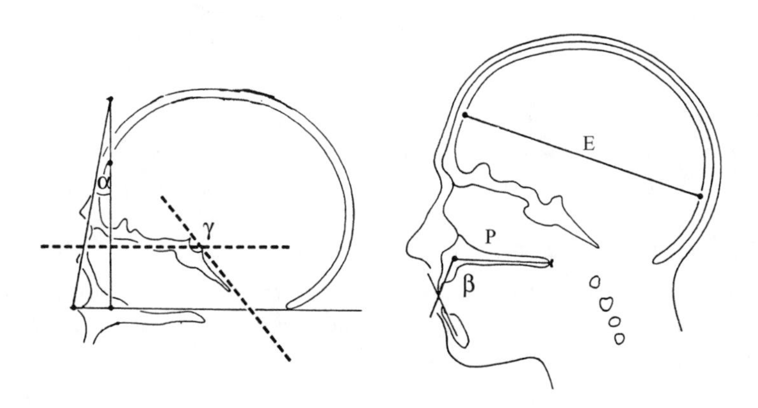

Fig. 133: Several skull parameters that are better preserved during human ontogeny (Flügel et al., 1983). The viscerocranial ratio P/E is the length of the hard palate divided by the greatest diameter of the brain cavity. α: prognathic angle, β: interincisive angle (the angle between the upper and lower incisors, γ: clivus angle.

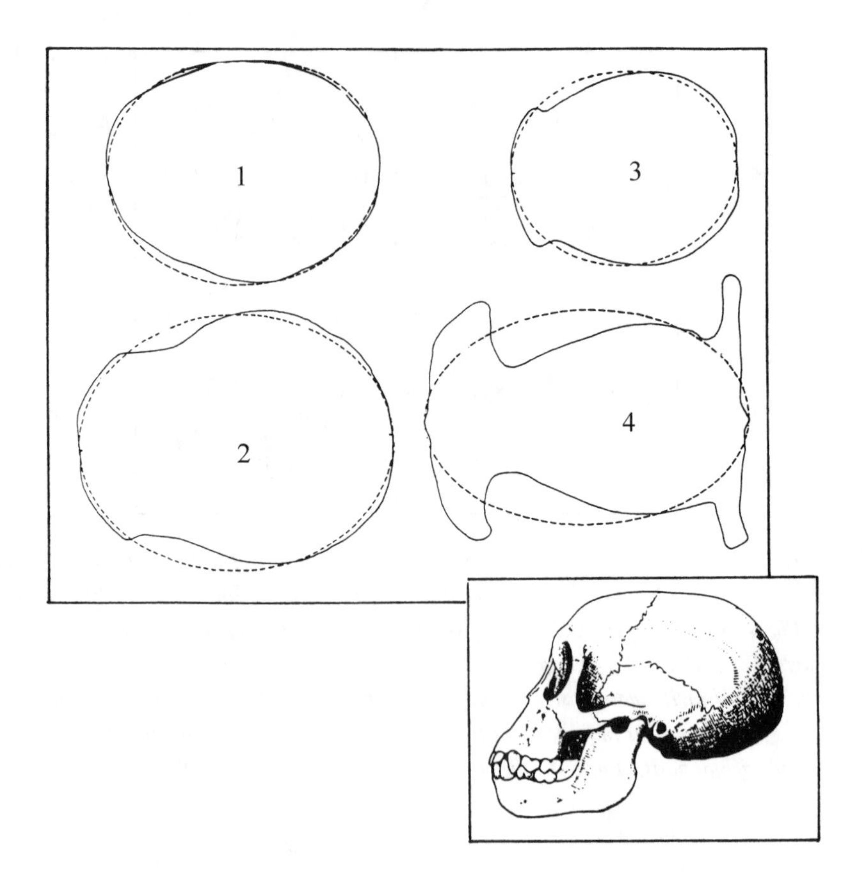

Fig. 134: Outline of the skull at its greatest circumference in (1) a four-year-old human, (2) an adult human, (3) a juvenile gorilla, (4) an adult gorilla. (The face is on the left.) The broken line indicates the ellipse whose long and short axes correspond to the greatest length and width of the skull (Ortmann, 1984). The shapes of juvenile human and gorilla skulls deviate only slightly from the elliptical form. The shape of the adult human skull is still nearly elliptical, and its deviations from the elliptical outline are similar to those of the juvenile gorilla. In contrast, the adult gorilla skull's deviations are in the same locations but much more advanced. This figure illustrates the principle expressed by Goethe—in the human being the animal element is present, but "is enhanced to serve higher purposes" and, "as far as both the eye and the intellect are concerned, it has been overshadowed." (See Figure 45, *p. 96.)*

Below, right: The skull of a juvenile gorilla (adapted from Selenka; cf. Figure 138, *p. 272). The canine teeth are still small, the skull has no brow ridges or sagittal crest, and the muzzle is still relatively undeveloped. The outer auditory canal still lies below the extension of the zygomatic arch (cf. also* Figure 149, *p. 291).*

> It must be emphasized that the skulls of adult haplorhines [tarsiers, platyrrhines, and catarrhines] either extend beyond or fall short of a true ellipse in the same locations as the human skull, but to a much greater extent. This is true of both the zygomatic arch and the lateral occipital region and indicates their relationships to the mastication muscles and neck muscles, respectively. Furthermore, in juvenile catarrhines the skull's deviation from a true ellipse is quantitatively equivalent to the deviation that develops over an entire human life span.

In animals, therefore, adult deviation from the universal initial skull form is directly related to the renunciation of uprightness (see Chapter 6). Animals need strong neck muscles in order to support the head in front of (rather than above) the body, and they develop strong chewing muscles because the mouth is forced to serve as a prehensile organ, which in turn is a consequence of the fact that the forelimbs are devoted to locomotion. In apes, therefore, the original shape of the head is designed for uprightness, and their quadruped stance is a secondary phenomenon.

It is especially illuminating to compare the chimpanzee (*Pan troglodytes*) to the pygmy chimpanzee or bonobo (*Pan paniscus*). The latter species, which is found only along the Congo River, is considered a neotenic variation of the common chimpanzee and has several more humanlike characteristics. For example, the foramen magnum of this species remains closer to the front of the head (Luboga and Wood, 1990). In comparison to the common chimpanzee, the pygmy chimpanzee deviates less from their shared juvenile gestalt, a phenomenon apparently due to retention of a number of humanlike traits.

There can be no doubt that Bolk's view of the human skull as a fetalized form remains basically correct, although it has been criticized from various perspectives (Devillers et al., 1990; Shea, 1989). *Figure 135* illustrates once again that an ape skull is initially very similar in shape to a human skull but is then animalized in ways that are sometimes quite disturbing. It is difficult to consider such illustrations without beginning to suspect that these animals have "renounced" their humanlike origins.

*Fig. 135: Three successive stages in the development of a macaque skull (*Macaca nigra*, from Sulawesi; cf.* Figures 18 *and* 19, *p. 39). The juvenile skull is very humanlike, with a relatively undeveloped muzzle and a pronounced forehead; the eye sockets lie below the brain cavity, and the outer end of the auditory canal lies below the zygomatic arch. In the adult, the development of the muzzle is extremely advanced by primate standards. The eye sockets are located in front of the brain cavity, the forehead has completely disappeared, and the auditory canal has shifted to above the extension of the zygomatic arch. A median sagittal crest, which is totally absent in the juvenile form, has appeared on the cranium (cf.* Figure 19, *p. 39) (A. Dabelov, 1928).*

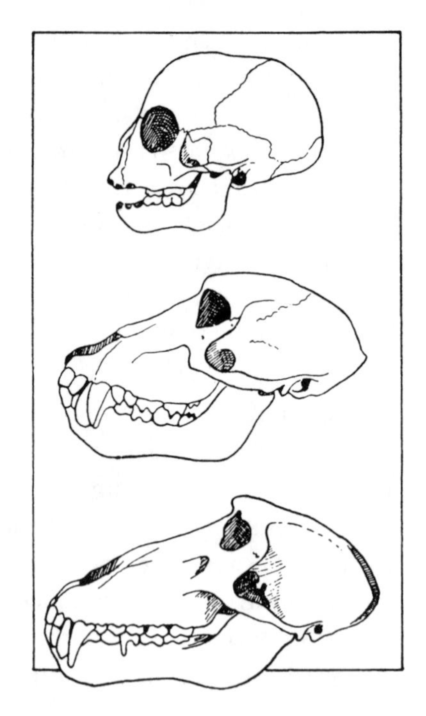

The Teeth

A smaller mouth goes hand in hand with a dental arch that is shorter in proportion to the length of the skull. The dental arch is approximately 25 percent shorter, on average, in monkeys and nearly 65 percent shorter in humans than it is in insectivores and lower primates. The tendency toward a shorter dental arch is accompanied by decreases in the number of teeth. Prehistoric mammals are presumed to have had approximately sixty teeth, reptiles even more. In several mammalian lines of evolution, the number of teeth tends to decrease. The evolutionary trend is clear. Martin (1990) writes, "Reduction in the length of the cheek-tooth row has doubtless occurred in the evolution of simian primates, an evolutionary development carried to an extreme in *Homo sapiens*."

Martin (1990) goes on to note that primates generally have relatively unspecialized teeth and that within the catarrhine group, humans and anthropoid apes have less specialized teeth than the lower catarrhines (Martin, 1990). According to Dahlberg (1968), retardation, resulting in lack of specialization, or paedomorphosis, offers in this case, too, "a clear explanation for the many previously unintelligible forms, patterns, and size reductions in hominid dentitions."

Thirty-two teeth remain in the human dental arch. In adults, both the upper and the lower jaw have four incisors, two canines, four premolars, and six molars (including the wisdom teeth). Molars are absent from the first dentition, which includes only 20 teeth (*Figure 136*).

The shape of the dental arch is very different in adult humans and adult anthropoid apes (*Figure 137*). In humans, the shape of the dental arch is always curved, although considerable racial and individual variation is possible. The measurements of the dental arch describe a rectangle known as the "basic rectangle of the mandible," or BRM, which was first defined by the great paleontologist Louis Leakey. The base of this rectangle connects the rearmost points on the third molars of the lower jaw; the opposite side runs through the bases of the incisors. The BRM index is defined as the rectangle's height as a percentage of its width; i.e., a square has a BRM index of 100. In humans, the BRM ranges from 80 to 116; that is, the human BRM is nearly square, as Leakey discovered.

In anthropoid apes, the dental arch is nearly rectangular rather than curved, and the BRM index is very high, even in comparison to much lower primates. The Eastern tarsier and several platyrrhines have BRM indices strikingly similar to those of humans. Gibbons, too, remain much closer to the human shape than do the large anthropoid apes. Lower primates also have a more curved dental arch, which anthropoid apes have abandoned. I maintain, therefore, that the curved shape and low BRM index of the human dental arch are primitive and prototypic characteristics related to the absence of a highly developed muzzle (as in all primate fetuses). The fact that the dental arch of some lower primates is more humanlike than that of the large anthropoid apes suggests that the dentition of the latter should be considered a more specialized pattern.

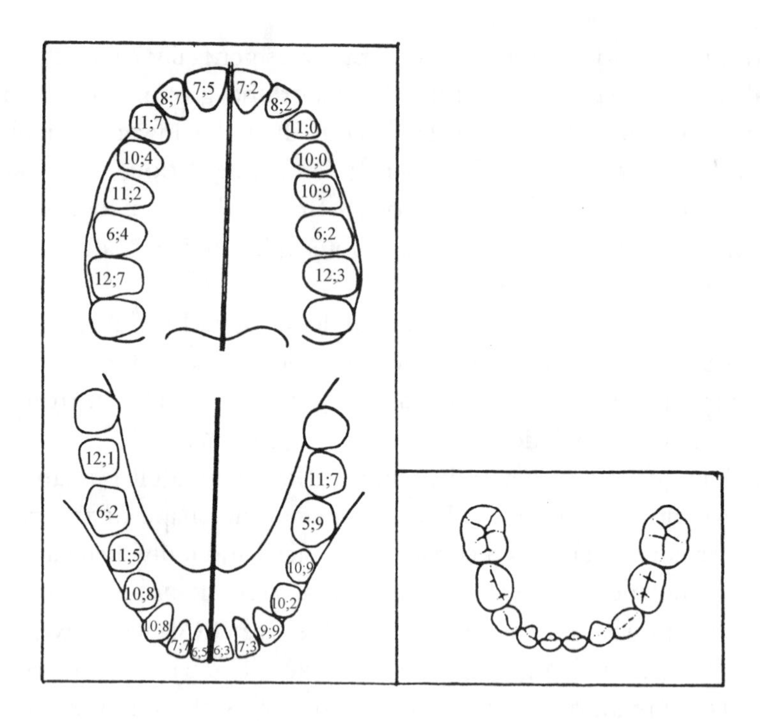

Fig. 136: left: The shape of the human dental arch, with the average date of emergence of adult teeth given in years and months. Above: upper jaw; below: lower jaw. The figures in the left half are for boys, those on the right for girls. In general, the change of teeth occurs somewhat earlier in girls. With the exception of the third molar, the wisdom tooth, which may erupt at any time between the ages of 18 and 80, the adult set of teeth is completed during the second seven-year period (Sinclair, 1973).

Right: Human baby teeth, lower jaw (Aiello and Dean, 1990). The first set of teeth has no molars, but the shape of the arch is already the same as it is in the adult.

There are obvious gender differences in the teeth of some monkeys. In these species, the BRM is higher among males, whose dental arch also tends to be more rectangular. The difference between female and male macaques is shown in *Figure 137* (j and k), the differences between female and male gorillas and chimpanzees in Figure 138. Among baboons—species with a very pronounced muzzle—Kinzey (1970) found average BRM indices of 232 for males and 212 for females. Montagu (1982) suggests that females of all species generally display more paedomorphic characteristics than males do. Females, for example, have less hair and live longer. The less developed muzzles of female apes and their correspondingly lower BRM indices are probably examples of the same phenomenon. With regard to apes, therefore, the old saying, "The woman is the future of the man," may be true (see also *Figure 138*). *Figures 136, 27, 37,* and *38*, however, illustrate that in certain respects males develop more slowly. And as *Figure 38* shows, earlier puberty in females means that they may also acquire typically human characteristics at an earlier age than males do. Differing rates of development in males and females certainly still conceal some significant anthropological riddles.

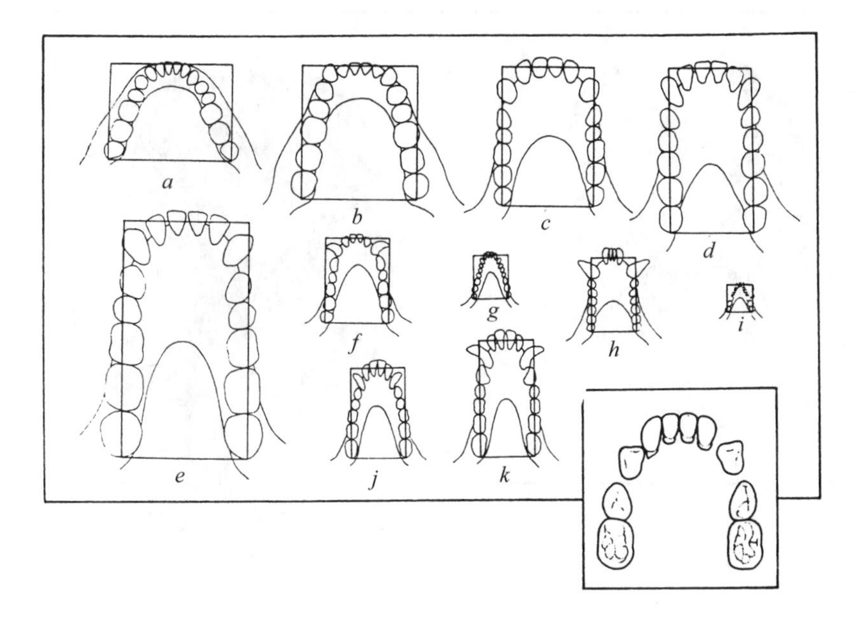

Fig. 137: The lower jaws (BRM in parentheses) of: (a) Homo sapiens, *(77); (b)* Homo sapiens, *(116); (c) chimpanzee (145); (d) orangutan (196); (e) gorilla (185); (f) gibbon (133); (g)* Callicebus *(120), (h)* Cacajao *(160); (I) Eastern tarsier (104), (j) macaque, female (166); (k) macaque, male (205). (a) and (b) are the most extreme human examples that researchers have discovered to date. The large anthropoid apes (c, d, e) have more or less rectangular dental arches with high BRM indices (Kinzey, 1970).*

Below, right: Baby teeth of the gorilla. The shape of the arch is still humanlike (Aiello and Dean, 1990).

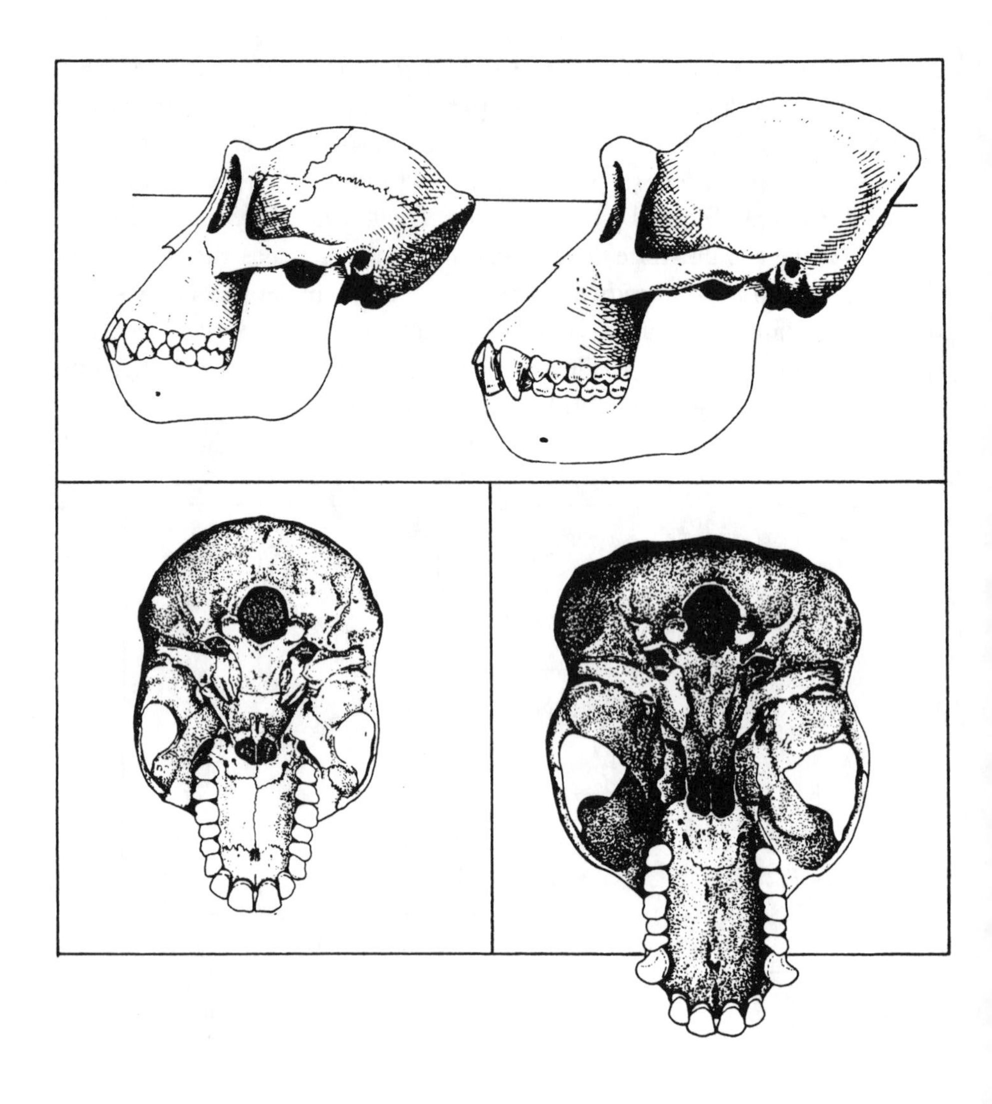

Fig. 138: Top: Skulls of female (left) and male (right) gorillas. The skulls of female animals are generally more paedomorphic, with smaller canine teeth and less protruding brow ridges and sagittal crests (Naef, 1926).

Bottom: Undersides of the skulls of female (left) and male (right) chimpanzees. In many respects, the female is more paedomorphic and thus more humanlike. The foramen magnum is still closer to the center of the skull, the teeth are shorter and arranged in an arch, and the canine teeth and diastemata (spaces between teeth) are less developed (Vandebroek, 1961).

Some Details of Dentition Patterns in Humans and Anthropoid Apes

In apes, the canine teeth are almost always disproportionately large and clearly longer than the other teeth; they are also definitely more developed in males than in females. In contrast, all human teeth are approximately the same length. In *The Descent of Man*, first published in 1871, Charles Darwin presented the hypothesis that the prehistoric ancestors of humans had large canine teeth. The more they used tools and weapons, the more superfluous these teeth supposedly became; as a result, human teeth gradually evolved into the present dentition pattern with shorter canines. Since the beginning of the twentieth century, Darwin's view has been almost universally accepted. The Piltdown skull, one of the greatest frauds in the history of paleontology, played an important role in this acceptance (Grigson, 1990; Weiner, 1955). Shortly before World War I, a so-called fossil consisting of fragments of a human cranium and a lower jawbone was discovered in the English village of Piltdown. In 1913, Teilhard de Chardin—whom many researchers believe to have been involved in the deception—discovered a large, apelike canine tooth at the same site. Although no part of the jawbone belonging to the tooth was found, it was generally assumed that the tooth, lower jaw, and partial cranium stemmed from a single individual. *Eoanthropus*, as the fossil was named, "proved" that at some point in time, humans with large brains and large, protruding canines had existed. The deception was resolved only after 1950, when the jawbone proved to be that of an orangutan. The Piltdown skull was a fraud. In the interim, the pseudofossil had illegitimately lent credence to Darwin's view for nearly half a century.

Other than the fraudulent Piltdown skull, no fossil finds prove that any prehistoric hominids ever had apelike canine teeth (Aiello and Dean, 1990; Kinzey, 1971). Although it is true that the canines of Neanderthals and *Homo erectus* were, on the average, larger than those of modern humans, so were their other teeth. Most adult male anthropoid apes have large, dagger-like canines, especially in the upper jaw. Consequently, the overwhelming majority of primatologists maintains that small human canine teeth are a derivative, specialized characteristic (Greenfield, 1992). Nonetheless, it cannot be denied that the protruding canines of apes constitute a deviation from the more uniform, generalized pattern of their first dentition. According to Greenfield, adult human canine teeth approximate the shape that is universally found in juvenile primates. In contrast, the protruding canines of male apes must be considered a one-sided transformation of that original, generalized shape. Because this transformation does not occur in humans, small human canines are a Type I characteristic.

The dentition pattern of anthropoid apes includes many specialized idiosyncrasies that are absent in humans. Two examples are:

1. In anthropoid apes, the first premolar of the lower jaw is specially adapted to sharpening the large upper canine. This specialization is absent in humans. It is interesting to note that many fossilized higher primates, such as *Gigantopithecus*, *Ramapithecus*, and *Sivapithecus*, are closer to humans than to anthropoid apes in this respect (Martin, 1990). The mysterious *Oreopithecus*, whose inclusion in the higher primate group remains contested, also has small canines and unspecialized first premolars.

2. The incisors are relatively small in humans and much broader in anthropoid apes. Furthermore, human incisors are more upright (angle β in *Figure 133*, p. 265). "Man gives the impression of having the incisors and the bones in which they are embedded pushed into the oral cavity. This flattens out the angle formed by the incisors when the mouth is closed" (Lenneberg, 1967). Here, too, the situation in anthropoid apes must be considered more specialized (Clark, 1967). And again, the human shape is found in the fossil primates listed above (Martin, 1990). It has also been observed that the incisors are more upright in juvenile apes (*Figure 143*, p. 280).

In summary, we can say that human teeth permanently retain the generalized developmental pattern also found in juvenile apes and higher primate fossils. In adult apes, development continues beyond this point, especially with regard to the canines, and their dentition pattern assumes a degree of specificity that must be considered a one-sided specialization rather than a logical further development of the juvenile configuration. Because this specialization is so common among higher primates, however, it is easily misinterpreted as a generalized phenomenon.

Teeth and Speech

The structure of the human jaw and teeth cannot be understood purely from the perspective of the activity of eating. As Kipp (1980) justifiably points out, although human teeth are very often interpreted as the teeth of an omnivore, this view is one-sided. While it is true that human teeth (and the human digestive system as a whole) are not typical of those of either carnivores or herbivores, the human mouth is also a speech organ. Kipp points out that most mammals have teeth of widely varying sizes (see *Figure 141*). In animals, dentition patterns with no gaps in the front row of teeth are rare, and because of the mouth's prehensile function, animal incisors almost never stand upright in the jawbone. In contrast, according to Kipp (1980), the human dentition pattern is "uniquely uniform with regard to the height of the teeth. The front row of teeth is completely closed and without gaps (the effect of a gap on speech becomes immediately apparent when a front tooth is missing). The characteristically human vertical orientation of the incisors in the jaw is essential to the formation of consonants." Lenneberg (1967), who also saw relationships between speech and specific attributes of the human dentition pattern, writes:

> One deviation in the shape of man's denture is the conspicuous absence of the enlarged canines which are so prominent in most males of most other primate forms. Owing to the great evenness in height and width of all teeth in man, the denture forms an unbroken palisade around the oral cavity. This is the essential prerequisite for the production of sounds such as f, v, s, sh, th and others.

The Mandible (Lower Jawbone)

General Remarks

The shape of the human lower jawbone (mandible) represents an extreme case in comparison to the mandibles of anthropoid apes (see *Figure 139*). In humans, the lower jaw is relatively small and weak, and its front portion, the so-called body, is very short, which of course is directly related to the lack of muzzle development in humans. The ramus, or upright branch of the mandible, ends in two extensions (the coronoid process in front and the condyloid process in back) that support the temporomandibular joint. In humans, the condyloid process is usually larger than the coronoid process, while in apes the opposite is true (Aiello and Dean, 1990).

One conspicuous feature of the human jaw is its protruding chin, so obviously different from the receding chin of almost all ape species and fossil hominids. The potential for chin development seems to be present in all animal species, appearing whenever the dental arch is artificially shortened (see *Figure 140*). In comparison to the shape of animal chins, the human dental arch looks as if it had been pulled backward even more than the base of the lower jaw.

As a rule, the mandible is already shorter in apes than it is in other mammals (see *Figure 141*). In most mammals, the dental arch is long and the series of teeth interrupted, but in apes the arch appears shorter and the spaces between teeth are generally smaller. Shortening the mandible and closing the spaces between teeth is an ancient tendency; a qualitative leap occurred as early as the transition from reptiles to mammals. The lower jaws of reptiles still include various bones that have been incorporated into the skull in humans and other mammals. Reptiles, for example, have only one ossicle in the middle ear, while mammals have three. The two additional mammalian ossicles are the metamorphosed equivalents of reptilian bones that are components of the so-called primary temporomandibular joint. From reptiles to mammals to humans, we can trace a process in which the lower jaw retreats toward the skull, with some of its parts being incorporated into the skull.

Facial proportions are also informative in this context. In the human newborn, the entire facial skull (including the mandible) is still small in proportion to the cranium (cf. *Figure 142*). In all mammals, the cranium initially has a developmental head start as a result of the cephalocaudal pattern of development (see Chapter 3). The infantile mandible also differs from the adult form in several respects. In the newborn, the chin does not yet protrude, and the ramus is still relatively short.

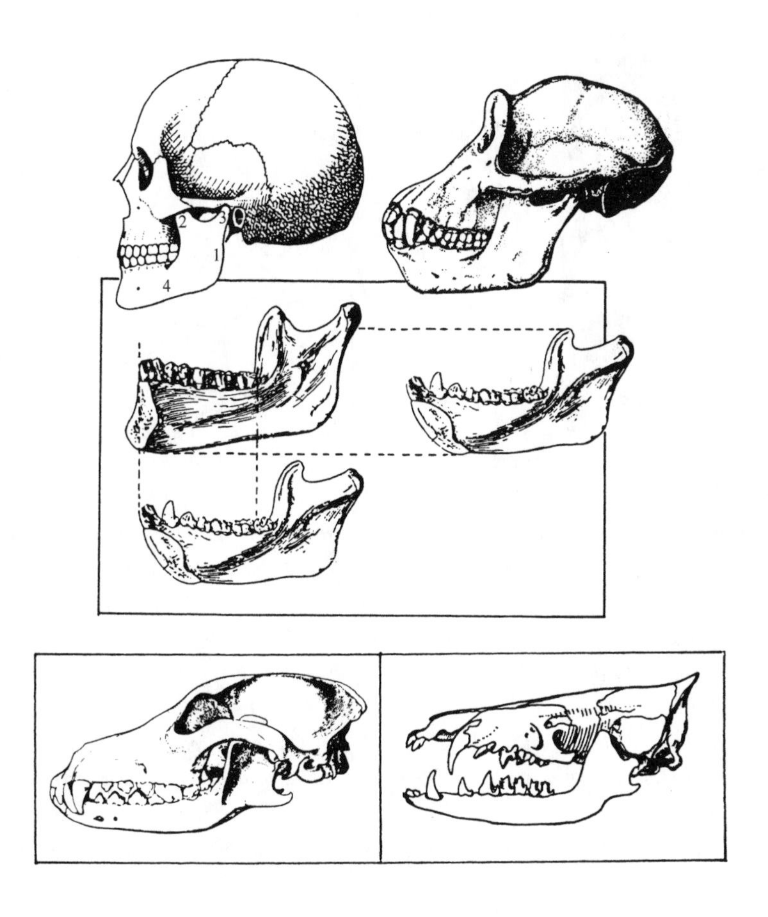

Fig. 139: The mandible in humans and chimpanzees.

Above: Skull of a human (left) and a chimpanzee (right). The human mandible is relatively small and short. The parts of the lower jaw are: (1) the upright branch or ramus, which has two processes, (2) the coronoid process and (3) the condyloid process. The front portion of the mandible is called the body (4). The coronoid process provides insertion sites for the jaw muscles; the condyloid process supports the joint surface.

Center: Half jawbones of a human (upper left) and a chimpanzee (upper right and lower left) of almost equal size. In the human, the ramus is proportionally taller and the row of teeth shorter; the condyloid process is taller and the chin protrudes rather than receding as it does in the ape.

Below: Skulls of two typical mammals: dog (left), hedgehog (right). In these species, the lower jaw is much longer, the chin recedes even more than it does in the average primate, and the location of the condyloid process is very low (Böker, 1935; Kipp, 1955; Vandebroek, 1961).

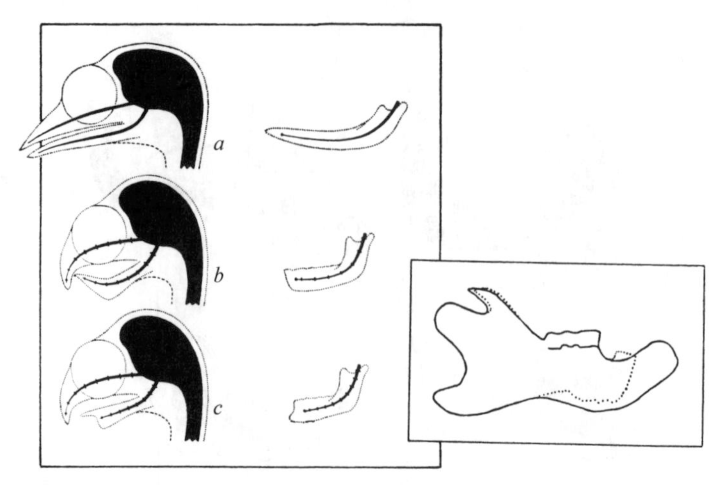

Fig. 140: Artificially induced chin development in animals.

Left: Abnormal chin development in a chick embryo after administering a teratogenic (malformation-causing) substance. (a): normal embryo (left) with the mandible of a lower mammal (right) for comparison; (b): slight malformation in the embryo, with a chimpanzee mandible; (c): severe malformation of the embryo produces a "chin" comparable to that of a human mandible. The black area indicates the location of the brain and main nerve fasciculi (Roth and Krkoska, 1978).

Right: mandible of a normal rat (solid line) and a rat whose incisors were extracted at an early stage of development (dotted line). Shortening the dental arch produces a chin of sorts (Riesenfeld, 1969).

These experiments seem to indicate that chin formation is not a specialization but a more or less necessary consequence of shortening the mandible by any means.

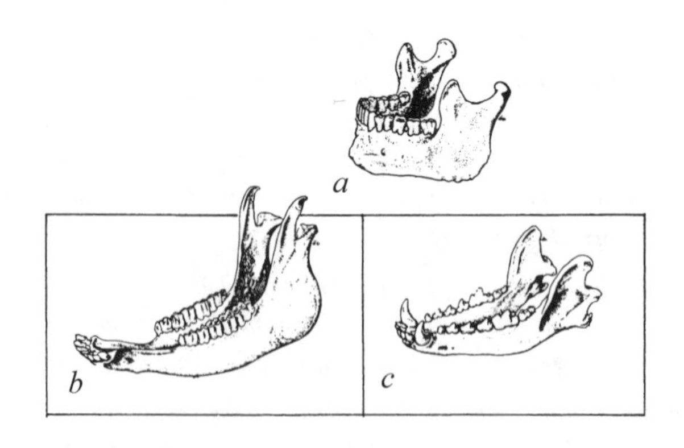

Fig. 141: The mandible in (a) human, (b) horse, and (c) dog.

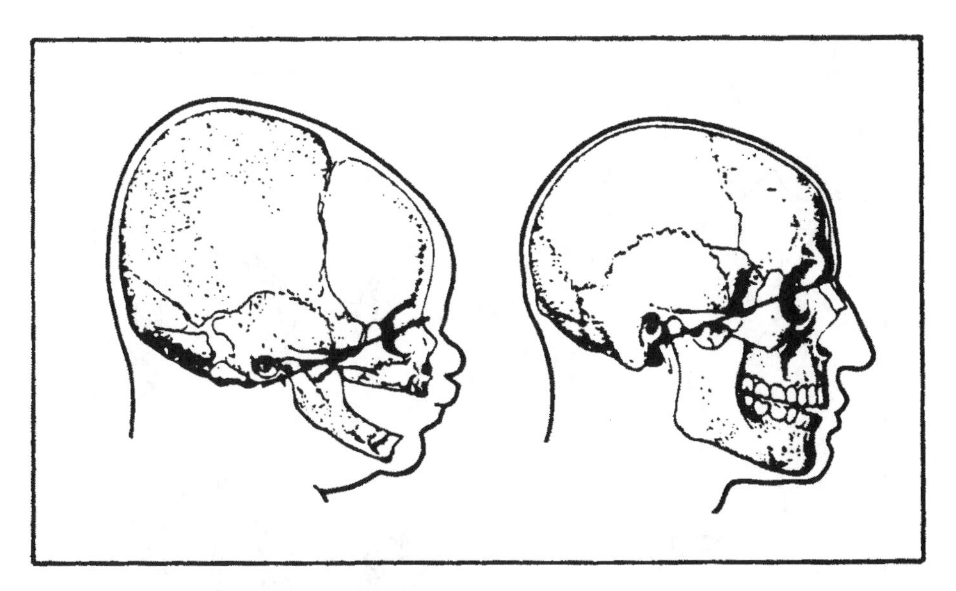

Fig. 142: Cranium and facial skull in the human newborn and adult. Most of the growth of the facial skull is vertical. The ramus (the upright branch of the lower jaw) is disproportionately larger in the adult (Sullivan, 1986).

The shortness of the human mandible, combined with its location below the skull—which requires a tall ramus—produces efficient leverage because the load arm of the lever remains short. From a purely mechanical perspective, the elongated jaws of animals are less efficient—an unavoidable situation, since animals use their mouths as prehensile organs, which in turn is related to the fact that their forelimbs are used for locomotion.

We already know that the dental arch has a typical curved shape in humans. In mathematical terms, this shape is best described as a catenary (the curve formed by a cord suspended between two fixed points). This catenary shape is already evident at a very early stage in fetal development (Burdi, 1968). In humans, it is preserved in both the first and second dentitions (see *Figure 143*). Scott's important study in comparative anatomy reveals that the catenary shape of the dental arch is evident in the very early fetal stages of all mammals (1957, 1967). Furthermore, the shape of the base of the mandible is also a catenary in most mammals (see *Figures 144* and *146*).

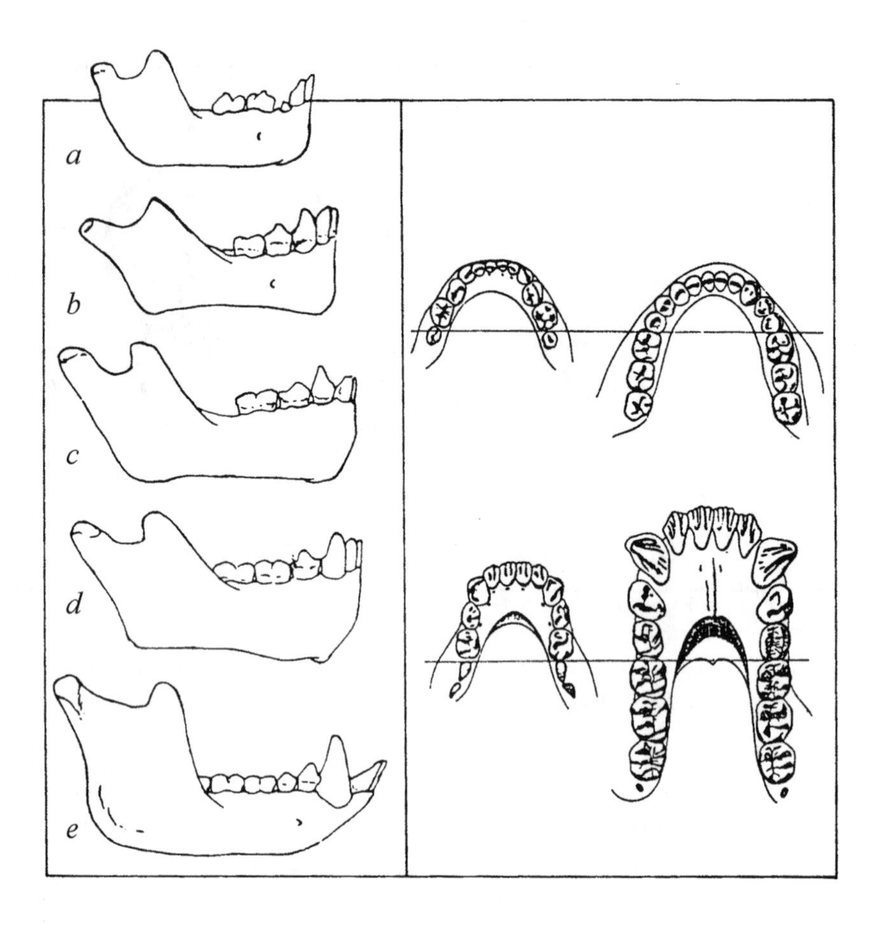

Fig. 143: The first dentition in humans and anthropoid apes.

Left: Five successive stages in the development of the jaw and teeth of a young chimpanzee. The chin of the juvenile anthropoid ape neither protrudes nor recedes; the lower jaw gradually assumes the mature diagonal shape. Throughout the first dentition (stages a-d), the canine is only slightly taller than the other teeth. In the second dentition (stage e), the canine is huge in comparison to the other teeth, and the incisors angle diagonally forward. The second dentition in humans retains the two juvenile characteristics of shorter canines and relatively vertical incisors (Bolk, 1924).

Right: First (left) and second (right) dentitions in humans (above) and orangutans (below). In humans, the front portion of the dental arch, where the baby teeth appear, does not increase in size after age six, but in the ape, juvenile proportions are completely abandoned. The size of the front of the arch continues to increase, and the originally somewhat arching, closed row of teeth is replaced by a rectangular pattern with diastemata (Kipp, 1980).

The lower border of the adult human mandible conforms even more closely to a catenary than the dental arches. In animals in which the dentition does not conform to a catenary curve such as ungulates, marsupials and rodents, the lower border of the mandible usually does so. This is also the case in the anthropoid apes. In the young chimpanzee, the dental arch conforms closely to the catenary form, but in older animals, while dentition is still in use, the arch has begun to develop the typical adult noncatenary form. While other than human dentitions do not as a rule conform to a catenary curve, the lower border of the mandible conforms with a remarkable constancy in the great majority of mammals (Scott, 1957).

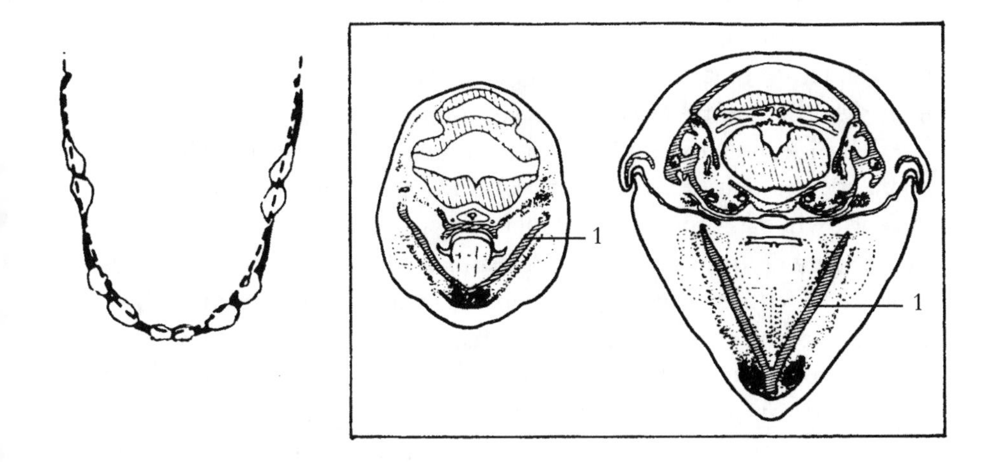

Fig. 144: The catenary as the basic shape of the dental arch.

Left: A catenary (broken line) drawn through the rudimentary dental arch of a twelve-week-old human fetus (Burdi 1968). In humans, this shape is preserved (see Figure 136*).*

Right: Cross-section of the head of a mouse embryo at 14 days (left) and 16 days (right). (The plane of the cross-section runs through the base of the mandible.) The shape of the rudimentary mandible (1) is still fairly close to a catenary in the younger embryo but has already become more pointed in the older one (Hinrichsen, 1959).

Scott points out that mammals have spaces (diastemata) between the incisors and the molars. The anterior portion of the dental arch assumes a prehensile function, while the posterior portion takes over the function of mastication. Scott comments:

> This division of the dentition into two functionally distinct parts occurs at the very beginning of mammalian evolution . . . In the higher primates it is much less marked, and in man it no longer exists . . . It is interesting to notice that in this as in certain other morphological features man retains the fetal conformation (1957).

A closed row of teeth in the shape of a catenary is actually a universal mammalian trait. Scott observes, for example, that the catenary is still clearly recognizable in juvenile apes and that the row of teeth is still fairly closed even in newborn rodents. The closed, catenary pattern of human dentition, therefore, is a Type I characteristic. The human mandible, whose structure determines the shape of the dental arch, must therefore also be an original, prototypic shape. This means that the mouth is not predestined to be a prehensile organ and that muzzle development in animals must be interpreted as a specialization.

The location of the canines in animals is especially interesting. In some species (lemurs, for example), the canines are much closer to the incisors, while in other species, including the great majority of higher primates, they are much closer to the molars. In some other instances, the canines are either completely absent or located exactly midway between the incisors and the molars. In humans, a consistently closed row of teeth is the general rule, but less pronounced forms of all animal dentition patterns occur as individual variations. "While in monkeys and anthropoid apes the canines obviously 'belong' to the cheek tooth series, there seem to be two types of human dentitions: a type in which the canines are more in line with the cheek teeth, and a type in which they are more in line with the incisors" (Scott, 1957).

Unlike monkeys (and also anthropoid apes with their obviously specialized dentition patterns), humans—as per Goethe's formulation (see Chapter 3)—have "overshadowed" the various patterns that occur in the animal kingdom. The shape of the human dental arch and mandible exemplifies the entire relationship between humans and animals. The shape found in humans is the original, prototypic form. In animals, this form is recognizable during the earliest stages of development but is later overwhelmed by specialization. The further evolution proceeds, the more the fundamental, humanlike gestalt emerges. Diluted versions or shadowy traces of animal specializations (including those typical of animals less closely related to humans) can be found in individual examples of the fully manifest human gestalt.

The Chin and the Change of Teeth

The protruding human chin, a formation that does not exist in any other primate, is not present during juvenile stages of development. The human fetus still has a somewhat rounded, slightly apelike chin. This roundness disappears postnatally, but babies' chins are still not truly pointed. Louis Bolk, who devoted himself extensively to this problem, initially concluded that this phenomenon appears to constitute an exception to the general rule that human characteristics develop through retardation. He eventually discovered that the human chin cannot be considered an independent trait but develops as the passive result of extreme retardation of the human dentition pattern.

Bolk discovered that in both humans and anthropoid apes, the fetus has a somewhat rounded chin whose shape later becomes angular during early childhood (see *Figure 143*). In the siamang, a gibbon relative, this angular shape can persist into adulthood (see *Figure 15*, p. 32). In humans, evolution takes another step in the same direction; during the second seven-year period (between the ages of six and thirteen, according to Bolk), the protruding chin develops (see *Figure 145*).

Bolk discovered that chin protrusion in humans (which appears simultaneously with the change of teeth; see *Figure 136*) is related to retardation of the dentition pattern. He also noted another strange fact that had been mentioned by several earlier investigators, namely, that the front portion of the dental arch, which supports the fully emerged baby teeth, stops growing as soon as the change of teeth begins. The permanent incisors, canines, and premolars must make do with the space that the corresponding baby teeth occupied. With regard to the shape and size of the dental arch, human development stops short at juvenile proportions.

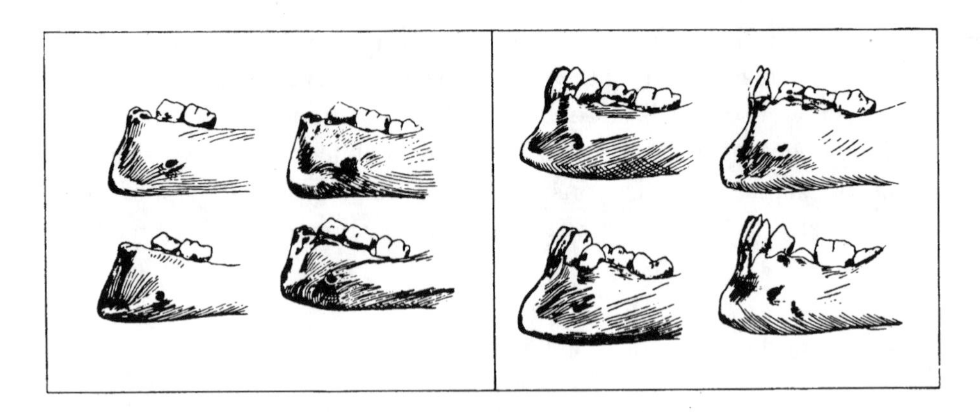

Fig. 145: Development of the protruding chin in humans.
Left: Mandibles of six- to seven-year-old children. The chin is still straight.
Right: Mandibles of eleven- to twelve-year-olds. The protruding, typically human chin is already present (Bolk, 1924).

This is not the case with anthropoid apes, whose dental arches continue to grow rapidly, assuming elongated rectangular shapes (see *Figure 143*). In apes, the dental arch abandons its original catenary shape and diastemata appear between the teeth. The change in shape of the base of the mandible, however, is somewhat more conservative in both anthropoid apes and other mammals; the catenary shape often remains recognizable on the underside of the mandible. The rapid growth of the dental arch in anthropoid apes causes gradual recession of the chin as the dental arch grows beyond the base of the mandible. The prototypic tendency, in which the base of the mandible gradually catches up with the dental arch, is reversed in apes.

In humans, however, this fetal tendency persists. The dental arch ceases to grow, while the base of the mandible continues to catch up for a short time, forming the human chin. Hence Bolk concluded that the human chin is essentially a negative trait that develops as a result of pronounced retardation of the dentition pattern in humans. And in fact, growth of the mandibular base in humans is also inhibited in comparison to the corresponding development in apes, although less inhibition occurs on this level than on the level of the dental arch. The retardation of the dental arch in comparison to the mandibular base is most evident in the chin, but its consequences are apparent in the entire outline of the jaw. As early as 1904, Weidenreich commented:

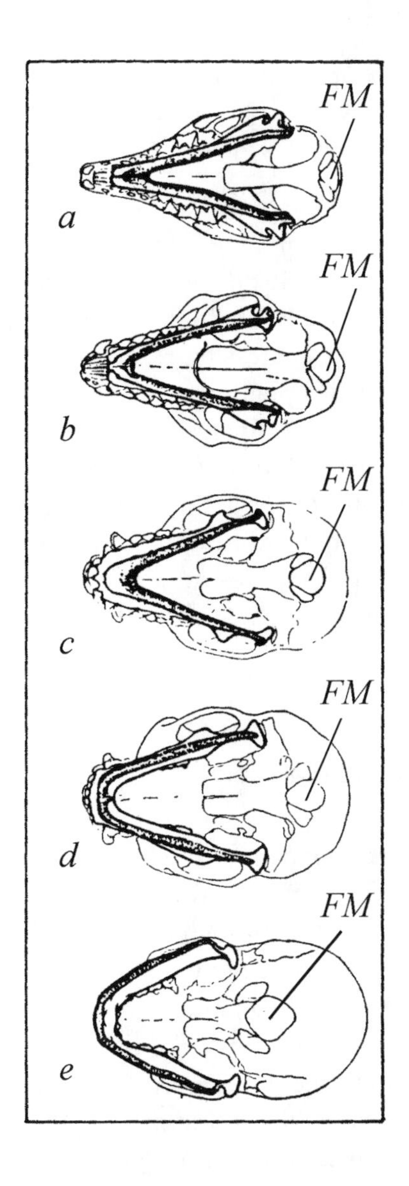

Fig. 146:

*View from below of the skull, mandible, and dental arch of primates. (a) tree shrew (*Tupaia*), (b) lemur, (c)* Cercopithecus*, (d) gibbon, and (e) human. The base of the mandible is indicated by a bold line and stippling. In all of these animals, the shape of the front portion of the base is a catenary. In tree shrews and lemurs, the suture line between the two halves of the mandible is not ossified, and the incisors are still nearly horizontal. In the higher primates (c-e), the teeth draw ever closer to the boundary of the base of the mandible; that is, they appear to move inward. In gibbons, this shift toward the back is already so advanced that some of the molars are visible on the inside of the mandible. In humans, all teeth are visible on the inside, and retardation of the dental arch relative to the base of the mandible is very apparent. Note also the forward shift of the foramen magnum (FM) in this series (DuBrul, 1958).*

The recession of the alveolar portion occurs throughout the lower jaw. In apes the teeth and alveoli [dental sockets] on the side of the jaw sit directly on the basal portion and the incisors extend far beyond it, but in humans the basal portion of the jaw protrudes to the side and to the front. As a result, the chin appears in front,

> while on the sides the dental sockets (especially those of the molars) are located on the interior of the jaw, above the basal portion, which is separated from them by a furrow.

This recession of the dental arch relative to the basal part of the jaw can be readily observed. The more humanlike a primate is, the less its teeth extend outside the base of the mandible. This tendency climaxes in adult humans, whose teeth are visible on the inside of the mandible as seen from below (see *Figure 146*). Note also that increasing recession of the dental arch in primates parallels the degree of uprightness of the incisors, which are still angled sharply forward in lower primates.

Kipp (1955) points out that cessation of growth of the dental arch is very important for human speech. Humans learn to speak during the first seven-year period. If the shape of the oral cavity were to change significantly after that time—as is the case in anthropoid apes—we would have to relearn all of the subtle motions of speech that we master in the early years of life.

The Retarded Growth Pattern of the Human Mandible

Bolk did not explain *why* growth of the human dental arch is retarded in comparison to growth of the mandibular base. Why is the retardation process seemingly more active on the level of the teeth than on the level of the chin? And why does the ramus grow much more between birth and adulthood than the body (the front part of the mandible) does?

To understand these phenomena, it is helpful to divide the mandible into three approximate zones – the dental arch (the upper portion of the body that supports the teeth), the basal portion (the lower part of the body), and the ramus. These three mandibular zones do not develop at the same time. Rudimentary teeth appear very early, and the parts of the mandible that surround the teeth are the first to ossify (see *Figure 147*). At the same early embryonic stage, the hind portion of the body is still cartilaginous, and the ramus is still entirely undeveloped. Through retardation of this staggered growth pattern within the mandible, the "latecomer" (the ramus) ultimately becomes disproportionally larger than the portion that began to develop first (the dental arch). This original developmental sequence (dental arch, mandibular base, and ramus) is more or less preserved in the course of further growth. During decisive phases of growth, the dental arch is always closer to its final dimensions than the base of the jaw, which in turn has a head start on the ramus.

Baughan et al. (1979) distinguish the following three growth patterns in the bones of the facial skull:

The cranial growth pattern

Bones that follow the cranial (head-directed) growth pattern achieve 90 percent or more of their final size by the age of six and undergo only a weak growth spurt during puberty. This growth pattern is evident in the bones surrounding the brain (including the base of the skull). The same pattern is apparent in the throat and also, in extreme form, in the front portion of the dental arch.

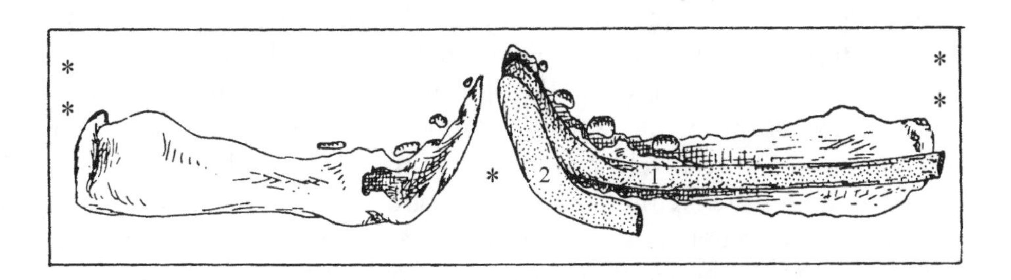

Fig. 147: Exterior (left) and interior (right) of the embryonic human mandible (= chin end; ** = joint end) (Bolk, 1924). The jaw develops along the exterior of a cartilaginous bar known as Meckel's cartilage, which functions as a scaffolding of sorts and later disappears. (Meckel's cartilage: (1) in the right half of the jaw; (2) in the left half of the jaw, shown cut away.) Initially, the mandible consists of connective tissue, which later ossifies. At the stage illustrated here, this process has already begun in the front portion of the jaw. Rudiments of baby teeth are already visible along the dental arch. At this stage, the ramus remains undeveloped. Where ossification has already begun, the dental arch has a developmental head start on the other parts of the lower jaw.*

The facial growth pattern

Bones that follow this growth pattern achieve almost 80 percent of their final size by age six. Their remaining growth is accomplished slowly, with a moderate growth spurt in puberty. This growth pattern applies to the base of the jaw and to the vertical dimension of the upper jaw. The chin comes about as a result of the difference between the facial growth rate (which includes the base of the jaw) and the cranial growth rate (which includes the dental arch).

The growth pattern of the ramus

According to Baughan et al., this growth pattern is related to that of the long bones of the limbs. During the second seven-year period, the growth of the ramus, as a percentage of its ultimate size, is significantly retarded relative to the growth of the base of the jaw (see *Figure 148*). The pubertal growth spurt is very pronounced, and the temporomandibular joint itself matures only around age twenty. As a whole, therefore, the growth of the ramus lags behind that of the mandibular base, which in turn lags behind that of the dental arch and palate.

The mandible's pattern of growth is comparable to that of the limbs. Just as a caudocephalic or distoproximal growth pattern is evident in the limbs, growth in the mandible moves from the teeth to the base of the jaw to the temporomandibular joint. (This growth pattern is also called distoproximal, since it also moves in the direction of the cranium.) Retardation of the limbs' distoproximal growth pattern results in relatively shorter fingers and toes and relatively longer parts of the body that lie closer to the torso (humerus and femur). Similarly, retardation of the corresponding pattern of growth in the mandible results in a protruding chin and a highly developed ramus; the location of the condyloid process is high (see *Figure 139*, upper left). Hence, retardation of the distoproximal growth pattern underlies all of these unique features of the human mandible. The relatively tall ramus and the elevated location of the head of the mandible (condyle) can therefore be considered hypermorphic phenomena.

The metamorphosis of the human mandible in comparison to the animal form allows the human head as a whole to retain a retarded shape. The tall ramus permits the facial skull to retain its original position below the cranium, and the fact that the dental arch remains short corresponds to the lack of muzzle development in humans.

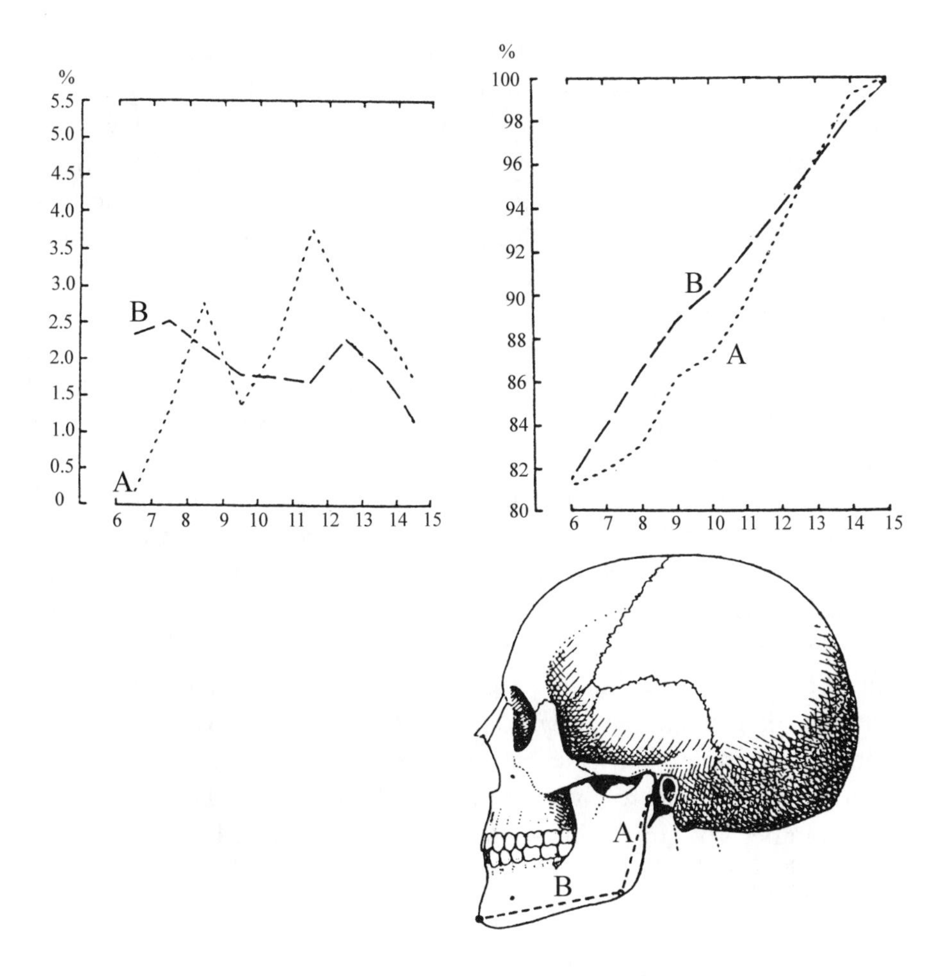

Fig. 148: The course of growth in the ramus (the data apply to girls; in boys, the pubertal growth spurt would undoubtedly begin later) (Baughan et al., 1979).

Upper left: rate of increase (expressed as a percentage of the size achieved at age 15) in distances (A) (between the condyle and the gonial angle) and (B) (between the mental protuberance and the gonial angle), as a function of age (in years).

Above right: lengths of distances (A) and (B), expressed as percentages of the size achieved at age 15.

Below: Distance (A) is a measure of the size of the ramus; distance (B) is a measure of the size of the base of the jaw. During the second seven-year period, the size of the ramus is retarded relative to the size of the mandibular base; after puberty, the relationship tends to reverse. The elementary growth spurt is slower in the ramus than in the base of the jaw, but this relationship is reversed in the pubertal growth spurt.

The altered proportions of the human mandible develop through retardation of a universal growth pattern rather than through specialization. Permanent retention of the catenary shape of the human lower jaw clearly demonstrates the prototypic character of this part of the body, which in turn is related to uprightness in several respects. The fact that the facial skull remains short allows the head to occupy a balanced position above the upright torso (see Chapter 6), while on the other hand muzzle development is unnecessary because an upright torso also leaves the arms free to serve the prehensile function. At the same time, the short human mandible—specifically, its retarded, upright, and closed row of teeth—is a prerequisite to developing the purely human function of speech.

The mandible is like a little limb within the head. With this in mind, it is interesting to make a closer comparison between the growth of the mandibular base and that of the ramus (*Figure 148*). According to data compiled by Baughan et al. (1979), the *elementary* growth spurt takes place earlier in the base of the jaw than in the ramus, while the *pubertal* growth spurt affects the ramus earlier than the base of the jaw. These facts suggest that the elementary growth spurt acts in a distoproximal direction (from the dental arch to the base of the jaw to the ramus), while the pubertal growth spurt follows the proximodistal pattern. This means that the phenomena of growth in the mandible closely parallel those observed in the limbs (see *Figure 37*, p. 80).

The Temporomandibular Joint

The temporomandibular joints of humans and apes are distinctly different. In humans, the head of the mandible rests in the mandibular fossa (see *Figure 149*). This depression in the temporal bone is bounded by two processes, the articular tubercle in front and the post-mandibular process in back. Apes do not have such a clear, S-shaped indentation; the articular tubercle is either less developed or absent, while the post-mandibular process is even more highly developed than it is in humans.

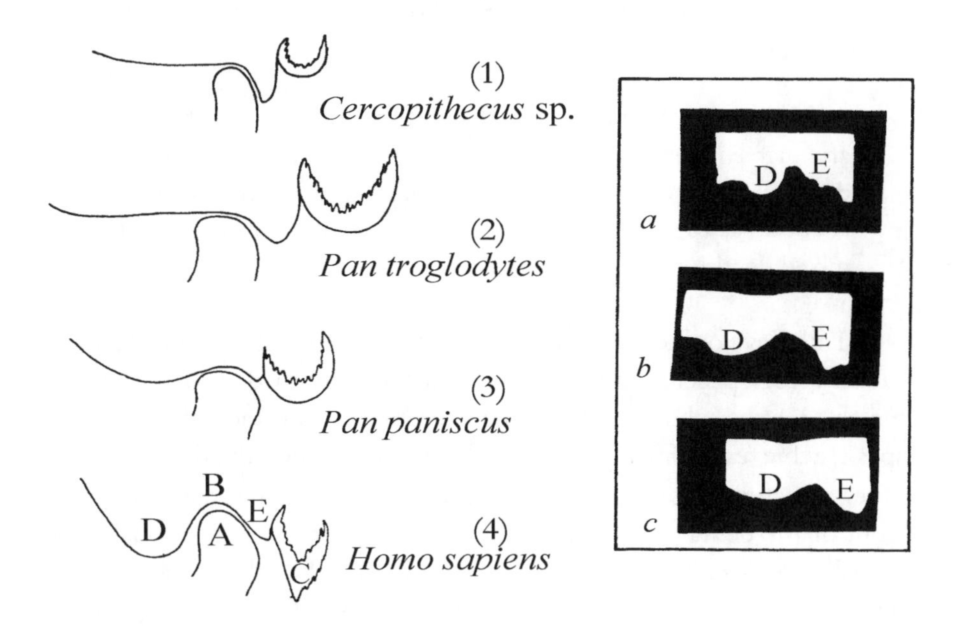

Fig. 149: The temporomandibular joint.

Left: schematic representation of the mandibular fossa in (1) Cercopithecus *sp. (a lower catarrhine), (2)* Pan troglodytes *(chimpanzee), (3)* Pan paniscus *(bonobo or pygmy chimpanzee), and (4)* Homo sapiens *(human). (A): head of the mandible, (B): mandibular fossa, (C): tympanic part of the auditory canal, (D): articular tubercle, (E): post-mandibular process. The mandibular fossa is beginning to develop in the more paedomorphic bonobo but is absent in the ordinary chimpanzee (Vandebroek, 1961). The apes are characterized by a highly developed post-mandibular process and a flattened or absent articular tubercle. Note also the low position of the auditory canal in humans.*

Right: Cross-section of the temporomandibular joint in: (a) a human, (b) a female gorilla, (c) a male gorilla (Aiello and Dean, 1990). In female gorillas, the shape of the skull is generally more highly paedomorphic and the mandibular fossa deeper than in males (see Figure 138, *p. 272).*

At birth, the human temporomandibular joint has not yet achieved its ultimate shape:

> At birth the temporal fossa is flat and there is little evidence of an articular eminence, but during the first three years the eminence develops to resemble the mature S-shape which is finally obtained at about six to seven years of age, in other words, at the start of the mixed dentition (Keith, 1982).

This development seems to indicate a relationship with the human capacity for speech. The S-shape, which begins to develop at approximately the age when the baby teeth erupt and the individual begins to speak, achieves its final form around age seven, when the development of other structures closely associated with speech (brain cavity, size of the front of the dental arch, etc.) comes to a conclusion. A relationship with speech is also suggested by the fact that the head of the mandible always remains in the mandibular fossa during speaking (Fuss, 1990). (When the mouth is opened wide, the head of the mandible slides forward over the articular tubercle.)

Thus it seems that, in humans, the initially ape-like temporomandibular joint with no mandibular fossa develops its final human form in the course of early childhood. This state of affairs is generally understood as a Haeckelian recapitulation in which an apelike ancestral configuration is repeated in human ontogeny (Petrovits, 1930). To date, little research has been done on the ontogeny of the temporomandibular joint in apes, but Hinton and Carlson (1983) made a remarkable discovery when they studied the growth of this joint in rhesus monkeys. The joint initially develops in a humanlike direction, assuming an obvious although not highly developed S-shape. After birth, as muzzle development continues, the joint reverts to the flat shape typical of adult monkeys.

> The slight concavity of the mandibular fossa in the neonate becomes gradually flatter with age in the rhesus monkey, in contrast to a continued deepening in humans . . . with the result that an eminence is often no longer recognizable in adult animals (Hinton and Carlson, 1983).

It would be interesting to study the ontogeny of the temporomandibular joint in anthropoid apes in greater detail. Presumably, in this case, too, initial development in the human direction would be succeeded by reversion to a flattened fossa. Remarkably, the mandibular fossa is much more apparent in the pygmy chimpanzee species than in common chimpanzees. Many researchers consider the pygmy chimpanzee a paedomorphic form of the common species (Shea, 1989). For example, the pygmy chimpanzee has a shorter muzzle, smaller canine teeth, and a more parabolic dental arch than the common species (Vandebroek, 1961). Less advanced animalization in this species probably explains its more human-like mandibular fossa. In female gorillas, the fossa also appears deeper than it is in males of the same species. In this case, too, the more paedomorphic form has the deeper fossa (Aiello and Dean, 1990).

The absence of the articular tubercle in monkeys is related to the mastication function. Both Vandebroek (1961) and Jones (1947) point out that the protruding canines of monkeys impede any lateral chewing motions of the lower jaw. When a human being chews, the sideways ("grinding") motion of the lower jaw is very apparent. In monkeys, however, the lower jaw moves mainly forward and back; hence the S-shaped, stabilizing mandibular fossa is replaced by a flat plane on which the head of the mandible can move back and forth. Aiello and Dean (1990) point out that the joint socket in monkeys is quite tight around the head of the mandible, which also makes sideways motion difficult. In humans, the shapes of the joint socket and the head of the mandible are less closely matched, so the mandible remains more mobile, especially with regard to lateral motion.

Hinton and Carlson (1983) maintain that the shape of the mature temporomandibular joint in monkeys develops under the influence of the pressure exerted on it during mastication. In humans, the bony wall of the mandibular fossa (*Figure 149*) is often paper-thin, but in monkeys it becomes thicker after birth as a result of accelerated ossification. In monkeys, the post-mandibular process is also much more highly developed and often produces an indentation on the back of the head of the mandible, which suggests that heavy pressure is exerted from front to back. Clearly, the temporomandibular joint is subject to less pressure in humans than it is in monkeys.

The human temporomandibular joint plays a decisive role in hearing one's own voice, which is very important for learning to speak. The mandible picks up the vibrations of the vocal chords and transmits them via the temporomandibular joint to the auditory canal and the bones of the skull (von Békésy, 1932, 1941). The fact that the head of the mandible lies in the mandibular fossa is one factor that facilitates transmission of sound. From this perspective, the specific shape of the human temporomandibular joint is directly related to the capacity for speech. The presence of a highly developed articular tubercle (*Figure 149*) not only stabilizes the position of the head of the mandible but also increases the surface area of the joint, which improves sound transmission through the joint. In humans, the external auditory canal is lower and closer to the temporomandibular joint than it is in apes (Clauser, 1971). Sound transmission is also improved by the fact that in humans (as in all catarrhines) the auditory organs are enclosed in a bony capsule. According to Clauser, the mastoid process, which is located directly behind the external auditory canal and is highly developed in humans, plays a role in sound transmission (see *Figure 150* and *151*).

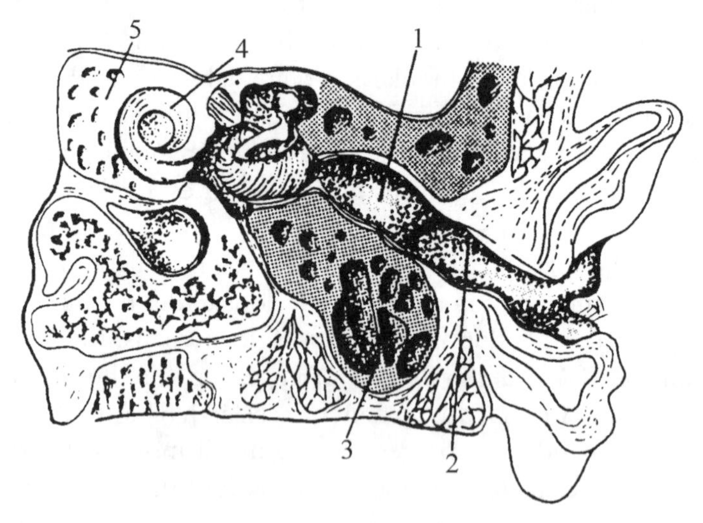

Fig. 150: The bones bordering on the inner portion of the human external auditory canal. (1) external auditory canal, bony portion, (2) cartilaginous portion, (3) mastoid process, (4) cochlea, (5) petrous portion of the temporal bone (Clauser, 1971).

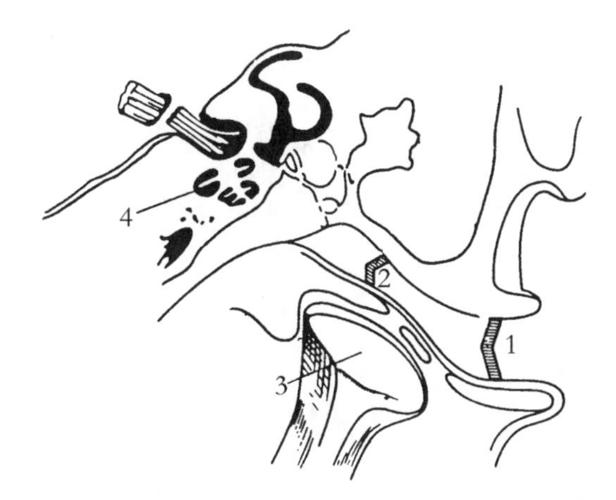

Fig. 151: An experiment by von Békésy (1941). One's own voice is heard more loudly when the auditory canal is closed near its outer end (1) but less loudly when the blockage is close to the eardrum (2); (3) head of the mandible, (4) cross-section through the cochlea (Clauser, 1971).

On the other hand, Mills (1978) demonstrates that there is also a physiological connection between the presence of the articular tubercle and certain characteristics of human dentition. In humans, the lower teeth turn inward (see *Figure 146,* p. 285). As we have seen, this phenomenon can be interpreted as a consequence of retardation of the universal pattern of mandibular growth. Because the human cranium is broad in proportion to the dental arch, the two temporomandibular joints are far apart, and the mandible becomes broader at the back. Contact between the upper and lower teeth is assured by the fact that human upper teeth are somewhat buccally oriented (i.e., outward-pointing) while the lower teeth are more lingually oriented (i.e., inward-pointing). In this case, maintaining contact between the upper and lower teeth during lateral motion of the mandible in one direction requires lowering the branch of the mandible on the other side (*Figure 152* illustrates this situation.) In humans, the head of one side of the mandible moves over the articular tubercle to produce this lowering. In apes, lateral mandibular motion is restricted because the canine teeth protrude far beyond the other teeth. In these animals, therefore, the mandible moves backward and forward to a much greater extent than it does in humans, and the articular tubercle serves no purpose. In addition, these animals' teeth are oriented differently, so the motion illustrated in *Figure 152* occurs less or not at all.

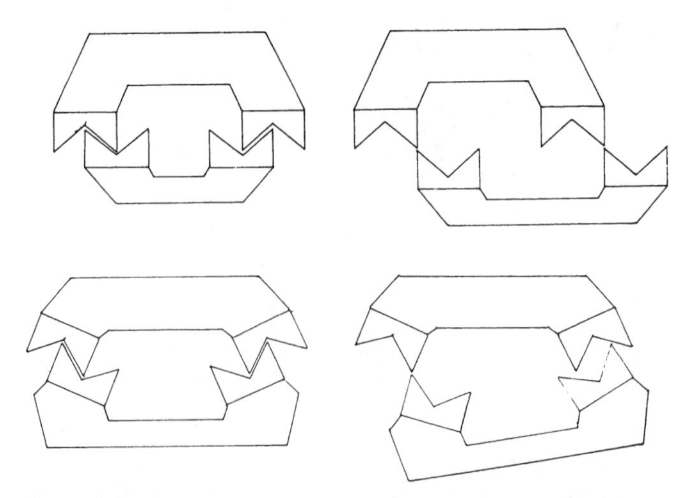

Fig. 152: Schematic representation of the role of tooth orientation in lateral movement of the lower jaw. Top: When both rows of teeth are roughly vertical and positioned above each other, the mandible can maintain tooth contact while moving sideways without tipping. Bottom: When the upper teeth point outward and the lower inward (because the base of the jaw has become relatively broad), the same movement is associated with tipping of the mandible. For a more realistic representation, see Mills (1978).

The function of the articular tubercle can undoubtedly be considered from other perspectives (see, for example, DuBrul, 1958 and 1976). A complex interrelationship of mutually complementary functions and mutually influential developments exists among the shapes of the mandible, dental arch, skull, and larynx. The perspectives we have mentioned thus far, however, demonstrate the various direct or indirect relationships between the specific shape of the human temporomandibular joint and typically human characteristics—specifically, the capacity for speech. The absence of large canines, for example, is a prerequisite to lateral mandibular motion (which is facilitated by the presence of the articular tubercle, according to Mills) but is also related to speech, which is facilitated by an even row of teeth.

It is remarkable, therefore, that the articular tubercle makes a transitory appearance in juvenile macaques and that its beginnings are also found in the adults of certain African anthropoid ape species. Thus the mandibular fossa seems to constitute a transition between Type I and

Type II characteristics. It is a Type I characteristic inasmuch as its beginnings appear at an early developmental phase in lower apes, only to disappear under the pressure of animal specialization (i.e., muzzle development). It is also a Type II characteristic, however, because it is not yet present during fetal development, remains visible in adulthood in the highest primates (especially pygmy chimpanzees and gorillas) and is related to certain specifically human attributes such as the capacity for speech.

The Cheeks

The opening of the human mouth—in comparison to that of land vertebrates in general—is conspicuously small. Reptiles and birds have no cheeks; their mouth openings extend all the way to the temporomandibular joints. As such, the cheek is a mammalian characteristic. Its development permits not only sucking but also extensive chewing. But in comparison to humans, other mammals still have relatively large mouth openings, a phenomenon that is surely related to muzzle development. The small human mouth is very important for speech. Hockett (1967) writes:

> For vocal-auditory communication *Homo sapiens* has an advantage over the other hominoids in the relatively small external opening of his oral cavity. This is a matter of physical acoustics that can be tested in any kitchen. Allowing for differences of material and shape, a container with a small mouth, such as a beer bottle, resonates better than one with a large mouth, such as a jelly glass.

Lenneberg (1967) states that a small opening is very important for the formation not only of vowels but also of some consonants such as p and b. The small volume of the mouth allows the pressure required for the formation of these sounds to build up more quickly.

Figure 153 shows the first rudiments of the face. At age five weeks, a centrally located, fissure-like indentation, the primitive mouth or stomodeum, has developed in the human embryo. Above it, around the two so-called olfactory placodes (which develop into the nasal pits and later into the actual olfactory organ), two quickly growing folds develop, the lateral nasal prominence, which develops into the nostrils, and the medial nasal prominence, which is located closer to the center and develops into the central portions of the nose and upper lip. In the five-week-old embryo, the eyes are still located on the sides of the head. Because the distance between the eyes increases relatively less than the size of the head as a whole, they later lie closer to the front. Under each eye are two protuberances, the maxillary process (which will later develop into the upper jaw) and below it the mandibular process (which will develop into the lower jaw). These two protuberances develop through involution of the same (the first) branchial arch (Starck, 1975).

As the head grows, the mouth opening becomes relatively smaller and—like the eyes—contracts into the center of the face. The nose splits off from the mouth. In the ten-week-old fetus, the upper lip has already developed. This process is much simpler in humans than in animals. In mammals—probably due to the beginnings of muzzle development—the stomodeum (mouth) is too large, and fusion of the mandibular and maxillary processes is necessary to make it smaller. In certain animals such as guinea pigs, the entire cheek develops as a result of this extra process. Many mammals—bats, for example—have combined cheeks and lips with a visible suture where the mandibular and maxillary processes fuse. This fusion can result in unique situations. In rabbits, for example, the process is unusual in that the outer skin closes along the inside of the cheek, producing the potential for hair growth there (Starck, 1975).

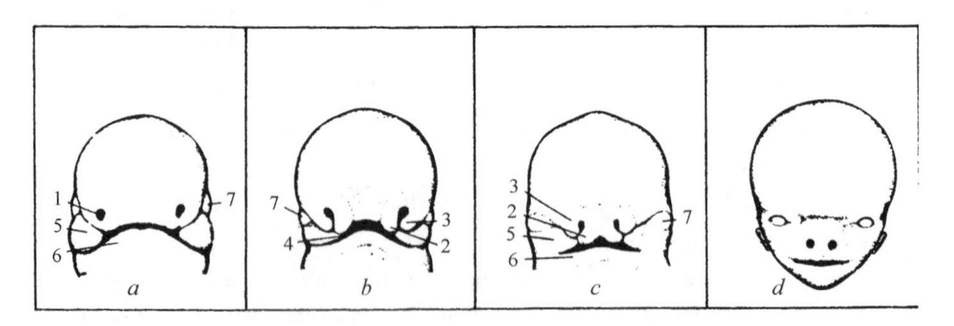

Fig. 153: Development of the human face in an embryo at age:

(a) five, (b) six, (c) seven, and (d) ten weeks. (1): olfactory placode or nasal pit, (2): medial nasal prominence, (3): lateral nasal prominence, (4): primitive mouth, (5): maxillary prominence, (6): mandibular prominence, (7): eye (Langman, 1966).

In humans, there is no question of the cheeks growing closed. The indentation between the two branches of the first branchial arch (the maxillary and mandibular processes) grows just wide enough to permit the development of the primitive mouth (Heine, 1979 and 1983; Starck, 1975). The developmental route taken by mammals (excessive separation followed by secondary unification) suggests a deviation (probably related to muzzle development) from the prototypic developmental plan. Muzzle development causes an exaggerated increase in the size of the mouth, which is partially balanced out in mammals through further extension of the cheeks. The compact human oral cavity with its small mouth is a Type I characteristic that prefigures the appearance of speech.

The Ossification Pattern of Facial Skull Sutures

The Conundrum of the Premaxillae (Ossa Incisiva)

Looking at the facial skull of an ape from the front, we notice two bones that appear to be completely absent in humans. In humans, the two maxillae (upper jawbones) are located on either side of the plane of symmetry of the face. They meet between the nasal opening and the mouth and extend to the zygomatic bone on the side and to the frontal bone above. The maxillae are separated by the nasal bone above the level of the nasal cavity. In apes, two smaller bones, the premaxillae (or ossa incisiva), lie between the two maxillae. The upper incisors are rooted in these bones. The premaxillae touch below the nasal cavity and extend upward to the nasal bone (see *Figure 154*). It looks as if we were dealing here with a difference between the human and the animal structural plan. Almost all mammals have premaxillae.

During the Middle Ages, it was universally assumed that premaxillae were also to be found in the human facial skull. This assumption rested on the authority of the Greek physician Galen (Claudius Galenus, 131-201 AD), whose description of the human skeleton was clearly based on ape bones. Andreas Vesalius incurred the enmity of Galen's supporters with his book *De Humani Corporis Fabrica*, published in 1543, in which he was the first to vehemently contest Galen's claim that there is a visible suture between the maxilla and the premaxilla in the human facial skull.

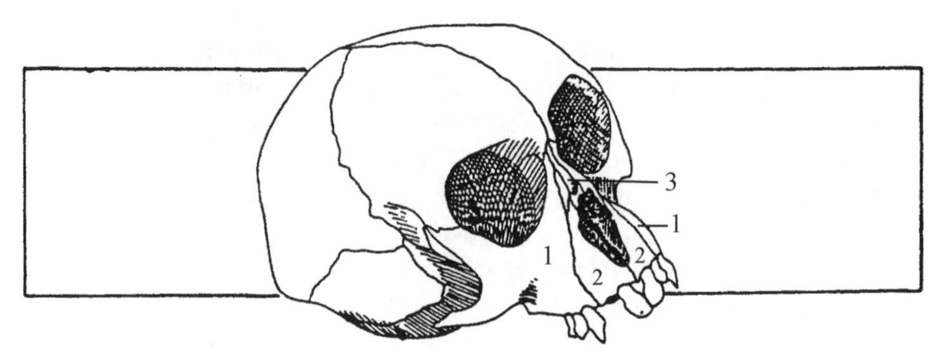

*Fig. 154: Skull of a young monkey (*Erythrocebus *sp.). (1): maxilla, (2): premaxilla, (3): nasal bone. Note the small nasomaxillary bone between the nasal bone and the maxilla. It is also sometimes seen in humans (Montagu, 1935).*

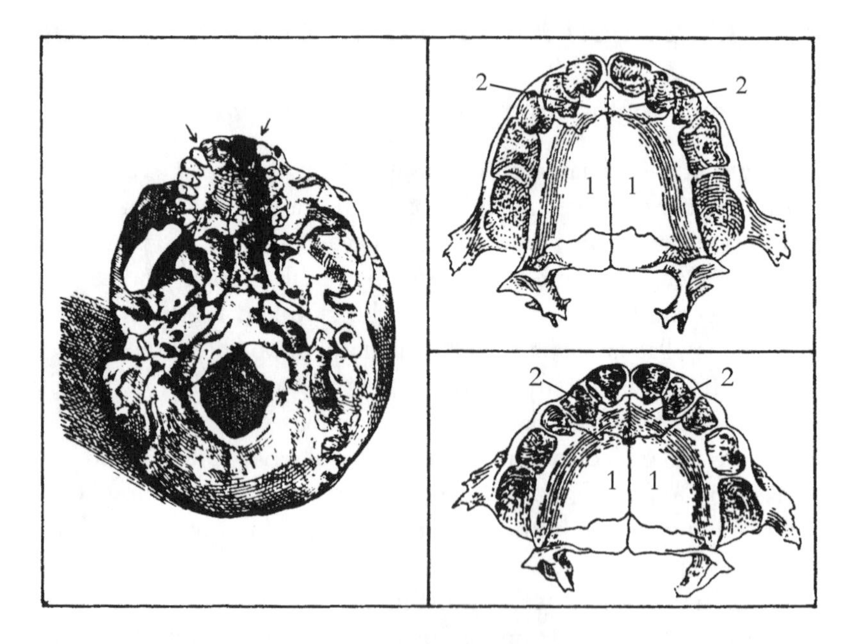

Fig. 155: The suture between the maxilla and the premaxilla in the human palate.

Left: Drawing by Andreas Vesalius.

Right: The palate of the human fetus at 7.6 months (top) and 9.9 months (bottom). The majority of the palate is formed by the two jawbones, in which all teeth are rooted (1). The premaxillae are visible at the front of the palate (2). The sutures delineate these bones as an "island" located directly behind the incisors and surrounded by the maxillae (Montagu, 1935).

The premaxillae, however, do indeed exist in humans. Vesalius had already determined that in many cases a suture is visible in the human palate in the place where the suture between the maxilla and the os incisivum is visible in apes (see *Figure 155*). Vicq D'Azyr, Goethe, and others also pointed to the structural similarity in the premaxillae of humans and apes.

Although the human premaxillae do develop during the early fetal period, in comparison to animals, humans are in a borderline situation from the very earliest stage of development. The human premaxillae are never truly independent but are connected to the maxillae from the very beginning (Wood et al., 1968). Even in the early fetal stage, no sutures between the human maxillae and premaxillae are visible between the nose and the mouth.

Bolk, who studied this issue in 1912, discovered that these sutures also disappear over the course of a lifetime in several ape species. The sutures may disappear even before birth in chimpanzees; they vanish during childhood in orangutans, later still in gorillas, and are only sometimes still visible in adult spider monkeys. In humans, however, no suture is ever visible on the side of the face. An independent premaxilla is an animal characteristic that, in Goethe's words, is "overshadowed" in humans.

As we said earlier, however, it is possible to discover a trace of the premaxilla in the human palate (see *Figure 155*). In one quarter of adult skulls and most juvenile skulls, a suture separating the maxillae (which make up most of the palate) from the premaxillae is still visible at the front of the mouth in the area of the upper incisors. Montagu (1935) studied this suture in human fetuses and made a remarkable discovery. Essentially, the suture that defines the premaxillae forms a closed circle and runs along the back of the incisors (see *Figure 155*). The incisors are actually rooted in part of the maxilla, so it is somewhat misleading to call the human premaxillae "ossa incisiva."

It becomes apparent that an unusual process, not observed in animals, takes place in humans. From the very beginning, the human premaxilla is "under attack" by the maxilla. At a very early stage in fetal development (at the end of the third month), outgrowths of the two maxillae cover the premaxillae on the outside of the face between the mouth and the nose. These two outgrowths meet in the upper lip. As a result, no sutures between the maxilla and premaxilla are ever visible on the outside of the face below the nose, because the maxilla completely covers the premaxilla. In apes, while the suture between the maxilla and premaxilla may be obscured by ossification, it is present nonetheless and runs right through to the outer surface of the skull. In humans, however, this situation never arises; the premaxilla is completely overgrown by the maxilla. This means of closing off the premaxilla is uniquely human. Montagu writes, "Man is in this respect unique . . . Each of the four incisor sockets is, in man, almost entirely formed by the maxilla, whereas in all other animals these sockets are clearly formed entirely by the premaxilla." He draws a direct connection between the enclosure of the human premaxilla and the fact that humans retain a flat facial skull:

> A brief consideration of the form of the face in the primates leaves little room for doubt that the chief factor instrumental in bringing about this unique change in the maxillo-premaxillary relationship in man has been the progressive and almost total reduction of the snout. The premaxilla, which in the lower mammals and primates forms an appreciable part of the pronounced snout, has in man been so completely retracted that it may be said to have been almost totally embraced by the face (1935).

Goethe (1786) also saw the rapid incorporation of the premaxillae as an expression of human resistance to muzzle development; he considered the preservation of a closed row of teeth (which is important for speech) to be one purpose of this resistance.

Suppression of muzzle development began with the first appearance of primates at the beginning of the Tertiary period and is associated with suppression of the premaxilla (Martin, 1990). At each level of evolution, however, a few "throwbacks" occur; for example, baboon species developed large muzzles anew at a later stage of evolution (see *Figure 56*, p. 116). Martin, like Bolk (1912), sees a direct relationship between shorter muzzles and shorter premaxillae in primates. Shorter premaxillae definitely contribute to shortening the muzzle.

In humans, the sutures between the maxillae and premaxillae ossify very rapidly, at least behind the upper lip. Essentially, these bones start to fuse from the very beginning. A tendency toward acceleration of this ossification process can already be observed in gibbons and gorillas, is more pronounced in chimpanzees and orangutans, and reaches an extreme in humans (Schultz, 1969). Only in humans is the tendency toward early ossification sufficiently swift and forceful to preserve the flatness of the facial skull. In anthropoid apes, the attempt at enclosing the premaxilla seems to fail, so that in the end considerable muzzle development may occur in these species.

The severely restricted development of the human premaxilla cannot be interpreted as an animal specialization. Rather, it represents the culmination of a tendency that reaches far back into the evolution of animal species. Schultz (1950b and 1951) points out that the shape of this bone, although highly variable both among and within animal species, is extraordinarily constant in humans. The premaxilla's potential for morphological variability is enormous; it can develop outgrowths in any direction. A very wide range of over- and underdevelopment is possible in chimpanzees and gorillas, and the time of onset of suture ossification also varies greatly in these species. Within the group of anthropoid apes, the variability is enormous. The ossa incisiva may completely surround the nasal opening, extend all the way to the zygomatic bone, or develop extensive lateral "wings" (Schultz, 1951). Great variability in a trait indicates that it is a relatively recent evolutionary phenomenon. Hence, the great stability of the form of the premaxilla in humans and its enormous variability in anthropoid apes suggests that the human configuration is prototypic.

The tendency toward shortening of the os incisivum and muzzle, which characterizes the evolution of the primate group as a whole, is a Type I characteristic. The related tendency toward earlier suture ossification and enclosure of the premaxilla is a Type II characteristic in that it is already evident in the highest primates but reveals its purpose only in humans, who achieve complete suppression of muzzle development, thus paving the way for speech.

Suture ossification around the premaxilla is *not* premature in humans; rather, it is delayed in animals, in whom specialization (muzzle development) causes deviation from the prototypic, humanlike morphology of the skull (*Figures 135*, p. 268 and *144*, p. 281). As we will see in a subsequent section, Bokian retardation produces the apparent acceleration of the ossification process around the premaxillae in humans.

Other Apparent Acceleration Phenomena in the Human Facial Skull

Although the lower jaw has no equivalent of the premaxilla, premature ossification is also observed there in higher primates and humans. The left and right halves of the mandible first appear as two separate bones (cf. *Figure 147*, p. 287). In many mammals (such as most prosimians, dogs, squirrels, rats, and cattle; see *Figure 146*, p. 285), these bones remain connected only by flexible cartilage throughout the animal's lifetime. Many animals actively take advantage of these bones' mobility relative to each other; certain kangaroos, for example, use their lower incisors like scissors when eating grass. In other mammals, ossification of this mandibular symphysis occurs after a certain point in development (Beecher, 1977 and 1979). The process is most rapid, however, in higher primates. Schultz (1956) has observed in macaques, gibbons, and orangutans that the two lower jawbones fuse completely by the time the first incisors of the first dentition erupt. In humans, the last traces of the suture disappear between the ages of one and two. In African anthropoid apes, this process takes longer, namely, until the first dentition is complete. Relative to the long growth period of humans, the chin suture disappears very quickly, but there is nothing unusual about its location in comparison to other higher primates.

Various mechanisms have been considered to explain the rapid fusion of the mandibles in humans and higher primates. Beecher (1977), for example, maintains that fusion of the two halves of the mandible permits transmission of force from one side to the other during chewing. (Most mammals chew on one side only.) If the two halves of the mandible are firmly connected, the chewing muscles on both sides can contribute the force needed to masticate food on one side of the mouth. This explanation, however, encounters difficulties, because some animals (such as cattle) chew forcefully on tough foods but do not have fused mandibles. Conversely, we can follow Martin (1990) in asking why lower primates do not have fused mandibles if fusion is so advantageous. Martin cites various authors who note that an entire complex of traits (overall larger size, a deeper mandible, a taller ramus, a shorter row of teeth, etc.) seems to be associated with ossification of the chin suture. As we saw earlier, all of these traits are effects of retardation. It seems logical, therefore, to consider rapid fusion of the mandibles in higher primates (which effectively shortens the muzzle and dental arch) a manifestation of the same retardation process. Bolk (1924) also studied this phenomenon and demonstrated that growth stagnation in the lower jaw can be observed in apes and especially in humans, beginning with the early fetal stages of development.

Less pronounced premature ossification is also observed in the upper portion of the primate facial skull. In the fetus, the forehead consists of two bones (the ossa frontalia; see *Figure 156*) that generally remain permanently separated in lower vertebrates (Bolk, 1917b). Ossification of the metopic (frontal) suture also occurs either late or not at all in most mammals. Within the primate group, however, we find a tendency toward increasingly early ossification. In lemurs, the metopic suture disappears in the middle of the juvenile period, but in higher primates, such as gibbons and orangutans, ossification begins prenatally and is completed shortly after birth.

In humans, too, the metopic suture ossifies earlier than the other cranial sutures. The situation in humans, however, is not uniform; in a certain percentage of individuals, this suture remains unossified into adulthood. (This phenomenon, known as *metopism*, is very common among Bolivian Indians.) In anthropoid apes, metopism is extremely rare and rapid ossification is the rule, indicating a tendency to preserve the fetal shape of the forehead and prevent the originally domed forehead from shifting to the front. When we trace the degree of acceleration of ossification along the axis of symmetry of the human face, we find that it is most intense at the level of the mouth opening. The key factor here is blocking the premaxillae, the "muzzle bones." In higher primates, the acceleration of ossification above and below the rows of teeth is not much weaker than it is in humans, but at the level of the mouth itself it is too weak to halt the development of the animal facial gestalt.

The ambivalent situation with regard to the human metopic suture is undoubtedly related to the fact that this suture is located in an area of transition between the facial skull and the cranium. The sutures of the cranium remain unossified much longer in humans (and also in some apes) than they do in most mammals, and still longer in humans than in anthropoid apes. Bolk (1926c) often cited this phenomenon as an example of human fetalization. Flügel, Schram, and Rohen (1983) relate this contrast directly to differences in skull morphology between humans and apes:

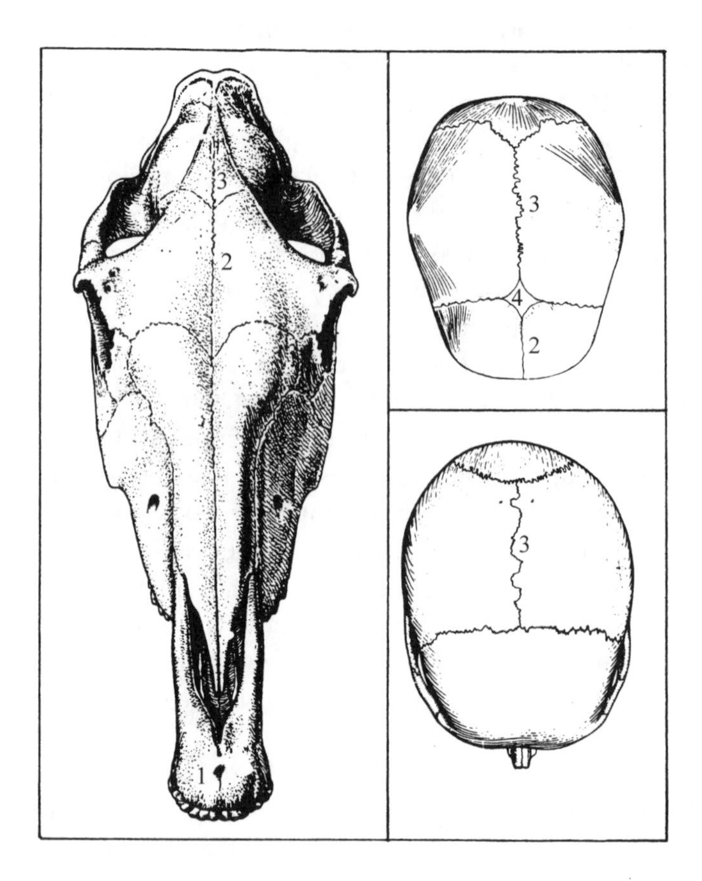

Fig. 156: Left: Skull of a horse, showing the ossa incisiva (1), metopic suture in the ossa frontalia (2), and sagittal suture (3).

Upper right: Skull of a newborn human. As in the mature horse and many other adult mammals, the metopic and sagittal sutures are not yet ossified and the large fontanel (4) has not yet closed.

Lower right: View of an adult human skull. The large fontanel and the metopic suture have ossified (Langman, 1966).

In the monkey species studied, the original form of the skull, which in the newborn resembles that of humans, changes from a more spherical to an elongated form because of the continuous prognathic growth of the chewing apparatus enabled by proliferating connective tissue in the intermaxillary sutures. In contrast to this, the sutures of the neurocranium close off early, restricting the postnatal growth of the neurocranium, so that in the monkey the volume of the neurocranial cavity remains relatively small.

In the human, completely different tendencies in the growth pattern of the two skull regions become evident. The sutures of the neurocranium remain open, whereas those of the facial bones close off early after birth. Thus prognathic growth is prevented and the clivius angle, as well as the viscerocranial quotient, remain constant during the entire postnatal development.

The Ossification Pattern of the Human Facial Skull as a Bokian Retardation Process

Relative to the corresponding events in animal development, suture ossification around the human mouth appears accelerated. Can this acceleration be understood as a retardation process, or is it a human specialization unlike any we have yet encountered?

The key to understanding this phenomenon is provided by the fact that within the skull as a whole, the jaws are clearly early ontogenetic developments. Not only in humans but also in all mammals, the upper and lower jawbones appear at a very early stage of development (Bruce, 1941). These bones are already present as such when the bones of the roof of the skull are not even present in the form of cartilaginous structures. As a result, the sutures around the mouth appear much earlier than those in the roof of the skull (where even newborns still have several fontanels).

If we see the various skull sutures as equivalent organs in the Bokian sense (see page 195), the development of the human face immediately makes sense. The skull sutures around the mouth are ontogenetic forerunners in comparison to the cranial sutures, and the frontal bones, located below the cranium, also appear fairly early. The Bokian principle of development predicts that in humans, ontogenetic forerunners will be qualitatively more developed—in this case, more ossified—than latecomers but will remain quantitatively smaller (see *Figure 103*, p. 195). This is exactly what we observe in the human skull. On the one hand, the forerunners (the sutures around the mouth) ossify very quickly; on the other hand, these sutures (like the bones they connect) are very small in comparison to those of the cranium. The fact that the human metopic suture ossifies more rapidly than the other cranial sutures can also be understood in the same way.

The situation in animals, however, cannot be interpreted in terms of retardation in the same sense. In animals, ossification around the mouth is suppressed to allow alteration of the prototypic, humanlike proportions and development of a real, pointed muzzle. In this instance, too, "developmental quanta" are removed from the generalized pattern of development and replaced by specialized ontogenetic steps (see *Figure 23*, page 55).

Additional Comments on the Seven-Year Rhythm in Skull Development

Bolk recognized the unique status of the seventh year in the life of a human being. As we have already learned, he observed that the growth of the front of the dental arch stops around the seventh year. He also maintained that the brain achieves its final volume during the seventh year:

> It can indeed be said that the size of the cerebral cavity almost ceases to increase after the seventh year. After that, the skull wall continues to thicken and develop relief formations, and very remarkable growth phenomena can be observed in the base of the skull, but the increase in size of the cranial space is completed in the seventh year (1915b).

Bolk discovered a remarkable phenomenon with regard to the seventh year. The so-called sagittal suture, which separates the parietal bones, is an extension of the line of the rapidly vanishing metopic suture (see *Figure 156*). Normally, the sagittal suture remains unossified into adulthood. By studying a large number of children's skulls, Bolk discovered that the probability of premature ossification of the sagittal suture increases around the seventh year of life:

> During the sixth and seventh years, the skull of the child passes through a critical period with regard to premature closure of the sagittal suture. During this brief time span, premature sagittal

> ossification occurs in approximately three percent of our children. Once they have outgrown this stage, the possibility of closure still occurring during childhood becomes slight—we might even say, impossible (1915b).

Because the cranial cavity achieves its final form around age seven, early ossification of the sagittal suture at this stage does not cause deformation. Ossification in early childhood, however, results in scaphocephaly (a wedge-shaped skull).

As a rule, the seven-year rhythm is clearly apparent in the development of the human skull (cf. *Figure 35*, p. 78). In the first seven-year period, the cranium, the front portion of the dental arch, and the S-shape of the temporomandibular joint achieve their final forms. In the second seven-year period, the chin develops (see *Figure 145*, p. 284) and the second dentition occurs (see *Figure 136*, p. 270). A cartilaginous zone of growth between the sphenoid bone and the basilar part (pars basilaris) of the occipital bone remains active into puberty in connection with the metamorphosis undergone by the facial skull during the second seven-year period (Sullivan, 1986). Ossification of this growth zone, in turn, is completed around age twenty and is considered a criterion of skull maturity (Bolk, 1915b). It should also be noted that the stage of most active suture ossification begins around the end of the fourth seven-year period.

The seven-year rhythm, which is also clearly recognizable in human growth (see Chapter 3) and human reproduction (see Chapter 8), goes hand in hand with extensive decoupling of the growth of different areas of the body. With regard to skull growth, Bolk comments:

> Remarkably, as soon as the cranium stops growing, the second phase of rapid growth of the facial skull commences, as does the second dentition. The difference in maturation times between the facial skull and the cranium is greater in the human species than in any other primate (1915b).

10

THE PHENOMENON OF SYNERGISTIC COMPOSITION

The further we progress in our investigation of human development, the less inclined we become to view it as the random product of dissociated material mechanisms. As we saw in the last chapter, the form of the human head results from the continued growth of the cranium on the one hand and suppression of muzzle development on the other. Both of these complementary processes work together synergistically to maintain the youthful, typically human form of the head. Upon further consideration, we come to see all aspects of human development as elements that together form a meaningful composition.

The Descent of the Larynx

The location of the larynx in adult humans is unique. In mammals, the larynx is located in the upper part of the throat, so that the epiglottis and the uvula touch. The same is also true of human newborns. In this situation, the larynx functions as a type of tube connecting the trachea with the nasal cavity. The resulting air passage bypasses the connection between the oral cavity and the esophagus (which is located behind the trachea), thus allowing babies (and mammals in general) to breathe and drink at the same time. In contrast, the larynx in human adults lies much farther down, producing an important separation between the uvula and the epiglottis (see *Figures 157* and *158*). This new configuration makes it possible for humans to breathe freely through either the mouth or the nose.

> With the retreat of the larynx...entirely new operations are possible. Man is the first animal able to close off the nasal tubes from the rest of the airway completely, easily, speedily and habitually. The [human] soft palate acts as a valve (DuBrul, 1958).

The unique descent of the human larynx results in the development of a broad and controllable speech organ, the human pharynx. At first glance, this appears to be a clear case of human specialization in preparation for the appearance of speech. Lieberman (1984), for example, insists that the human larynx contradicts the idea that the human species is paedomorphic.

> Adult human beings diverge from newborn infants as much as, if not more than, other adult primates diverge from their newborns. Human newborn infants conform to the general principle noted by Darwin: they are closer to the newborn forms of nonhuman primates; they reveal our common ancestry. Human newborn infants, in particular, retain the supra-laryngeal airways and associated skeletal morphology that occur in living nonhuman primates.

Nonetheless, the descent of the human larynx is not a specialization. It should be noted that laryngeal descent begins early in life, during embryonic and fetal development, and appears to continue throughout the entire life span. "There is a gradual descent through the embryo and fetus and child," writes Negus (1949). Wind (1970) confirms the discovery that the larynx descends rapidly during the first years of life and then continues to descend, but more slowly. The range of individual variation in later laryngeal descent seems to be quite broad (Flügel and Rohen, 1991).

The deep location of the human larynx is the logical continuation of a movement that is common in the early stages of animal development. In monkeys, considerable laryngeal descent generally occurs during the first months or years of life, but it stops earlier than it does in humans

(Flügel and Rohen, 1991). In higher primates in particular, a consistent decrease in the overlap between the epiglottis and the soft palate has been confirmed. DuBrul (1958), for example, discovered increasing separation between the epiglottis and soft palate within the primate series consisting of tree shrews, lemurs, Diana monkeys, gibbons, and humans. He found that among lemurs (which are lower primates), the overlap is already smaller than it is in tree shrews (classified by some as lower primates). In the Diana monkey (a lower catarrhine), the presence of the uvula—an extension of the soft palate that is largely undeveloped in lower primates—guarantees the overlap. DuBrul (1958) writes this about the gibbon: "It is difficult to judge whether, during life, the palate and epiglottis usually touch in the rest position. In dissected specimens they seem free of each other." Wind (1970) made similar observations about chimpanzees: "Topographically the chimpanzee shows one difference from the monkey: its laryngeal aperture projects less far into the pharynx, so its epiglottis does not reach the palate and has an antevelar position. Consequently there is a small gap between the two structures, resembling the human condition." And Negus (1949) writes, "The higher apes—the gorilla, chimpanzee, and orangutan—also show this separation [of the epiglottis and the soft palate] but in all of them the pharynx is small." Laryngeal descent can be considered a universal tendency that is already recognizable in catarrhines in general and quite obvious in that group's highest members, the anthropoid apes (Bernstein, 1923). In other words, laryngeal descent is greater in more retarded species. However, the unique potential it creates—namely, speech—becomes a reality only in humans. Thus the deep location of the adult human larynx is an interesting example both of hypermorphosis and of a Type II characteristic.

The fact that the human larynx is a strangely unspecialized organ attracted the attention of earlier researchers:

> In the structure of its skeletal parts, the human larynx is so prototypic that its development suggests (as Gegenbaur rightly pointed out) that human beings branched off from the vertebrate trunk at a very early stage. This primitiveness, which is also characteristic of other human traits, is all the more remarkable in the human larynx in that, as an organ of speech, it possesses a degree of perfection that is absent in animals (Goerttler, 1954).

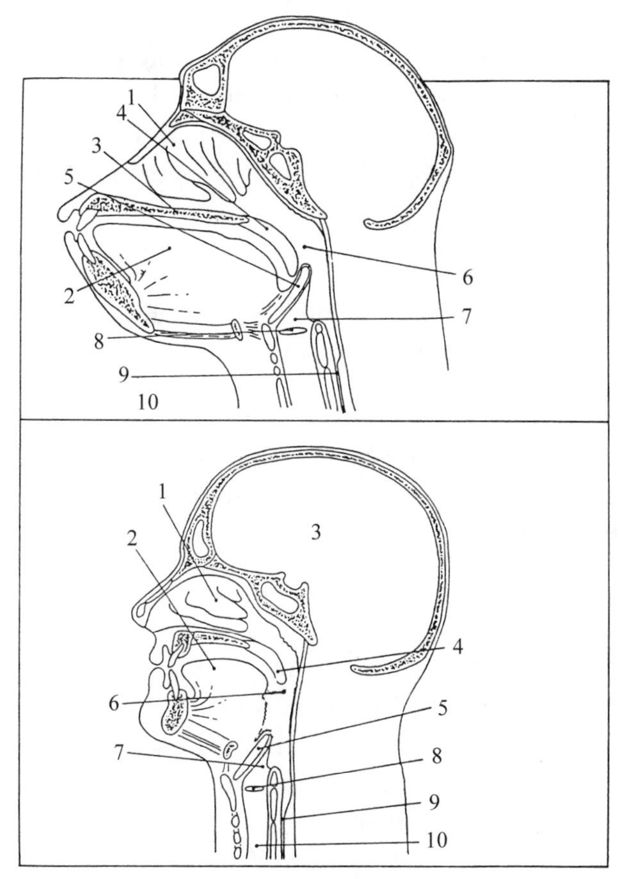

Fig. 157: Nasal, oral, and pharyngeal cavities in the chimpanzee (top) and human (below).

(1) nasal cavity, (2) tongue, (3) hard palate, (4) soft palate (5) epiglottis, (6) pharynx, (7) larynx, (8) vocal cords, (9) esophagus, (10) trachea (Laitman, 1986).

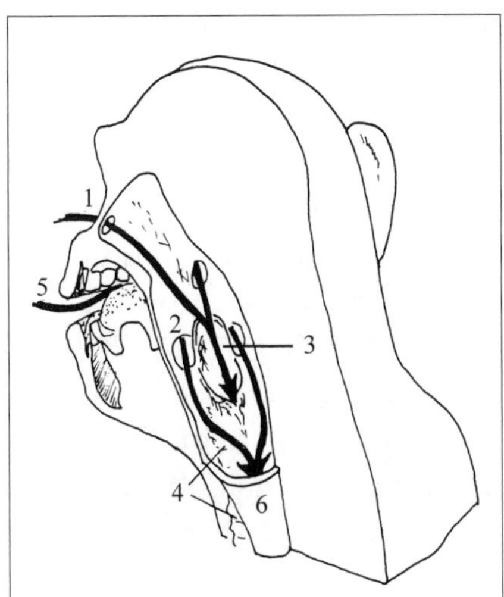

Fig. 158: Simultaneous breathing and drinking in a mammal. The stream of air (1) passes over the soft palate (2) and epiglottis (3) to the trachea (4). The liquid (5) reaches the esophagus (6) by flowing around on both sides of the epiglottis (Laitman, 1986).

In this connection, Wind (1970) comments, "[A] neotenous character of man may possibly be shown in his larynx, since compared with other mammalian larynges it has, from a morphological point of view, a rather generalized or non-specialized appearance." Kipp (1955) points to two important differences between laryngeal structure in humans and other primates:

- Vocal cords are muscular in humans but membranous in lower primates. Those of anthropoid apes occupy an intermediate position.
- Lower primates and anthropoid apes have multiple air sacks adjacent to the larynx. These so-called laryngeal sacs are almost nonexistent in humans.

Negus (1949) considered only the first of these two traits a sign of human primitiveness. According to Negus, muscular human vocal cords do not pass through an apelike stage in their ontogeny and therefore also did not evolve from the membrane-like vocal cords of apes. Negus did assume, however, that the ancestors of human beings had larger laryngeal sacs and that these have regressed in modern humans. In contrast, Wind, a later author, maintained that the common ancestors of humans and apes did not have the large pneumatic sacs that Negus attributed to them. From Wind's perspective, the unfulfilled potential of the human larynx to develop laryngeal sacs could be seen as another example of an animal tendency that is "overshadowed" in humans (see Chapter 4).

In connection with the first of the two above-mentioned traits (muscular human vocal cords), Goerttler (1950) also mentions that part of the human larynx is derived from heart tissue, a process that does not occur in other mammals; see also Schilling (1952). The larynx and the heart are in close proximity during early embryological stages, and according to these two researchers, part of the human vocal apparatus develops from tissue initially belonging to the heart region. Anthropoid apes have not been studied with this in mind; more research is needed here.

In summary, there appears to be little or no reason for viewing the morphology of the human larynx as specialized. Its most prominent distinguishing feature—the depth of its descent—is a clear example of hypermorphosis.

Laryngeal Descent and the Fetalization of the Human Skull

Laryngeal descent in humans is directly related to specific retardation phenomena in the human skull. In humans, the front of the dental arch stops growing relatively early, and important sutures around the mouth ossify quickly (Chapter 9), suppressing muzzle development. Thus human retention of the fetal form is more pronounced. The tongue grows more slowly in humans than in animals, and its back end is lower and located farther back, forming the movable anterior wall of the pharynx—a very important consideration in shaping speech sounds (see *Figure 159*). Because the larynx lies directly behind the tongue, the fact that the tongue shifts toward the back necessitates a downward shift in the larynx, which does in fact occur as a logically consistent continuation of prenatal laryngeal descent.

In animals, laryngeal descent ceases much earlier, and at the same time the flatness of the face is abandoned. The muzzle grows larger; the tongue extends forward but remains completely within the oral cavity. In humans, both the fetal structure (a flat facial skull) and the fetal tendency (laryngeal descent) persist. In animals, however, a double reversal occurs. Laryngeal descent ceases and the muzzle replaces the flat facial skull. Human retention of the flat facial skull is a Type I characteristic, while the depth of laryngeal descent is a Type II characteristic. These two traits, however, are inextricably linked. They form a coherent whole, and both serve the specifically human function of speech. From the perspective of the human line of evolution, there is no fundamental difference between Type I and Type II characteristics. The distinction is fluid and depends in part on which animal species is being compared to humans. Type I characteristics generally relate to broad, early-manifesting trends in the emergence of the human gestalt, while Type II characteristics relate to later, more subtle developments that elaborate on and give meaning to the earlier tendencies.

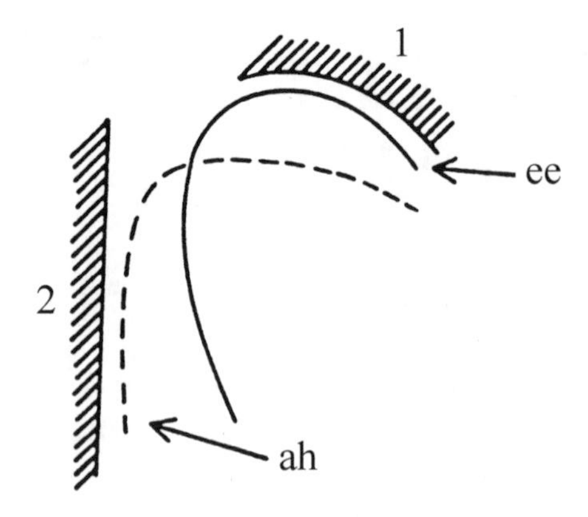

Fig. 159: Diagram of the position of the tongue relative to the palate (1) and the rear wall of the pharynx (2) in the formation of the two vowels "ah" and "ee." (Lieberman, 1984).

The Brain

The human head is relatively small, accounting for approximately 5.4 percent of the weight of the entire body. According to Schultz's calculations, the corresponding figures are 5.8 percent for chimpanzees and orangutans and 7.25 percent for gibbons (1950a). For lower catarrhines, the figures are higher due to the fact that smaller animals always have larger brains (and therefore also larger heads) in proportion to total body weight. The head is also always very large in proportion to the body during the early embryonic stage (see *Figure 29*, p. 66). The disproportionally low weight of the head in the human adult is due to retardation of the cephalocaudal growth pattern (see *Figures 25* and *26*, p. 62).

The heads of monkeys display several specializations that do not occur in humans. Brain growth ceases early in monkeys and continues longer in humans. As we have already mentioned, monkeys develop a protruding muzzle, a feature that is totally absent in humans. Muzzle development accounts for the disproportionally greater weight of the monkey head. On the whole, the shape of the human skull is retarded. Its high forehead, rounded top, and relatively undeveloped jaws are characteristic of all mammals during the fetal stage. This universal skull form suggests that the brain is disproportionally large in mammal fetuses. In the adult human skull, retaining a relatively large brain is directly related to the retention of various other fetal characteristics (see F*igure 160*, for example).

In all primates, the brain grows very rapidly during the prenatal phase of life. At this stage, there is no noticeable difference between the brain volume of a human being and that of a monkey. Schultz (1950a) supplies the following figures:

	Body weight	*Brain volume*
human fetus	550 g	12.5 cm^3
newborn spider monkey	510 g	12.5 cm^3
chimpanzee fetus	570 g	14.1 cm^3
newborn macaque	400 g	13.0 cm^3

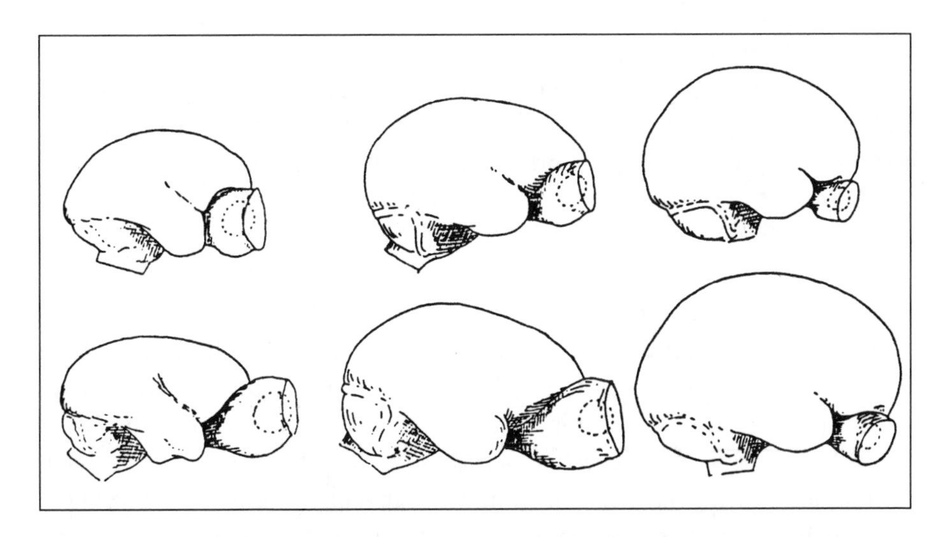

Fig. 160: Relative positions of the brain, eye socket, and eyeball (dotted line). Juvenile (upper row) and adult (lower row) specimens of (left to right): macaque, siamang, and human. In the juvenile specimens, the eye socket lies directly below the forebrain. This configuration is preserved in the adult human, but in the adult macaque the eyeball is located in front of the brain. This example illustrates how the large size of the human brain permits the retention of fetal skull proportions (Bolk, 1923b).

In humans, the fetal pattern of rapid brain growth persists for several years after birth, but in monkeys brain growth ceases much earlier. As a result, the difference between adult monkeys and humans is quite large. Brain weight accounts for 2.0 percent of a human adult's body weight. This figure sinks to 0.92 percent in chimpanzees, 0.76 percent in orangutans, and 0.4 percent in gorillas. In gibbons, which are much lighter than either humans or large anthropoid apes, the percentage (1.94) is only slightly smaller than it is in humans.

The weights of some specific organs are roughly constant percentages of total body weight in all mammal species (Stahl, 1965). For example, the relative weight of the heart in all mammals is 0.5 percent, and all mammals, regardless of size, have approximately 70 ml of blood per kilogram of body weight. With regard to other organs, the relationship is less simple. Larger mammals, for example, have disproportionally smaller kidneys. In a series of increasingly heavy mammal species, the relative increase in kidney weight is less than the increase in total body weight. On average, a mammal that is ten times heavier has kidneys that are only seven times heavier.

Even with regard to organs such as the kidneys, however, the formula describing the relationship between the mass of the organ and total body mass is a simple exponential equation. For example:

$$\text{kidney mass} = 0.0212 \times (\text{total body mass})^{0.85}$$

If we know an animal's total body weight, we can calculate the weight of the kidneys with great precision. Clearly, there is something resembling a universal mammalian structural plan that determines the relative weights of a number of important organs. This plan also applies to humans.

Like the kidneys, the brain increases in weight more slowly than the body as a whole. On average, an animal that is three times as heavy has a brain that is only twice as heavy. Larger mammals generally have smaller brains relative to the size of the body as a whole, while very small mammals may have relatively larger brains than humans. This same tendency is apparent in the ontogeny of each mammal species. The brains or kidneys of young mammals that are still small are larger relative to total body size than are those of adults of the same species.

Therefore, in order to determine whether a certain mammal species (or the human species) has unusually large or small kidneys or a disproportionately heavy brain, we must compare the weight of that organ to the average weight of the same organ in mammals of the same size. Such studies reveal that primate brains are consistently heavier than expected and that within the primate group, relative brain weight tends to increase regularly as the species become more humanlike (see *Figure 161a*). In lower primates, the brain is two to three times as large as the average brain in mammals of the same size, and in humans it is up to eight or nine times as large.

The increase in relative brain weight in the ascending phylogenetic sequence of primates is accompanied by increased retention of a number of fetal skull characteristics. The fetal pattern of rapid brain growth comes to an end shortly before birth in lower primates, at birth in anthropoid apes, and only between the ages of two and three in humans (see *Figure 161b*). The brain of a newborn macaque has already achieved 60 percent of its final size, while the corresponding figure is 46 percent for chimpanzees but only 25 percent for humans.

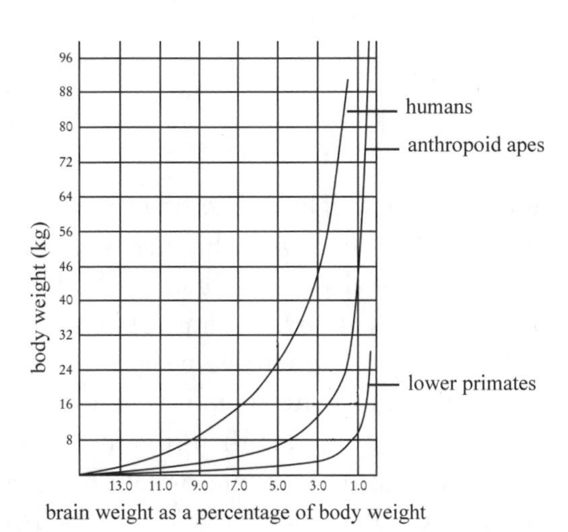

Fig. 161a: Graphic representation of brain weight as a percentage of total body weight. The three curves illustrate the differences among humans, anthropoid apes, and lower primates (Schultz, 1950a).

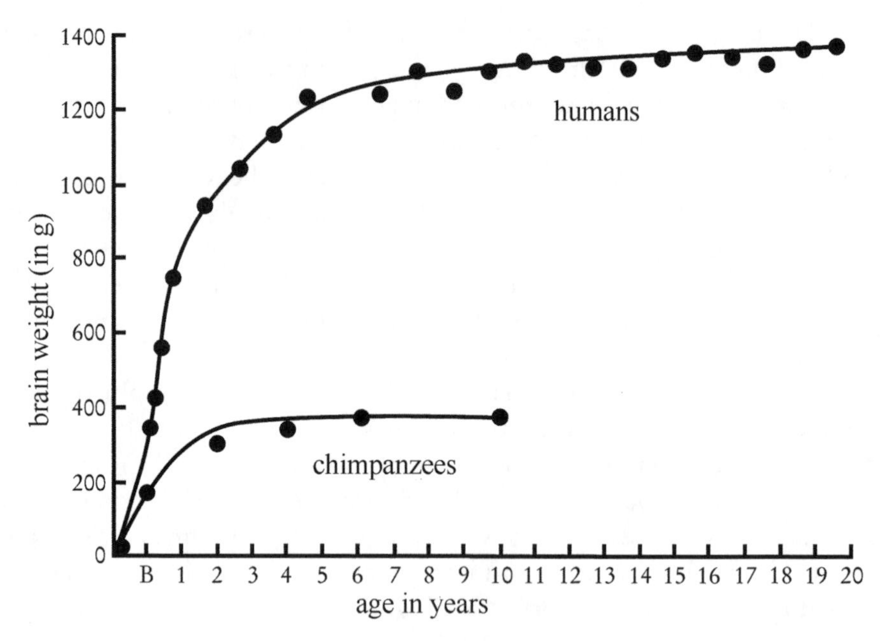

Fig. 161b: Brain weight in humans and chimpanzees as a function of age. The rapid prenatal growth rate persists in humans for several years, while chimpanzees abandon it shortly after birth (Passingham, 1982).

To explain the great weight of the brain in adult humans, we might maintain, as Bolk did, that fetal growth patterns persist longer in humans and that therefore our final ratio of brain to body weight more closely resembles the fetal state of affairs. (A similar phenomenon may be evident in the case of the kidneys. The relative weight of the kidneys not only is much higher in newborn primates than it is in adults of the same species but is also higher in adult humans than it is in monkeys; see Straus and Arcadi, 1958.) This simple Bolkian interpretation, however, is invalid. We know that ontogenetic "latecomers" tend to grow disproportionally large in a highly retarded organism, and in fact the kidneys do appear relatively late. The nervous system, however, begins to develop at an early ontogenetic stage. Hence, the relatively greater development of the human brain requires a different explanation.

Retardation Effects within the Brain

The round shape of the adult human skull, as we have already indicated, can be considered a Type I characteristic or fetalized trait. And yet the brain seems to be an organ that begins growing early. It also occupies most of the volume of the head, which only later begins to decrease in size relative to the rest of the body. Increasing retardation results in relatively small sizes among organs that get a head start. Why, then, does the more highly retarded human species have a relatively larger brain than the less retarded anthropoid apes? It becomes evident that the answer lies in the very different growth patterns of various parts of the brain. Some parts do indeed get off to a head start, undergoing substantial development in the early fetal stage, while other parts begin to develop much later and catch up to a very great extent. Retardation further enhances this latter pattern.

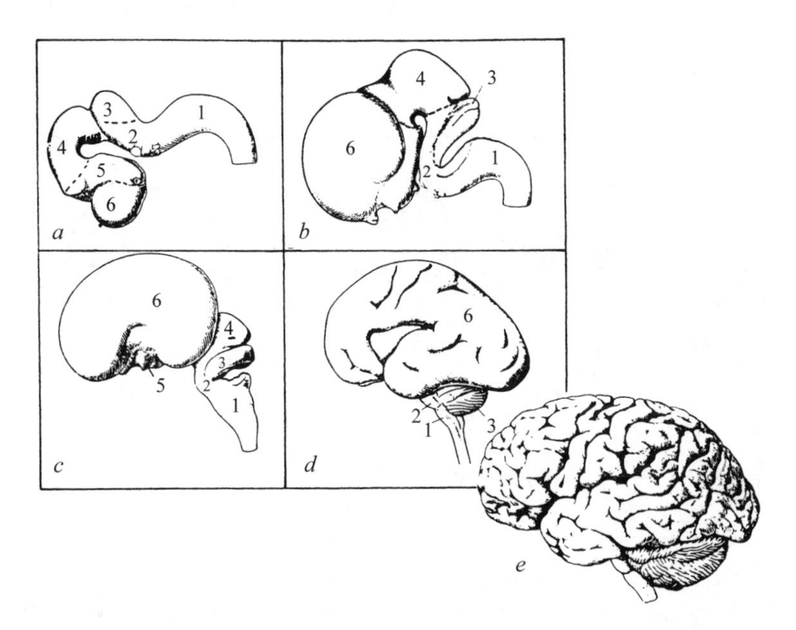

Fig. 162: Stages in the development of the human brain (Kahle et al., 1975). (a) embryo at 10 mm torso length; (b) embryo at 27 mm torso length; (c) fetus at 53 mm torso length; (d) 33 cm-long fetus; (e) adult. (1): medulla oblongata; (2): pons; (3): cerebellum; (4): mesencephalon (midbrain); (5): diencephalon; (6): telencephalon.

The nervous system takes shape through a process of involution. At the beginning of the third week of gestation, two longitudinal neural folds arise on the dorsal side of the embryonic disc, flanking the so-called neural groove. These folds grow toward each other and fuse, creating the neural tube (see *Figure 168*, p. 329). The fusion process begins in the middle of the embryo and proceeds in both directions (caudal and cranial). At the end of the fifth week, the neural tube is completely closed and takes the shape of a thin-walled canal that is differentiated into a series of successive vesicles (see *Figure 163*). Moving up from the spinal cord, the sequence is: medulla oblongata, pons, cerebellum (the three parts of the hindbrain); mesencephalon (the midbrain); diencephalon and telencephalon (the two parts of the forebrain) (see *Figure 162*). On the diencephalon, the primordium of the eye soon becomes visible. All of these vesicles, which are more or less individually visible little segments of the neural tube, are ring-shaped in cross section; the forebrain forms the closed end of the tube. The vesicles are connected by the fluid-filled canal produced by the involution process.

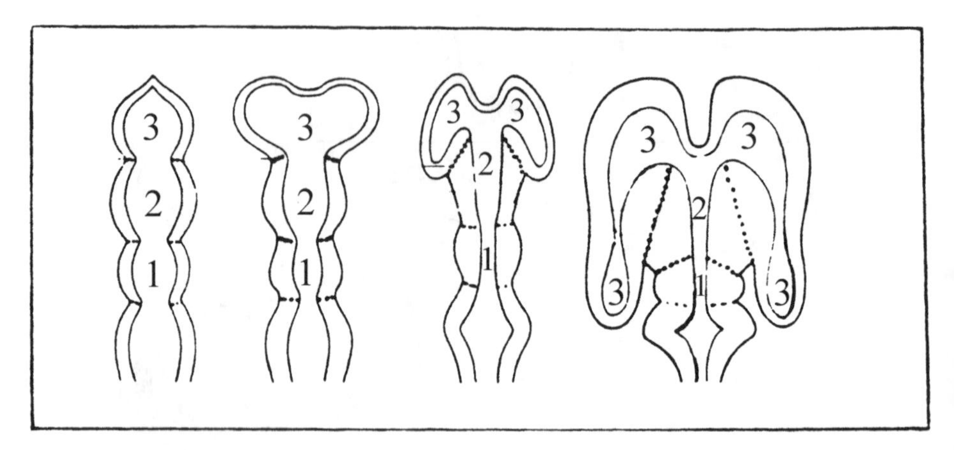

Fig. 163: Stages in the development of the human brain. Section through the upper portion of the neural tube in four successive stages. (1): hindbrain vesicle; (2): midbrain vesicle; (3): forebrain vesicle. The side walls of the brain vesicles are initially thin and undifferentiated. Later, they become thicker and develop into actual nerve tissue. This thickening process begins later in the forebrain, which is initially fairly small but eventually grows into two lobes that expand to the rear and cover other parts of the brain (Kahle et al., 1975).

From this point onward, the growth patterns of the individual brain vesicles vary widely. In the second fetal month, the area from the midbrain to the medulla oblongata grows faster than the forebrain. Initially, the walls of all the vesicles are made up of fairly generalized neuroepithelial cells, which are capable of replication. Only later do they specialize into actual nerve cells (neurons), which, in general, can no longer replicate. This specialization occurs much earlier in the brainstem than in the forebrain. In the course of the second month, individual nerves have become apparent in the brainstem, while the forebrain is still not much more than an undifferentiated vesicle.

The development of actual nerve tissue is accompanied by a thickening of the wall of the neural tube. This thickening begins in the more caudal parts of the brain and then proceeds to the midbrain, and forebrain in succession (see *Figure 163*). The forebrain vesicle is still thin-walled and undifferentiated at a stage when the other parts of the brain have obviously begun to develop. Although the development of the forebrain is delayed in comparison to that of other parts, the forebrain retains the original capacity for growth for a longer period. Bok (1924) explained the greater final size of the forebrain relative to the other parts of the brain as the result of retardation of the forebrain. Not a single differentiated nerve cell is to be found in the forebrain vesicle when the final differentiation of nerve tissue is already in full swing in the spinal cord:

> In what will become the cerebral cortex, the cells then divide rapidly. Only when an enclosed vesicle has developed does neuron differentiation also commence at this location. In this instance, a purely quantitative increase in relative volume comes about through retardation of neuron differentiation.

From the third month onward, the forebrain does in fact begin to catch up with the other parts of the brain, wrapping around the other portions as it expands.

In comparison to the underlying parts of the brain, the forebrain is a "late bloomer" that grows more rapidly than parts that began their growth earlier. In a Bokian sense, retardation of development leads to a relative increase in the size of organs that begin growth late. The extraordinary preponderance of the forebrain in the human adult is an example of hypermorphosis. Humans have highly developed forebrains and highly developed lower limbs for one and the same reason, namely, development as a whole is retarded and therefore late starters have ample opportunity to make up for their initial delay.

Figure 164 illustrates this state of affairs with regard to the lower body's continued growth in comparison to the spinal cord. In embryonic development, the spinal cord is one of the first organs to emerge in rudimentary form; the neural tube forms before the vertebral column. The spinal cord initially grows faster than most other organs, producing the typically curved shape of the embryo (see *Figure 87*, p. 167). Later, as the other organs make up for their initial delay, this extreme curvature disappears and the spinal cord lags behind, receding ever further into the vertebral column.

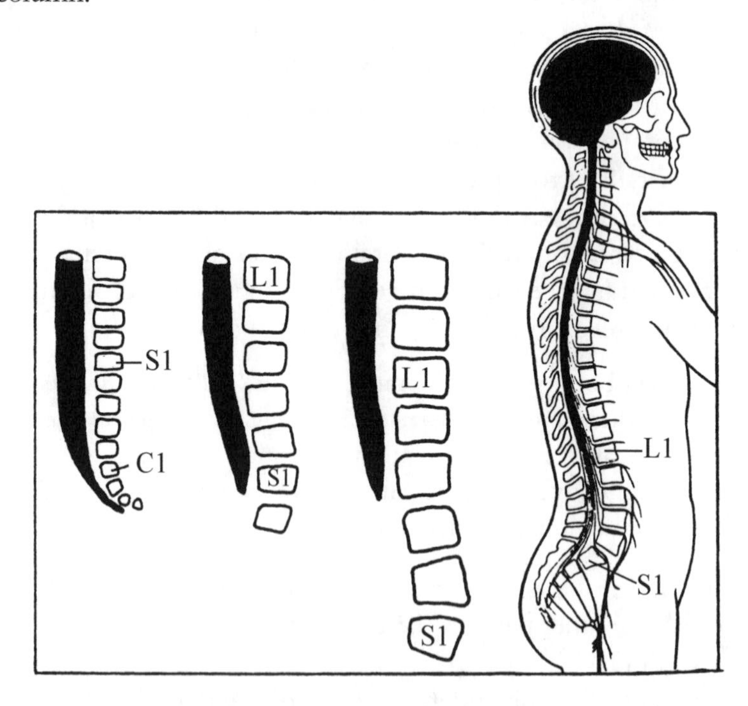

Fig. 164: The growth of the spinal cord (black) lags behind that of the vertebral column and lower body. Position of the spinal cord relative to the vertebrae in (left to right): fetus at three months, fetus at five months, newborn, adult. (C1): first coccygeal vertebra; (S1): first sacral vertebra; (L1): first lumbar vertebra (Kahle et al., 1975; Langman, 1966).

Within the ascending phylogenetic series of vertebrates, there is a clear trend toward increasing preponderance of the forebrain over the other parts of the central nervous system (see *Figure 165*). In amphibians and reptiles—classes of vertebrates that appeared on Earth before mammals—the forebrain is not especially large. Even in primitive mammals, however, the forebrain begins to overgrow the brain stem. This process climaxes in human beings.

In mammals, an analogous development occurs within the forebrain itself, whose surface (the cerebral cortex or neopallium) is reorganized to include a new structure, the neocortex. In higher mammals, the neocortex constitutes an increasingly large portion of the cortex at the expense of the phylogenetically older paleocortex and archicortex, which are suppressed. The neocortex occupies not quite one third of the forebrain in primitive insectivores such as hedgehogs. In humans, it occupies 98 percent, while the paleocortex and archicortex are totally marginalized and driven into the depths of the brain (see *Figure 166*). In embryonic development as in phylogeny, the neocortex appears later than the other parts of the forebrain; as the latest-emerging part of the brain, it catches up with the older parts, a process retardation greatly enhances in humans.

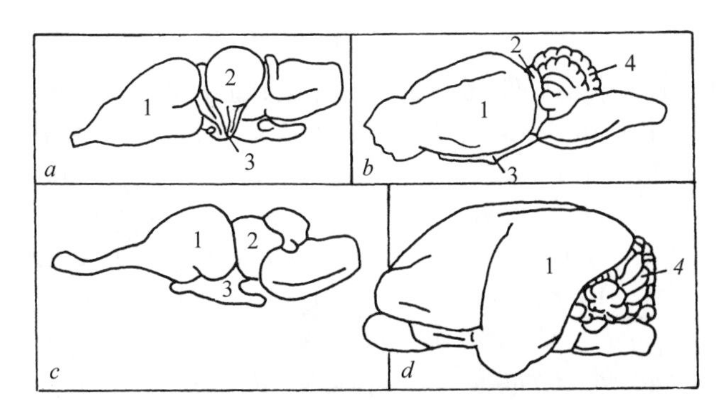

*Fig. 165: Telencephalization among terrestrial vertebrates. Brains of: (a) frog (amphibian); (b) crocodile (reptile); (c) hedgehog (primitive mammal); (d) bushbaby (*Galago *sp., a prosimian). (1): telencephalon (forebrain); (2): mesencephalon (midbrain); (3): diencephalon; (4): cerebellum. The more highly evolved the vertebrate, the more significant the forebrain becomes and the more it overgrows the other parts of the brain. This telencephalization trend climaxes in humans (Kahle et al., 1975).*

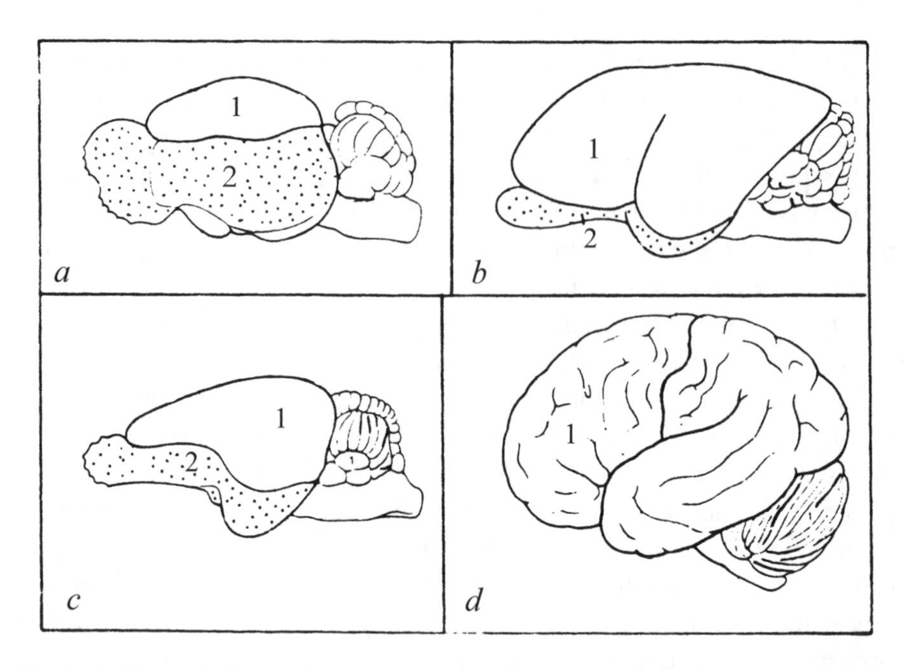

Fig. 166: Telencephalization in mammal species. Brains of: (a) hedgehog; (b) tree shrew; (c) lemur; (d) human. 1: neocortex; 2: paleocortex (dotted). In higher mammals, the paleocortex is increasingly overgrown by the neocortex; in humans, it is no longer visible from outside (Kahle et al., 1975).

The phenomenon of increasing preponderance of the forebrain (and within it, of the neocortex) is called *telencephalization*. This trend had its beginnings in the distant evolutionary past. The fact that human prenatal development recapitulates the broad outlines of phylogenetic development can be seen as a confirmation of Haeckel's fundamental biogenetic law (see page 33). Alternatively, we could say that *lower vertebrates abandoned the human pattern of development at earlier stages of evolution to follow specialized paths, while human beings maintained a consistent direction that involved increasingly differentiated development of the most recently emerged parts of the brain.* From this perspective, although the human brain cannot be understood as a simple enlargement of an ape brain, for example, it is nonetheless the unmistakable climax of a process that can be traced far back into the animal kingdom. As a general rule, the percentage of total brain volume occupied by the neocortex seems to increase in primates as the brain itself becomes larger (see *Figure 167*). For example, the neocortex is larger in proportion to the medulla oblongata in humans than it is in anthropoid apes. It is also disproportionally larger in anthropoid apes than in lower catarrhines and platyrrhines and in the latter group than in prosimians. In highly retarded primates, the disproportionate expansion of this last-developing part of the brain is roughly exponential (Passingham, 1982). (Cf. *Figure 167*; see also Finlay and Darlington, 1995.)

Like the organs of speech, the expanded neocortex is a characteristic that emerges ever more clearly in the course of evolution, achieving its climax and revealing its full potential only in humans. Telencephalization (or neo-encephalization) points not only to the emergence of qualitatively new characteristics or specializations but first and foremost to changes in the relative sizes of various parts of the brain through retardation of its overall pattern of growth. When new structures emerge (such as the neocortex in mammals; see *Figure 166*), they appear in the cephalic end of the brain, the ontogenetic latecomer. The new structure is then the one that later undergoes the greatest increase in size in more highly retarded organisms. Thus the human forebrain, which is extremely well developed and completely dominated by the neocortex, can be considered a Type II characteristic and an example of hypermorphosis.

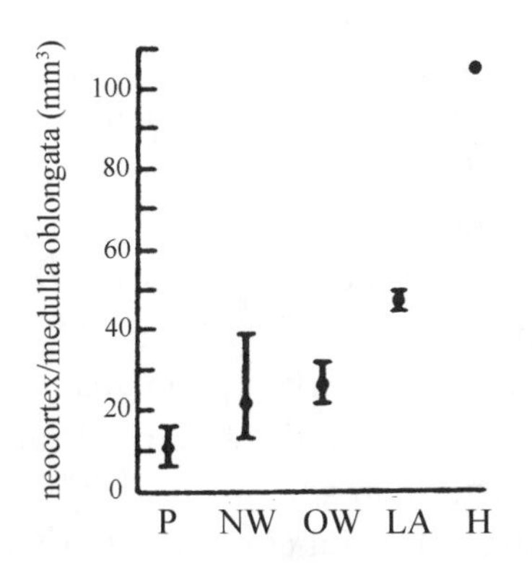

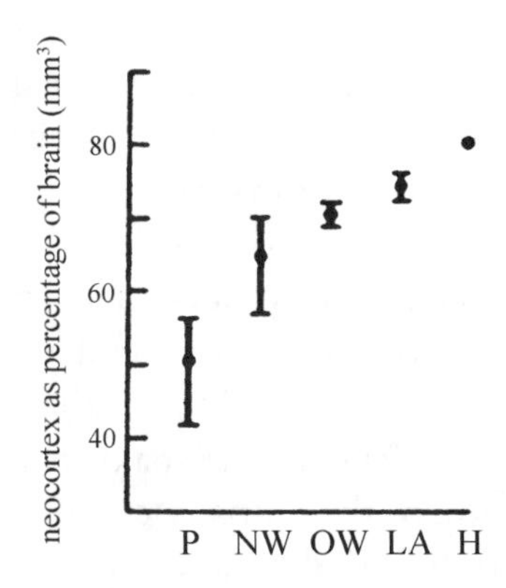

Fig. 167: Left: Volume of the neocortex in relationship to the volume of the medulla oblongata. Right: Volume of the neocortex as a percentage of total brain volume.

(P): prosimians; (NW): New World monkeys (platyrrhines); (OW): Old World (lower) monkeys; (LA): large anthropoid apes; (H): humans (Passingham, 1982).

The Forebrain as a Retarded Organ. Bokian Retardation

We saw that from the earliest stages of development, the forebrain lags behind the other parts of the brain to some extent. This delay occurs as early as the closure of the neural tube, which begins in the middle and then moves upward, forming the forebrain vesicle last. We also saw that nerve cells appear later in the forebrain than in other parts of the brain.

Even as a fully developed organ, the forebrain remains more fetalized in many respects than the rest of the central nervous system. Bok (1924 and 1926) gives several examples of this phenomenon. The original, fluid-filled central canal within the neural tube initially persists in the brain as the so-called ventricle system but is gradually fully integrated into the spinal cord. Similarly, Bok points to the distribution of gray and white matter in the brain and spinal cord (see *Figure 168*). Gray matter is an accumulation of neurons with long processes (axons) that allow them to make contact with distant cells. White matter consists of bundles of axons but contains no nerve cell bodies. In the spinal cord, gray matter lies in the center with white matter arranged around it. In the forebrain, the situation is reversed; the gray cortex surrounds the white matter. In the brain stem, we can observe a transitional situation. Bok points out that during the first stages of embryonic development, the gray matter in the spinal cord also tends to be arranged on the outside. Embryonic nerve cells (neuroblasts) extend their first processes in the direction of the center of the spinal cord rather than toward the outside; these processes later disappear.

Thus the brain supplies the prototypic example of Bokian retardation. In comparison to the spinal cord, the forebrain is an ontogenetic latecomer. Bok's principle predicts that in retarded organisms such as humans, such latecomer structures will ultimately become not only disproportionally larger but also more fetalized than structures that got off to an earlier start. In fact, this is exactly what we can observe in humans.

There are no major qualitative differences between the external structure of the human neocortex and the neocortex of an ape or a cow, which all follow the same general mammalian pattern (Rockel et al., 1980). One example is the convolutions of the brain, which appear on the surface of the forebrain beginning with the sixth fetal month (see *Figure 162*). These convolutions are not a specifically human feature; they develop according to laws that govern the mammalian brain in general. On the whole, the surface of the forebrain tends to grow in the same proportion as the brain's volume. But when an object simply increases in size and retains the same shape, its surface area increases less than its volume. (When the edge of a cube is doubled, its surface becomes four times as large and its volume eight times as large.) Hence, simply in order to allow the brain's surface area to grow at the same rate as its volume, larger mammalian brains must develop more convolutions. In this respect, there is nothing exceptional about the human pattern of convolutions. It has the complexity we would expect of a primate with a brain volume of nearly 1,400 cm^3. The elephant brain, with a volume of 4,150 cm^3, is more densely convoluted, while the brain of a horse displays less elaborate convolutions.

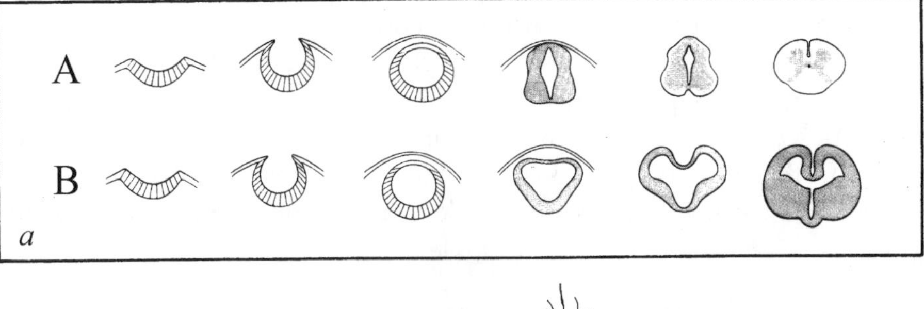

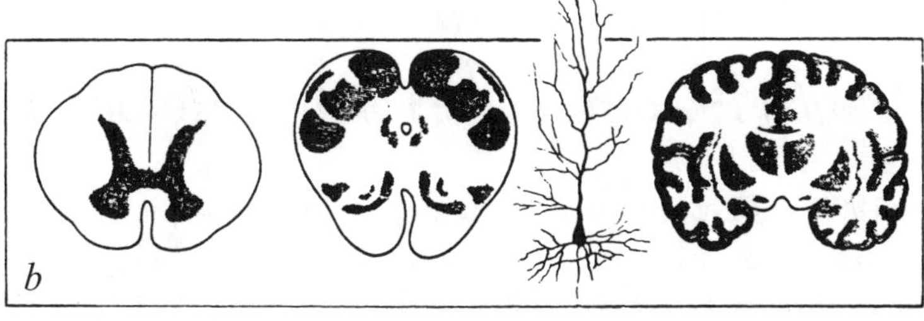

Fig.168: a: Diagram of the embryonic development of the human spinal cord (A) and brain (B). First, the fluid-filled neural tube closes off (cf. Figure 174, *p. 338**). The tube persists but is integrated into the spinal cord, where an outer layer of white matter also develops. Neither of these developments occurs in the brain (Kahle et al., 1975).*

b: Cross-sections of (left to right): the human spinal cord, brain stem, and forebrain, showing the distribution of white and gray matter. Gray matter lies in the center of the spinal cord but on the outside of the forebrain (cortex); the brain stem represents an intermediate situation. Also shown is a nerve cell from the cortex, with its cell body and long processes (Kahle et al., 1975).

In the human brain, just as in the human larynx, we encounter the complementary interaction or synergistic composition of reciprocally determined Type I and Type II characteristics. On the one hand, the human brain cavity retains its fetal proportions, and the overall form of the human skull is a Type I characteristic. On the other hand, the large size of the human brain must be interpreted as a Type II characteristic that comes about through a hypermorphic process in which the last parts to develop achieve the proportionally greatest sizes. These two characteristics constitute a structural unity. As a consequence of the great size of the human brain, several fetal phenomena are retained, such as the location of the brain above the eye sockets (see *Figure 160*, p. 317) or the low location of the external end of the auditory canal in relationship to the zygomatic bone (see *Figure 135*, p. 268). These fetal characteristics not only anticipate complete telencephalization, as it were, but also form a unity with it. When brain growth ceases prematurely, these anticipatory Type I characteristics also disappear.

As we have seen in so many of its aspects, human ontogeny allows for a further development of fetal characteristics while resisting the tendency to specialize. This same developmental principle is evident in the process of animal evolution as a whole.

Cranial Flexure and Profound Retardation

In 1923, Bolk published his views on the prototypic character of the orthognathic (i.e., muzzleless) skull. In the human being (unlike most mammals), the nasal cavity lies below the cranium, rather than in front of it as in the "prognathic" skulls of most animal species. Of course, muzzle development occurs in the latter type of skull.

The base of the skull from the foramen magnum to the bridge of the nose consists of three bony elements. The first, which lies directly in front of the foramen magnum, is an integral part of the occipital bone, which also forms the back of the skull. In humans, this front portion of the occipital bone angles fairly sharply upward and to the front. The next bone, the sphenoid, forms the front part of the temples as well as forming part of the base of the skull. The ethmoid bone abuts the sphenoid bone (see *Figure 169*).

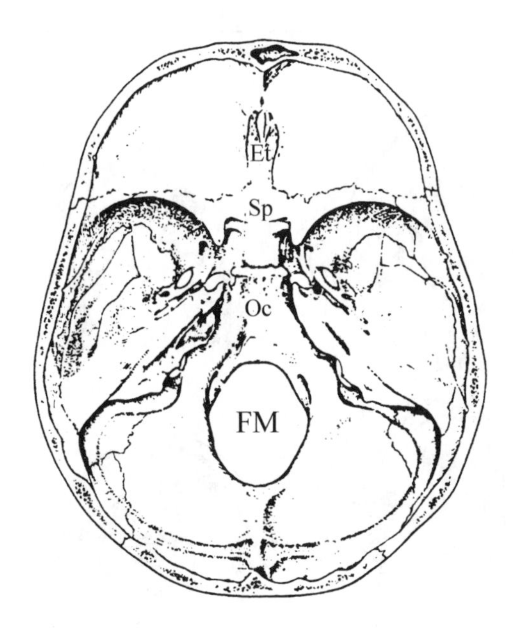

Fig. 169: View from above of the interior base of the skull (the surface on which the brain rests). The face is at the top. FM: foramen magnum; Oc: pars basilaris of the occipital bone; Sp: sphenoid bone; Et: ethmoid bone (Kahle et al., 1975).

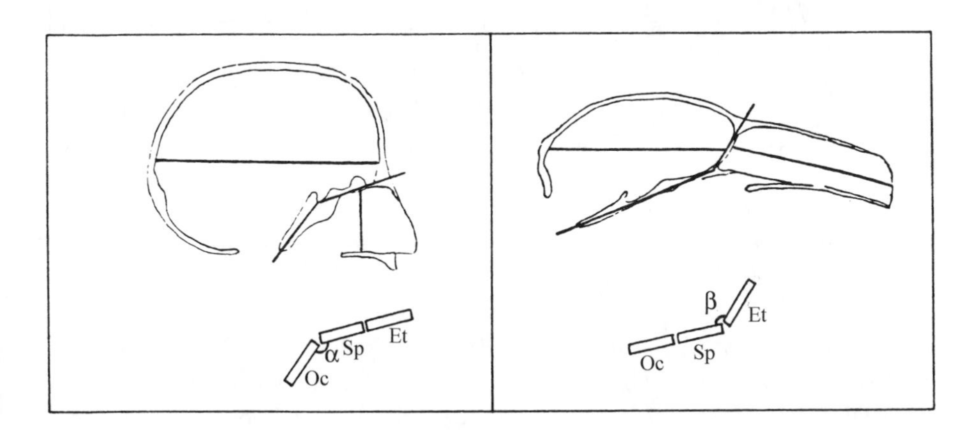

Fig. 170: Top: schematic section through the human head (left) and the head of a lemur (right). Bold lines indicate the sphenoid angle α and the ethmoid angle β (Bolk, 1923a).

Bottom: Schematic representation of the orientation of the occipital bone (Oc), the sphenoid bone (Sp), and the ethmoid bone (Et) in humans and lemurs.

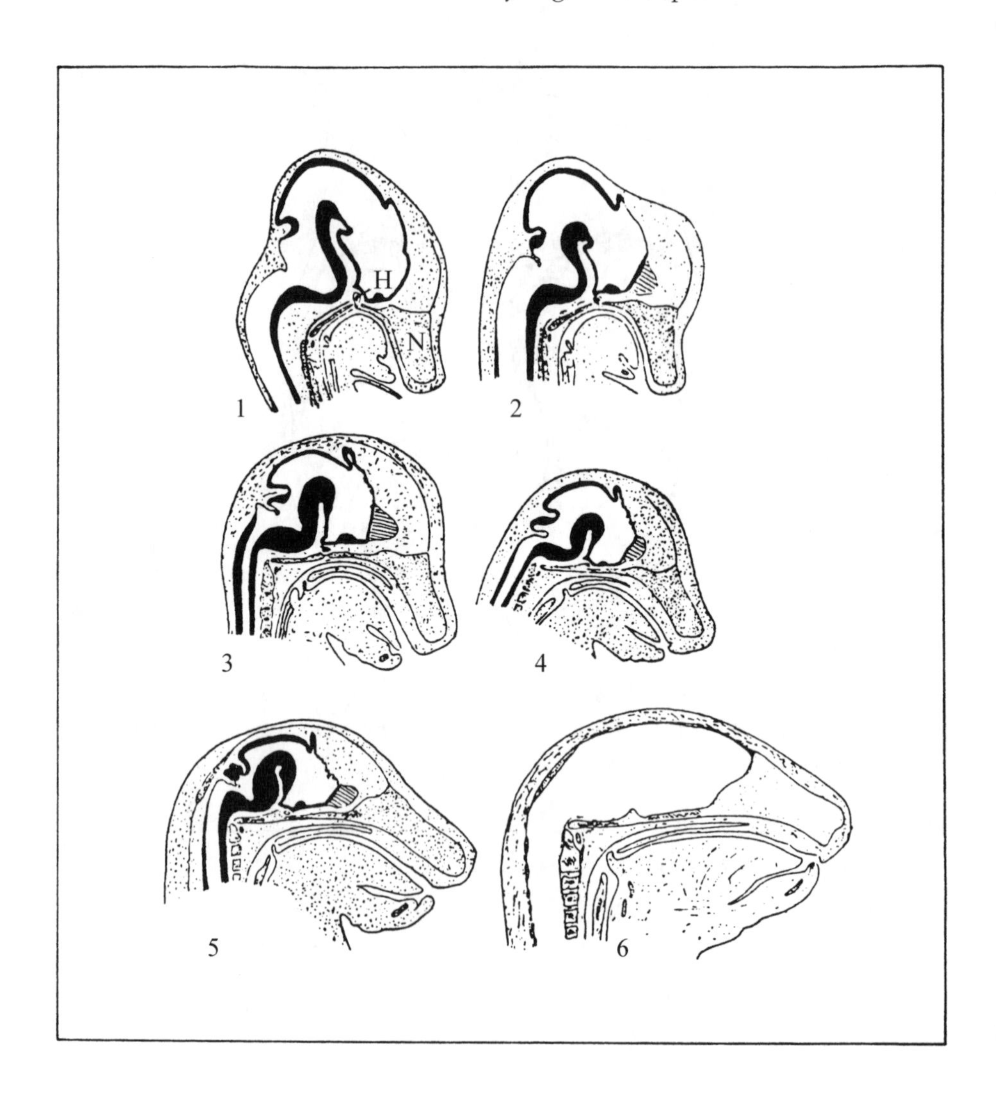

Fig. 171: Development of a rat skull (Bolk 1923a; cf. Figure 132, *p. 265). (1): an 11.5-mm embryo. The hypophysis (H) is located near the point of the sphenoid angle, and the nasal cavity (N) is oriented downward. (2): a 13.5-mm embryo. The sphenoid angle is flatter, but the nasal cavity still points downward. (3): a 20-mm embryo. The sphenoid angle has disappeared, and the nasal cavity begins to rotate. (4): a 25-mm embryo. (5): a 35-mm embryo. The ethmoid angle is already clearly evident. (6): a 43-mm embryo. The ethmoid angle has achieved its final size, and the nasal cavity is now located in front of the cranium.* Note *also that the foramen magnum has shifted to the rear and that the oral cavity and tongue have elongated.*

Crudely speaking, in the orthognathic skull the angle formed by the occipital bone and the other two bones (which lie almost in the same plane) is obtuse and open below (*Figure 170*). (For a critique of details of this statement, see, for example, Gould, 1977b.) This downward-facing angle is called the sphenoid angle (α in *Figure 170*).

In the prognathic skull, the sphenoid angle is almost straight (that is, the occipital and sphenoid bones now lie in the same plane), but the sphenoid bone and the ethmoid bone form an upward-facing obtuse angle. Figure 170 gives an example of this angle, called the ethmoid angle β in the skull of a lemur. Obviously, this orientation pushes the nasal cavity forward so that it lies in front of the cranium, and the animal develops a muzzle.

The foramen magnum, as we learned in a previous chapter, is located below the developing skull in the primate fetus but shifts toward the rear in the course of maturation. Schultz discovered that the foramen magnum in nonprimates is located at the rear of the skull from the very beginning, and Bolk discovered a directly related phenomenon. According to Bolk, all monkeys, as well as humans, have orthognathic skulls. Although many monkeys develop muzzles of sorts as the result of the strong forward thrust of the growth of the base of the nose and the jaw, such muzzles are different in character from other mammalian muzzles. Bolk points out that this is why monkeys look so humanlike. Their eyes are close together above the muzzle (monkeys have stereoscopic vision), whereas in true prognathic skulls the eyes are located more or less on the sides of the face.

Bolk made a fundamental discovery, namely, that during the embryonic period in mammals, a sphenoid angle (which as such is typical of the orthognathic skull) develops first. Later, this angle flattens out, and the ethmoid angle that shapes the prognathic skull develops. Bolk found this to be the case not only in mammals but also in reptiles (cf. *Figures 171* and *172*). He concluded that the orthognathic shape of the human skull is a persistent fetal characteristic. The sphenoid angle in the base of the human skull is "an essential character of . . . the primordial cranium of vertebrates in general" (1923a).

Figure 132 (page 265) is a reproduction of Bolk's comparison of fetal and adult forms in humans and dogs. We can see that the early fetal stage is very similar in both cases. The fetal flexures and angles of the skull are very well preserved in the human adult, but two changes occur in the dog.

In a direct adaptation to the dog's horizontal posture, the angle between the vertebral column and the base of the skull widens, and the head is reoriented so that the line of sight again becomes horizontal. (In quadrupeds, this generally means that the line of sight is roughly parallel to the vertebrae of the torso.) In this respect, the dog's original embryonic shape is still oriented toward uprightness. As we also see, the dog's double nasal cavity is located in front of the cranium. *Figures 171* and *172* illustrate the same phenomenon in rats, moles, and chickens.

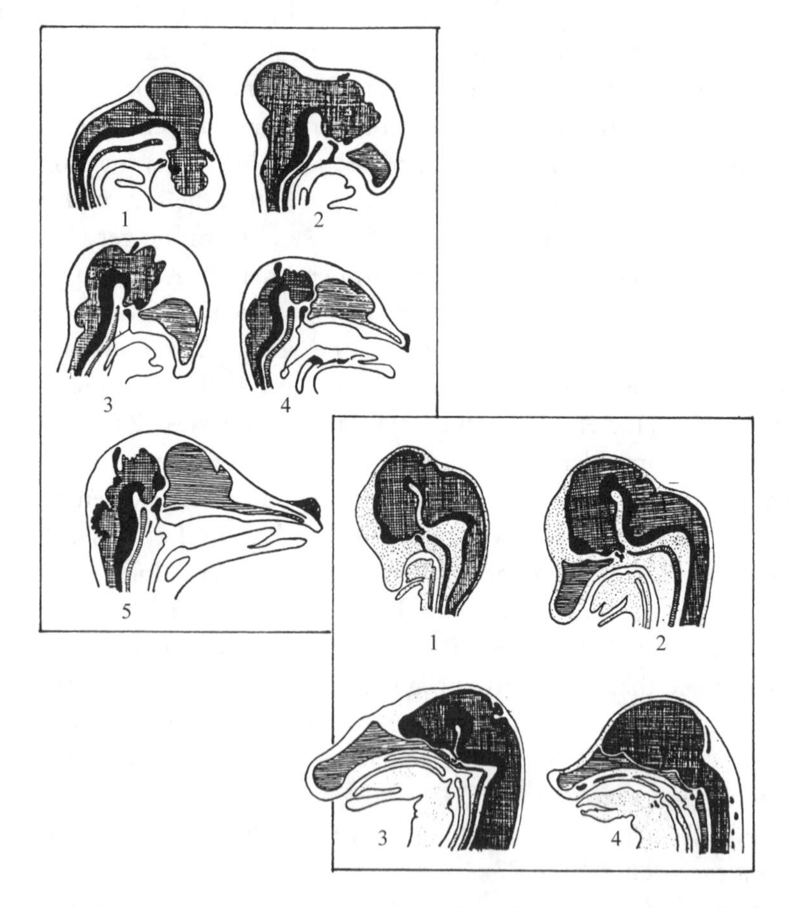

Fig. 172: Left: Development of the head in a chicken. The embryo at (1) 4.66 days; (2) 5.5 days; (3) 6.83 days; (4) 9 days; (5) 10 days. Right: Development of the head in a mole. The embryo at (1) 8 mm; (2) 10 mm; (3) 18 mm; (4) 24 mm (Dabelow, 1928). Crosshatching indicates the location of the brain, horizontal shading the location of the nasal cavity (cf. Figure 171*).*

The Tunicates as Overspecialized Vertebrates

The ancestry of vertebrates (animals with spinal columns: fish, amphibians, reptiles, mammals, and birds) is disputed, and alternative theories have been formulated (for an overview, see Jefferies, 1986 or Gee, 1996). Garstang's (1928) view, however, remains the most generally accepted. Garstang sees the tunicates as the closest relatives of vertebrates and vertebrates as the descendants of morphologically retarded tunicates.

Let's look at a typical tunicate or sea squirt. Several characteristics of the larval stage suggest a close relationship with vertebrates (see *Figure 173*). The sea squirt larva is a free-swimming animal that does not feed. It remains in this stage for only a few days while it searches for a place to attach itself, which it does by means of several adhesive papillae at the anterior end. The larva's brain vesicle includes a light-sensitive eyespot and is very conspicuous; its extension is the nerve cord, a rudimentary spinal cord. Located below the dorsal nerve cord is the notochord, a formation that is also present in the embryonic stage of vertebrates, where it functions as a predecessor of the spinal column.

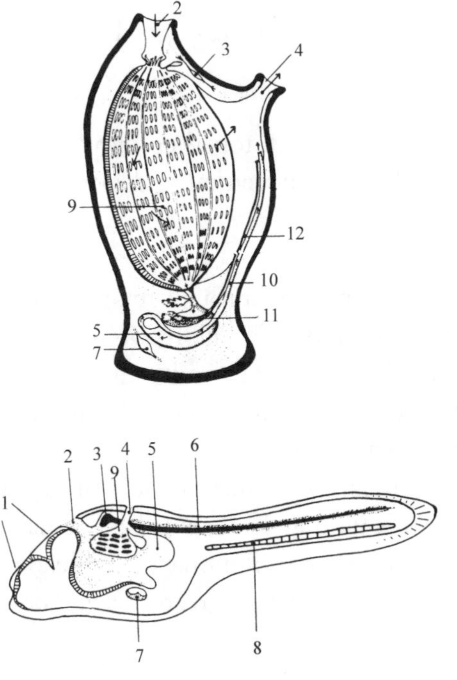

Fig. 173: Diagrams of the adult (upper) and larval (lower) forms of a sea squirt (Ascidiacea). (1) adhesive papillae (larval); (2) oral siphon; (3) brain vesicle (larva) or ganglion (adult); (4) atrial siphon (for elimination); (5) stomach; (6) nerve chord; (7) heart; (8) notochord; (9) pharynx with gill slits; (10) hindgut; (11) reproductive organs; (12) oviduct. The left and right sides of the adult animal as pictured here correspond to the ventral and dorsal sides of the larva, respectively (Buizer, 1983).

As soon as the larva becomes sessile, it undergoes a complete metamorphosis. The notochord disappears, the nervous system regresses and is replaced by a new system with no nerve cord at all, and the animal no longer has a recognizable back. In its new form, the sea squirt is a sessile, hermaphroditic filter-feeder. Water is sucked in through the oral siphon at the top of the body, and the pharyngeal gill slits filter out floating organic matter, which is immediately used as food. A smaller siphon located on one side is used for elimination. The pharynx, which contains the animal's filtering system, occupies much of the body's internal space. The stomach and the heart, which were relatively centrally located in the larva, shift to the "foot" in the adult.

In many respects, the sea squirt larva is already a specialized organism, although it cannot survive long in its free-swimming form. Nevertheless, it has a number of characteristics that clearly recall the overall structural plan of vertebrates:

- an elongated shape with distinct head and tail ends
- a notochord (cartilaginous rod)
- a brain (very rudimentary) and dorsal, hollow nerve cord
- mobility
- a centrally located heart and stomach, which together with the location of the nervous system distinguish ventral and dorsal sides.

Garstang postulates that vertebrates emerged through retardation of primitive ancestral tunicates that lived hundreds of millions of years ago. These animals retained the above-mentioned juvenile traits and avoided the extreme specialization of the sessile, filter-feeding stage. Vertebrates are related to tunicates in the same way that humans are related to apes. In each case, the initial fetal form is very similar, but while one line of evolution (the one leading to vertebrates and ultimately to humans) remains more similar to the initial fetal form, the other evolutionary line (leading to tunicates or chimpanzees, respectively) is subject to greater ontogenetic compression and specialization.

This view is supported by specific observations. For example, one group of tunicates (the Larvacea) moves freely through the water even in adulthood and undergoes less extensive metamorphosis of the larval form. The Larvacea may be examples of highly retarded tunicates.

Conversely, Gould (1977b) suggests there is no reason to conclude that the first vertebrates developed out of the larval form of an already highly specialized species. In his view, an adult animal similar to one of the unspecialized larval sea squirts may very well have been the ancestor of the vertebrates. Further evolution of this unspecialized form would have led, on the one hand, to sessile sea squirts (through extreme developmental compression and specialization) and on the other hand to vertebrates, who retained the fundamental traits of the prototypic gestalt and manifested them ever more explicitly. Both hypotheses, however, see vertebrates as organisms that are retarded in comparison to tunicates.

The Echinoderms

In 1845, Johannes Müller discovered that a microscopic marine animal, which had been given the generic name *Pluteus*, developed into a sea urchin, while other similar, glasslike little animals appeared to develop into starfish, sea cucumbers, and other echinoderms. The common structure of echinoderm larvae is simple, consisting of an elongated vesicle surrounding a bent intestinal tract (*Figure 174*). It is interesting to note that the ends of the intestinal tract display the same curvature that Bolk recognized as a universal characteristic of vertebrates. Only humans retain this curvature in adulthood. In echinoderm larvae, the oral orifice is roughly perpendicular to the main axis of the body—a characteristic also present in adult humans, where it is directly related to retention of the fetal cranial flexure.

According to Garstang, the shape of tunicate larvae evolved from that of echinoderm larvae. Echinoderm larvae have ciliated bands that allow these tiny animals to move. When these bands lie parallel to each other along the dorsal side and fuse over the intervening groove, a neural canal may form in the same way that it does during the embryonic development of all vertebrates (see *Figure 174*). Several detailed arguments support this theory.

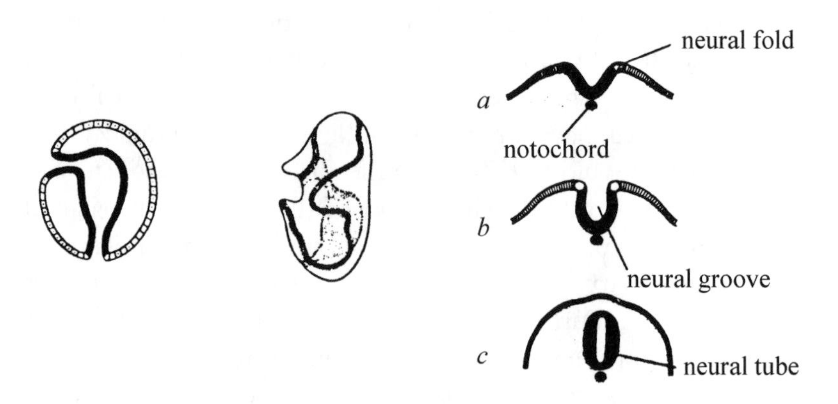

Fig. 174: Left: Diagrammatic longitudinal section through the ciliated larva of a marine invertebrate, with the intestinal wall indicated in black. This pattern is universal among the larvae of echinoderms, mollusks, some annelids, and acorn worms (Portmann, 1962).

Center: Larva of a sea cucumber (an example of an echinoderm). The intestinal tract displays the same ancient orientation that also appears in the cranial flexure and pelvic curvature of the vertebrate embryo. The illustration shows both the intestinal tract and the ciliated band that winds around both sides of the body.

Right: Diagram of the development of the spinal cord in the human embryo. (a) On the dorsal side above the notochord, a groove flanked by the neural folds develops. (b) The neural folds bend toward each other over the neural groove. (c) Fusion of the neural folds. The neural groove closes, forming the neural tube from which the spinal cord develops.

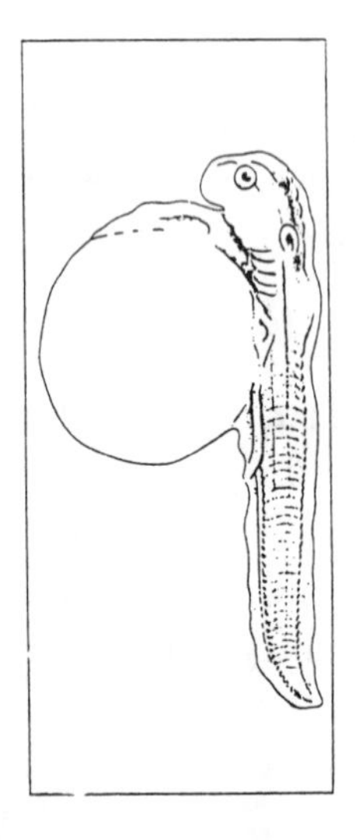

Fig. 175: A catfish embryo (Devillers et al., 1990). In fish species, as in other vertebrates, the head initially bends toward the torso, but the body straightens much earlier in fish than in higher terrestrial vertebrates.

Profound Retardation

Figure 176 illustrates one way of looking at the insights gained by Bolk and Garstang. The column on the left represents the evolutionary "mainstream," which displays minimal specialization but longer retention of traits from earlier phases. This sequence, rather than being seen as an actual line of evolution (since of course embryos or larvae are not descended from each other), should be understood in the sense intended by von Baer (page 33); that is, the traits that appear first are more general and common, while more specific characteristics appear only later. The unique aspect of the vertical line leading to humans is that while each new trait is significant for future human evolution, it is also fully integrated into the structural plan common to the state of evolution at which it appears. Such traits therefore correspond to the generalized components of evolution represented by the straight line in *Figure 45* (page 96).

Figure 176 depicts lines of animal specialization as branching off to the right. (In each instance, both fetal or juvenile and adult forms are shown.) These lines of evolution constitute deviations from the mainstream. The earlier the separation, the more extensive the consequences.

The mainstream evolves from a very simple form that is best described as that of an unspecialized marine invertebrate larva. Garstang (1928) called this form a "generalized dipleurula." It is characterized by:

- pervasive bilateral symmetry
- neural crests
- an alimentary canal with a recognizable angle in its initial portion.

The specialized line that evolves from this form leads to the echinoderms and transforms the neural crests into the larva's organs of locomotion. The larva later undergoes a dramatic metamorphosis to become an adult echinoderm—a starfish, for example. In this process, all of the characteristics of the initial form, including bilateral symmetry, are lost.

According to Garstang, the next form in the mainstream can be characterized as an unspecialized tunicate larva that retains and further develops traits from the previous stage. The neural crests merge into a rudimentary neural tube; the notochord emerges as a new element. The appearance of the notochord is in line with the overall structural plan of bilateral symmetry, but now the organism has a distinct dorsal side. The notochord exists not only in sea squirt larvae but also in the early embryonic stages of all vertebrates, where it serves as the precursor of the spine.

At this stage, the tunicates separate from the mainstream. Unlike echinoderms, they retain external bilateral symmetry, but they lose all characteristics related to the newly formed dorsal organs and elongated body structure. The ventral tilt of the "head" also disappears.

Vertebrates continue along the route that has been laid out. The nervous system develops to a much greater extent. The relative positions of the notochord, spinal cord, and digestive tract, as well as the forward-tilted head, are retained. Bilateral symmetry and an elongated body form, with an obvious dorsal side, finally become permanent at the adult stage, so animal specializations are less pronounced. The common initial form is expounded upon and becomes more explicit; new fixed traits (such as the four limbs of terrestrial vertebrates) evolve. Of course, the potential for limbs is not included in the earliest forms, but it is incorporated into the preexisting generalized structural plan without requiring reconstruction of the latter. In other words, a further potential of the generalized plan becomes explicit.

At this stage, additional characteristics of the original structural plan disappear from the adult animal forms that branch off from the mainstream. The most significant is the disappearance of the cranial flexure. In vertebrate embryos (such as the rabbit embryo, according to Haeckel), the oral cavity is still almost perpendicular to the spine, a trait that disappears in adult vertebrates. Retardation increases throughout the phylogenetic sequence of terrestrial vertebrates. Here, too, much more research is needed, but there is reason to say that amphibians are retarded in comparison to fish species, while mammals are retarded in comparison to amphibians and reptiles (see, for example, *Figure 101* above, page 190 and *Figure 112*, page 213).

The higher primates, the last species to abandon the evolutionary mainstream, are the least specialized of all animals. Their fetal and juvenile stages include characteristics that did not emerge until late in the evolutionary mainstream. These Type I characteristics, which indicate that apes are descended from humanlike forms, are lost in adult specimens as a result of specialization processes. Most of these traits, such as the central position of the foramen magnum or the balanced form of the head, are structurally meaningful only in humans.

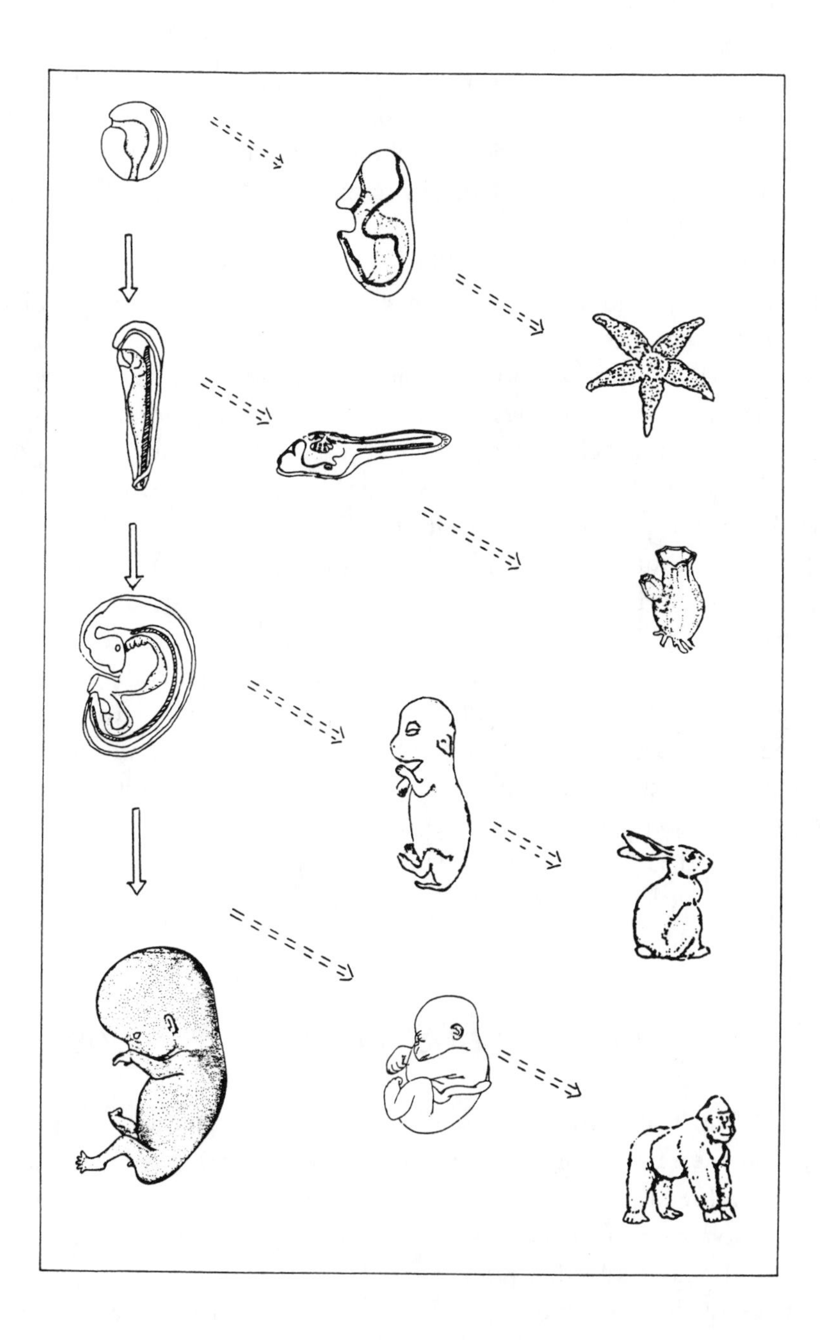

Fig. 176: Diagram of the main line of evolution (beginning with the larva of an aquatic animal and ending with a human fetus) and its side branches.

What we have called Type II characteristics were the last traits to appear in the evolutionary mainstream. Consequently, such traits appear only in the highest animals and even there in diluted form. They are much more closely associated with typically human capabilities (uprightness, the use of tools, speech, etc.).

Humans consistently retain all of the characteristics of the prototypic structural plan, including an oral cavity that is more or less perpendicular to the direction of the back. Human beings are the logical end of the evolutionary mainstream. By and large, *human evolution proceeds in a straight line. It does not specialize; it simply becomes more explicit*. Throughout the mainstream of evolution, earlier characteristics are preserved and elaborated upon in increasingly detailed traits that are logically consistent with preexisting ones. From this perspective, evolution is human from the very beginning. The cranial flexure, for example, already points to upright body posture. Increasingly subtle details (the central location of the foramen magnum, the shape of the hands and feet, etc.) complement the first general traits, which are retained. As the human gestalt evolves, it is not so much transformed as more clearly delineated, like a shape emerging from the fog. For the most part, Bolk restricted his studies to primates and other mammals; that is, to the animals most closely related to humans. Clearly, however, the phenomenon of retardation predates mammals and extends far into the lower animal kingdom. This extension can be called "profound retardation."

The Phenomenon of Synergistic Composition

The most important differences between humans and animals are based on the retardation process. There is no *a priori* reason to assume that general retardation of the chordate, vertebrate, or mammalian gestalt should produce a viable organism, let alone an exceptional one. And yet there can be no doubt that human abilities, in spite of their unspecialized character, qualitatively transcend the abilities of animal species. The retardation process seems to uncover potentials in humans that had lain dormant in the prototypic mammalian or primate gestalt and even in the prototypic vertebrate or chordate form. The many different consequences of retardation do not add up to a disjointed whole. Together, they provide an overwhelming abundance of possibilities. To use a mechanical analogy, they appear to be an arbitrary assortment of individually conceived structural parts. On closer inspection, however, they fit together in some inexplicable way to form a functionally superior motor. The different consequences of retardation effects, Type I and Type II characteristics, manifestations of fetalization and hypermorphosis, all come together harmoniously in the human body and derive meaning from each other; that is, they form a composition, a whole that is greater than the sum of its parts. I will call this *the phenomenon of synergistic composition*.

Darwin was familiar with the metaphor of an assortment of individual building blocks that seem to belong together. He formulated the specific evolutionary mechanism of *preadaptation* to explain this unavoidable phenomenon:

> Although an organ may not have been originally formed for some special purpose, if it now serves this end we are justified in saying that it is specially contrived for it. On the same principle, if a man were to make a machine for some special purpose, but were to use old wheels, springs, and pulleys, only slightly altered, the whole machine, with all its parts, might be said to be specially contrived for that purpose. Thus throughout nature almost every part of each living being has probably served, in a slightly modified condition, for diverse purposes, and has acted in the living machinery of many ancient and distinct specific forms (Darwin, 1862; also quoted in Gould, 1980).

Preadaptation means that an organ that initially developed to serve one specific function later demonstrated its suitability with regard to a totally different function. Let's look at insect wings as an example. According to Darwin's theory, wings would have to have evolved slowly. However, it is obvious that incompletely developed wings are not suitable for flying but are a disadvantage and a burden to the insect. In actual fact, truly superfluous organs such as the eyes of cave-dwelling species often regress and disappear rapidly. We must therefore assume that useless, underdeveloped wings would quickly disappear without having a chance to develop into complete wings. If we assume, however, that these precursors of true wings initially served a different function, the picture looks quite different. For example, they may have served either as thermoregulatory surfaces that allowed the insect to warm up quickly in the morning sunlight or as organs for gliding over the surface of the water—that is, they may have been precursors of actual organs of flight while serving a totally different function. These precursors were not wings; they played a different role and performed a task that even tiny, partial wings could accomplish. Only after they had been developed to serve this other function did they then prove adapted to flying, continuing their evolution in the direction of this new function. Hence the precursors of wings were *preadapted* to flying although they did not develop to serve that purpose.

The Darwinian concept of preadaptation presupposes that a very concrete original function (such as thermoregulation) precedes an equally concrete function (such as flying). For this reason, preadaptation cannot explain the phenomenon of synergistic composition in humans. Human characteristics, rather than emerging from a number of very concrete but different specializations, derive from one and the same ontogenetic principle. If the Darwinian principle of preadaptation were applied to explaining these characteristics, the obvious conclusion would be that the pattern of development common to all vertebrates was "preadapted" to anthropogenesis. This, however, is simply a reformulation of the same enigma, namely, that the human species is not just an arbitrary animal species. *Human beings explicitly manifest what remains unspecialized and universal in animals, and this universal, unspecialized kernel, which is the true inner architect of the human organism, reveals itself as the bearer of extraordinary potentials (the capacities for uprightness, speech, culture, etc.).*

The larynx and brain offer interesting examples that illustrate the phenomenon of synergistic composition. The deep descent of the human larynx, a Type II characteristic, is intimately linked to retention of the flat fetal skull (a Type I characteristic). The human cranium also retains its juvenile spherical form (a Type I characteristic); structurally related to it, however, is the Type II characteristic of telencephalization. Thus we can say that the juvenile shape of the primate skull is the natural precursor of both the descended larynx and the highly developed brain in humans. And these last two characteristics are both involved, in a mutually complementary way, in the emergence of speech, which depends not only on the specific structure of the larynx but also on a highly developed nervous system capable of serving as the instrument of the ability to think. Type I characteristics, therefore, point to Type II characteristics, which in turn derive their meaning from the typically human attributes of which speech is the centerpiece.

Retardation also extends the juvenile period, giving us additional time for learning, which we need if we are to effectively acquire speech as a means of gaining access to our entire sociocultural heritage. Of course, the structure of the mouth must also be suited to speaking. Here, too, retardation plays a key role. Muzzle development is eliminated, and the dimensions of the first set of teeth (a closed row of almost uniform height, the size of the anterior dental arch, etc.) are preserved. The mouth itself must be freed from the need to serve the prehensile function, which is taken over by the hands. And in fact our hands, simply as a result of their retarded and unspecialized character, are extraordinarily suited to that function.

The upright posture of the human body frees the hands to serve the prehensile function. Uprightness, in turn, is made possible by retardation of the foot (retention of the fetal position of the big toe; hypermorphosis of the big toe and heel), by the great length of the lower limbs (also due to hypermorphosis), and by the preservation of fetal primate characteristics such as the central position of the foramen magnum on the underside of the skull. Uprightness is also made possible by the flattened shape of the human ribcage (hypermorphosis), which in turn is prefigured by the fetal structure of the arterial ramifications of the aortic arch. The flattened shape of the human thorax also extends the effective radius of the arms, which makes the hands still more suited to the prehensile function.

Uprightness is also prefigured in the connection (which is permanent in humans) between the pericardium and the diaphragm. As a result of uprightness, respiratory rhythms are emancipated from the constraints of locomotion. This initial emancipation is furthered by the fact that hairless skin (a Type I characteristic) and a highly developed system of eccrine sweat glands (a Type II characteristic) combine to form a highly effective cooling system that no longer depends on respiration. This dual emancipation of the lungs creates the opportunity for them to serve the function of speech.

As a result of hypermorphosis, the vertebral column develops cephalocaudally, in the direction of the sacrum, and the vertebrae become increasingly heavy toward the base of the spine, a pattern that also supports uprightness. The resulting heavy sacrum is structurally suited not only to carrying the extra burden of weight imposed by uprightness but also to accommodating a wider birth canal. The latter modification is related to the large circumference of the newborn's cranium, which in turn is related to the capacity for speech.

Thus the different consequences of retardation in humans are mutually complementary and result in a net gain of typically human capabilities (speech, uprightness, use of tools, etc.). This, however, is the essence of the phenomenon of synergistic composition (see *Figure 177*).[1] Gould (1987) characterized this phenomenon as follows:

> We have evolved our massive brains largely by the evolutionary process of neoteny: the slowing down of developmental rates and the consequent retention into adulthood of traits that mark the 'juvenile' stages of our ancestors. We retain the rapid fetal growth rate of neurons well beyond birth (when the brain of most mammals is nearly complete) and end our growth with the bulbous cranium and relatively large brain so characteristic of juvenile primates. Neoteny also slows down our maturation and gives us a long period of flexible childhood learning . . . As primates we belong to one of the few groups of mammals sufficiently unspecialized in bodily form to retain the morphological capacity for exploiting a broad range of environments and modes of life. A bat has committed its forelimbs to flight, a horse to running, and a whale to balancing and paddling. Culture and intelligence at a human level may

1. Some scientists consider humans as a product of "mosaic evolution," with hypermorphosis and paedomorphosis affecting traits that are randomly distributed from an ontogenetic point of view. I believe this to be an error. As I see it, hypermorphosis in humans selectively affects ontogenetic latecomers. Hypermorphosis in ontogenetic latecomers (in humans) replaces specialization (in animals). The concepts of hypermorphosis and fetalization as I use them encompass the fact that fetalization denotes the conservation of early traits, and hypermorphosis the development of latecoming traits.

> have required the evolution of a free forelimb and a generalized hand endowed with the capacity to manufacture and manipulate tools . . . Only the morphologically unspecialized among mammals have not made inflexible commitments to particular modes of life that preclude this prerequisite for intelligence.

Elsewhere, Gould suggests that a Darwinian explanation of human retardation is still conceivable:

> What is the adaptive significance of retarded development itself? The answer to this question probably lies in our social evolution. We are preeminently a learning animal. We are not particularly strong, swift, or well designed; we do not reproduce rapidly. Our advantage lies in our brain, with its remarkable capacity for learning by experience. To enhance our learning, we have lengthened our childhood by delaying sexual maturation with its adolescent yearning for independence. Our children are tied for longer periods to their parents, thus increasing their own time of learning and strengthening family ties as well (1977b).

For two reasons, this type of Darwinian explanation remains wholly unsatisfactory. First, extending the juvenile learning period is significant only if other prerequisites to increased learning activity—a larger brain, versatile and unspecialized hands—are also in place. The phenomenon of synergistic composition (that is, the fact that the process of retardation meets *all* of these conditions *at once*) remains completely unexplained. Each individual manifestation of the retardation process entails an entire complex of developments that complement each other in surprising ways. In and of itself, the complementary and cumulative character of retardation effects suggests that they were already prefigured in the prototypic structural plan for the animal body. Because the physical appearance of these effects occurs only at the end of primate evolution, when the retardation tendency has asserted itself fully, they cannot be explained as the result of a physical process of selection.

Gould himself presents a second reason why our retarded development cannot be explained as a result of selection for a longer learning period. He points out that human retardation has its roots in the distant past:

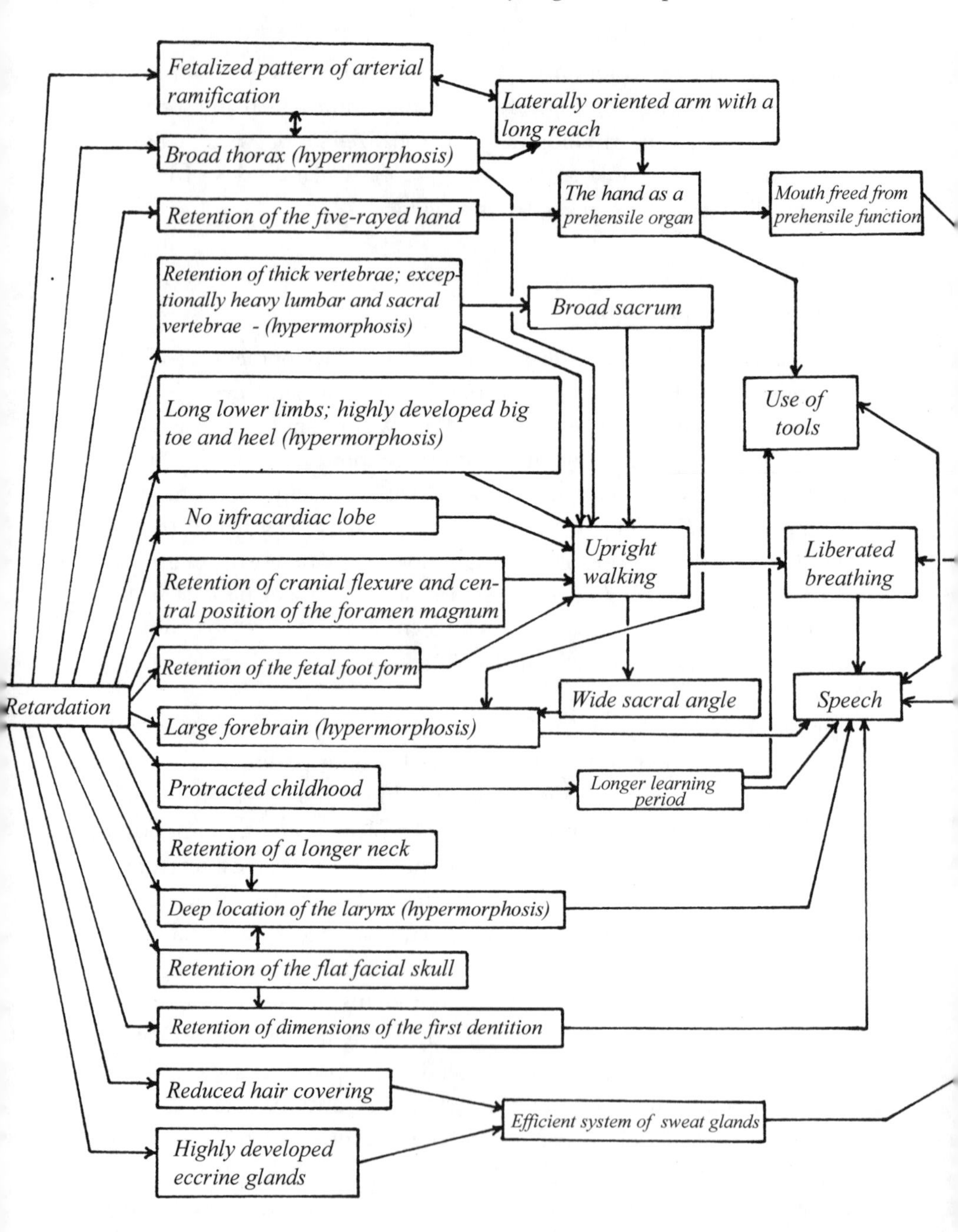

Fig. 177: Diagram of the interconnections among the different consequences of paedomorphosis in humans.

> Primates in general are retarded with respect to most other mammals of comparable body size. The trend continues throughout the evolution of primates. Apes are generally larger, mature more slowly, and live longer than monkeys and prosimians. The course and tempo of our lives has slowed up even more dramatically (1977b).

The retarded and unspecialized character of primates became evident as early as the early Tertiary period. *Tetonius homunculus*, a primate of that time, had a brain that was very small in comparison to the brains of modern mammals of comparable size (Jerison, 1979). In comparison to mammals alive at that time, however, this early primate had a brain volume three times the size of the average Tertiary mammalian brain. In this context, Gould notes:

> Primates have been ahead right from the start; our large brain is only an exaggeration of a pattern set at the beginning of the age of mammals. But why did such a large brain evolve in a group of small, primitive, tree-dwelling mammals, more similar to rats and shrews than to mammals conventionally judged as more advanced? . . . We simply do not know the answer to one of the most important questions we can ask (1977b).

The selection mechanism Gould proposes (promoting learning activity by extending the learning period) certainly cannot explain the large brain and overall retarded character of the first primates. If it could, the same mechanism would have had to be all the more obvious in almost all modern mammals, which have much larger brains than those early, long-extinct primates. Extreme retardation in humans is merely the keystone in this single developmental principle of retardation in primates, which asserted itself with increasing force as evolution progressed. Actually, the theme of retardation can be traced back to times prior to the emergence of primates. In all likelihood, the oldest retardation characteristic is retention of the embryonic cranial flexure (see *Figure 176*, p. 341).

The phenomenon of synergistic composition compellingly suggests that Bolk's "intrinsic evolutionary factor" is purposefully at work in animal evolution. The goal of the constant, increasing retardation of primates throughout evolution becomes visible only at the end, when human beings appear. Human speech turns out to be the focal point of an entire spectrum of retardation effects that form an intentional whole. The appearance of speech in the physical world was the goal to which evolution aspired. Through human beings, the Word appeared directly in matter for the first time.

Afterword: Is Darwinism Incomplete?

The present book clearly constitutes not only a critique of Darwinism, but a presentation of what I believe is a more complete view of evolution. I do not reject the possibility of Darwinian mechanisms playing a certain role in animal evolution. But I do claim that the facts do not sustain the idea, dominant among evolutionary theoreticians, that humans are basically a product of Darwinian mechanisms.

The particular abilities of *Homo sapiens* correspond to anatomical structures and developmental tendencies that appeared long before these abilities became operative. Moreover, from their first appearance, these structures constitute an integrated whole, just as sounds are integrated into a musical composition. Therefore, it makes no sense to explain phenomena such as the forward position of the foramen magnum, or the arterial branching pattern in a monkey fetus, as mere preadaptations. Saying that the generalized vertebrate or mammalian anatomy is preadapted to anthropogenesis merely restates the same baffling mystery.

In order to explain the emergence of humanity, we have to invoke some directive force or factor. In other words, something like an Aristotelian final cause must be at work in human evolution.

But doesn't such a proposition fly in the face of science and of rational thinking in general? Or are there general and rational grounds for doubting the completeness of materialistic and Darwinian explanations? I think there are reasons for profound doubt.

Darwinism, as a materialistic theory, essentially views living beings as extremely complex mechanisms consisting of atoms, molecules, etc., all of which obey the laws of physics and chemistry. Whether these laws are completely deterministic, or allow for some randomness, is immaterial. The important point is that there is no place for consciousness in this picture. Study the laws of physics and chemistry. Nowhere is there a formula, a law, or an equation predicting the emergence of such a thing as consciousness. If we examine a living human brain with a scalpel, with a nuclear magnetic resonance apparatus, or by any other means, we can observe or register physical processes such as the flow of tiny electrical currents, but we can never observe consciousness as such. The methods of anatomical or physical observation are simply not suited to observing consciousness.

Dennett (1991) and many other theoreticians conclude that consciousness simply does not exist. However, closer consideration soon reveals that this so-called explanation explains nothing at all, because a non-illusory experience of consciousness itself is a prerequisite of becoming aware of the "illusion" of consciousness!

But is consciousness merely an epiphenomenon, perhaps real (though certainly unexplained), yet without any consequences for the world of objects and the course of external events? Direct observation of our own lives shows that this is not the case. Whenever we make a decision based on rational and/or moral considerations, and act accordingly, it is the inner "content" of our deliberations, rather than physics or chemistry, that determines how we will influence the material world.

Materialists will retort that such thought contents are epiphenomenal, that physical and chemical processes in the human brain are the real links in the chain of causality underlying our acts. At this point, materialism becomes ideology. Not only does this statement contradict our direct experience; it contradicts itself. In fact, every distinction between truth and error would become meaningless if we were to reduce the content of thinking to underlying physical and chemical processes. We would consider something true only because certain physical processes in the brain create the impression or conviction of "truth." But such physical processes are determined, not by truth, but by the laws of physics. And if our own thinking were completely determined by such processes, we would be incapable of distinguishing whether something is "really" true or merely seems true because physical processes in our nervous system create that impression. Truth would be inaccessible and every quest for truth pointless.

For Darwinists, an additional problem arises here. If every process in the physical world, including consciousness, could be reduced to purely physical or chemical causes, then consciousness itself would have no survival value. If a gazelle's flight from a lion could be explained by the physical image of the lion in the gazelle's eye and the subsequent physical processing of this image by the gazelle's brain (which in turn sends electrical commands to the leg muscles and so on), it would be pointless for the gazelle to possess consciousness. Its entire body would function equally well without it.

There are many other logical problems with materialism (Verhulst, 1994a). One central problem results from the fact that meaning cannot be reduced to structure. As such, a material datum never "means" anything; that is, we cannot deduce the meaning associated with an object from its physical constitution. Consider a traffic sign, for example. The sign's generally accepted meaning cannot be deduced from its physical properties. If extraterrestrials were to commandeer such a sign and examine it in the laboratories on their planet, they would never be able to deduce its meaning. The sign's meaning transcends its physical structure. Physical structures do not, of themselves, exude "meaning," and this is true not only of signs but also of the structures in the human brain. Inasmuch as our brain structures operate according to the laws of physics and chemistry, they cannot indicate one possible meaning while excluding others. Without additional information, we cannot possibly deduce a person's thoughts from our knowledge (however detailed it may be) of his or her physical constitution or brain structure. Whatever structure or process we might observe, it could always "mean" almost anything. Hence, the statement that the contents of our thoughts are derived from our brain structure is mere ideology. I do not mean to deny that the brain plays a role as a necessary prerequisite to ordinary human thinking. Of course it does. Just as we need a radio in order to hear a program on the air, we need a brain in order to have rational thoughts. But it does not follow that the structure of the radio explains or produces the content of the program, or that the structure of the brain explains or produces the content of our thoughts.

These considerations lead to an important conclusion. If we accept the relevance of truth and logical thinking, then whenever we follow the chain of physical causality from our rational or moral acts into the nervous system, we eventually arrive at a point where it becomes impossible, as a matter of principle, to analyze these mental processes in terms of physical and chemical laws (Steiner, 1995, Chapter 9). Once again, is this reasonable? Do the physical sciences themselves suggest the existence of a limit beyond which we cannot proceed, where the track of physical causality dissolves, so to speak?[1] Quantum physics supplies the most fundamental theoretical framework for such issues. We must understand, however, that the formal structure of quantum theory presupposes but is unable to explain the existence of "form." It is interesting to note that a theoretician as early as Bolk already based his conclusions on a nonreductionist view of life. And he considered uniqueness of form the fundamental attribute of life. Here are two quotes:

> Life is not some complicated physical-chemical process, but a phenomenon in its own right. Life processes are accompanied by physical and chemical phenomena. Similarly, electrical discharges are accompanied by acoustical effects and emission of light; but electricity cannot be reduced to sound-producing light or light-emitting sound. In the same way, biological processes are accompanied by phenomena that can be understood and measured by the means and methods of physics and chemistry, but it does not follow that life itself is a physical-chemical phenomenon (Bolk, 1918).

> Life is not a property of matter; it is a property of form. The concept of a "living molecule" or "life molecule" is self-contradictory. There are no building blocks of life. Life is an attribute of form, and that form has two fundamental properties: it is unique, and it is transitory. Each form is different from all past and future forms; because it lives, it exists for one indivisible moment. Life is characterized by perpetual change; it is the continuous alteration of form. Because our methods are based on stability and on pinning down points in time, they do not permit us to grasp the essence of life (Bolk, 1918).

These Bolkian viewpoints seem hardly compatible with contemporary biological orthodoxy. The vast majority of biologists assumes as a matter of course that biology and chemistry can be reduced to physical principles and that biological form is dictated by the DNA molecule and not an ontologically autonomous aspect of reality. Strangely enough, many scientists hold that physical forms emerge through the mere clustering of "quantum" particles (atoms and the like), although the latter are formless in principle.[2]

Surprisingly, however—although difficult to reconcile with contemporary biology—Bolk's iconoclastic concept of life as a "property of form" is not unreasonable from the perspective of modern physics. Even the widespread idea that stereochemistry can be reduced (at least in principle) to quantum mechanics is incorrect (Woolley, 1978, 1985, 1991). For example, quantum mechanics cannot account for structural differences among isomers (for instance, between cyclopropane and propene or the two enantiomers of analine).[3]

Rather, the quantum mechanical treatment of a phenomenon produces a statistical prediction. By its very nature, this type of prediction makes sense only with regard to reproducible phenomena.[4] As Josephson (1988) and Bohr (1933, 1937, 1952) have indicated, irreproducible phenomena inevitably occur in living organisms because amplification mechanisms are so all-pervasive in biological structures (for example, in sense organs). In such cases, we cannot make statistical predictions because the phenomenon we are investigating cannot be considered an element in a well-defined statistical ensemble and is unique in the very specific sense that it can never be equated with any other phenomenon with any degree of certainty. As I mentioned earlier, Bolk viewed uniqueness of form as a fundamental characteristic of life. Form as such is part of all natural phenomena; each phenomenon is nothing more than a change of "form" (in the generalized Aristotelian sense of the term). Uniqueness of form is the hallmark of life.

Of course, we cannot use the methods of physics to prove that such unique phenomena occur in living beings. For the purposes of our argument, however, it is sufficient to know that this class of phenomena is possible and plausible.

We can now identify the ideological component of materialism very precisely. The unproven and implausible supposition that all phenomena in the material world are reproducible and open to quantitative physical analysis constitutes the core of the materialistic ideology that dominates contemporary science.

When neither classical physics nor quantum mechanics apply, however, the course of events is neither deterministic nor statistical, leaving room for the operation of factors that cannot be deduced from physical theory. These factors may have a teleological character that is revealed by logical connections or structural correspondences between phenomena that appear unrelated from a physical perspective. For example, logical connections may link human deeds; on a larger scale, structural connections can unify a phylogenetic tree of various classes of organisms. For example, the human form is prefigured by the phylogenetically much older developmental mode of mammals—a fact that points to a human-centered formative principle working as a guiding factor in mammalian evolution.

From this viewpoint, an organism's biochemical structures (DNA, proteins) are not its ultimate foundations but simply anchors cast by the biological gestalt in arriving at a stable material manifestation. These structures can be expected to be more related in organisms that more closely resemble each other but need not be interpretable in terms of a unique and unambiguous family tree, as Darwinism predicts. Molecular cladistics is indeed pervaded by parallelisms, back-mutations and the like. Data on myoglobins, for example, indicate that humans are more closely related to gibbons than to orangutans, birds more closely related to mammals than to crocodiles (Romero-Herrera et al., 1976, Bishop and Friday, 1987). These findings conflict strongly with other morphological, paleontological, and biochemical data. The *Homo-Pan-Gorilla* triad is a case in point: morphological indicators, such as adaptations that favor knuckle-walking, suggest a *Pan-Gorilla* clade, but DNA hybridization suggests a *Homo-Pan* clade. A comparison of an 896-nucleotide sequence of mitochondrial DNA in great apes and humans revealed not only eleven substitutions uniquely shared by humans and chimpanzees but also eight uniquely shared by chimpanzees and gorillas (Brown et al., 1982). "Parallel change" of this scope seems to preclude describing the human-chimp-gorilla relationship in terms of a conventional phylogenetic tree. And in fact, it appears that evolution often proceeds in multiple layers that are difficult to reconcile with certain aspects of Darwinism and that suggest instead a variety of factors acting across the same cluster of species (Grehan and Ainsworth, 1985; Heads, 1985; Schwabe, 1994; Went, 1972; Williamson, 1992).

Phylogenetic methods that attempt to shoehorn biological reality into linear schemes seem to produce nothing more than "least impossible" evolutionary trees of uncertain scientific value. That is not to say that evolution did not take place. However, the phenomena suggest that evolution is not directed solely by factors like natural selection, which operate separately along each evolutionary line. It seems that other factors are involved, operating together over large areas encompassing different evolutionary branches.

In recent years, the argument of "intelligent design" has been vigorously debated. This argument is as old as Darwinism itself, but recent authors have given it new relevance. William Dembski (1998) has added the claim that it can be quantified, and Michael Behe (1998) has given credence to the movement in the field of biochemistry. Basically, this argument states that biological systems are much too complex to have evolved through the Darwinian mechanisms of variation and natural selection. In other words, the Darwinian scenario implies the occurrence of highly improbable events. This approach has great potential merit, but its conclusions should be interpreted carefully. Neo-Darwinism, as a materialistic theory, views living organisms as clusters of atoms and molecules that interact according to the laws of physics and chemistry. If we take this materialistic picture as our starting point, we may well conclude that in this context the spontaneous emergence of life is too improbable to be believable. We must refrain from jumping to the conclusion, however, that some external force or Great Designer miraculously intervened to assemble atoms and molecules in a way that allowed evolution to continue. What we ought to question is the basic view of living beings as mere conglomerations of the particles of conventional physics.

The research included in this book indicates that there is an overarching lawfulness inherent in the development and evolution of organisms, and it shows in detail how this formative lawfulness is expressed in various organs. Such developmental principles as retardation and hypermorphosis are guided by the overarching principle of synergistic composition. The intelligence at work in evolution is clearly not imposed upon life from without but active within the evolutionary process itself.

Implications for Human Evolution

Paleontologists continue to discover new proto-human fossils, the most ancient of which have been found only in a few regions in Africa. At the time of this writing, the most recent finds were *Orrorin tugenensis*, discovered by a French-Kenyan team, and *Ardipithecus ramidus kadabba*, found by the Ethiopian researcher Yohannes Haile-Selassie.

There is a huge discrepancy between the actual findings and the stories that develop around them. Recently, *Time* (July 23, 2001, pp. 56-61) published a human evolutionary tree that is fairly representative of how paleontological data are processed and diffused among the educated public at large. Although this tree certainly does not represent scientific unanimity, it is an interesting example of how pseudo-scientific ideology can overwhelm the facts. The actual data are limited to a number of fossilized bones and very incomplete skeletons, together with some datings of earth layers. We have no data at all on the evolutionary connections among the creatures that were the sources of these fossils, and we do not have enough data on their outer appearance, mode of locomotion, etc. to produce accurate drawings of them. Their skin color, degree of hairiness, gait, and the like remain unknown. Nonetheless, the cover of *Time* shows a complete face of Ardipithecus ramidus kadabba glaring at the reader from among the leaves. The caption reads: "How apes became human." In the article, pseudo-realistic drawings of proto-humans, ostensibly based on actual data, are then shown marching along a number of parallel evolutionary paths. The resulting evolutionary tree, rather than reflecting actual knowledge, consists largely of unsupported hypotheses. Such images are icons of Darwinian ideology and are treacherous from a scientific viewpoint.

The evolutionary picture proposed by *Time* is implausible even from the perspective of elementary logic. At least two aspects render it highly suspect:

First, there is the curious fact that the evolutionary pathways leading to the chimpanzee and the gorilla are completely devoid of fossils. The evolutionary branches leading to humans and African apes are supposed to have separated approximately six to eight million years ago. During the period immediately following that split—that is, from six to three million years ago—we find six species or subspecies on the human branch: *Orrorin tugenensis*, *Ardipithecus ramidus kadabba*, *Ardipithecus ramidus ramidus*, *Australopithecus anamensis*, *Australopithecus afarensis*, and *Kenyanthropus platyops*. On the chimpanzee branch and the gorilla branch, we find exactly nothing. This disparity is too great to be accidental; there is only about one chance in 300 that random processes could produce such a lopsided result. There is no reason to believe that many more ancestors of humans than of chimps and gorillas were alive at that time. Furthermore, the *Time* article explicitly states (p. 50) that the ancestors of humans and African apes are thought to have lived in similar habitats (implying that their chances of fossilizing were roughly equal). Clearly there is a propensity among researchers to place their findings upon the line leading to humans.

At this point, of course, researchers will protest that all the fossils they discovered were clearly bipeds and therefore human. By implication, however, they eliminate a humanlike, bipedal creature as the possible ancestor of chimps and gorillas. Here we touch upon the ideological core of Darwinian human paleontology: a humanlike ancestor of an animal-like species is excluded *a priori*. A humanlike gestalt is rejected as a proto-form for animals, because that would indicate "anthropocentric" thinking and stand Darwinian evolution on its head. Consequently, all bipeds (or presumed bipeds) are located on or near the evolutionary path leading to *Homo sapiens*, and their fossils are "humanized" accordingly. For example, although the cranial capacity of australopithecines does not exceed that of chimpanzees, estimates of australopithecine brain volume have been inflated for the past several decades, making these creatures appear more humanlike and less apelike (Falk, 1998; Conroy et al., 1998).

Second, as in many other comparable evolutionary trees, there is a discrepancy between the upper and lower parts of the tree published in *Time*. Side branches with fossils placed on it appear only in more recent epochs, whereas the older part of the tree consists of a single trunk. Now, the probability of a proto-human fossil being on the main trunk should be independent of time. Yet the older fossils are all located on the main trunk leading directly to modern humans. Only at the point where several types of humanlike creatures exist simultaneously does the tree begin to branch. Any representation such as this is highly suspect on statistical grounds, of course, since there is no *a priori* reason for the more ancient fossils to predominate on the main trunk while the more recent fossils are concentrated on side branches. In other words, the evolutionary "tree" should look more like an evolutionary "bush."

Moreover, it should be noted that these mainstream representations of human and ape evolution always imply far-reaching parallelisms that do not receive the benefit of a great deal of discussion. For example, gorillas and chimps share a number of important traits (such as thin tooth enamel and adaptations related to knuckle-walking) that are completely absent in humans. Chimps and gorillas are thought to have no exclusive common ancestor, so the fact that they share these traits must mean one of two things: either the traits evolved independently in the chimpanzee and gorilla lines of descent, or they were present in the common ancestor of humans and African apes and were later eliminated in the human line. Similar considerations apply when we include the orangutan in the picture. Again, this ape shares a number of traits only with the African apes. The most impressive example is its high pelvic blades, which contrast with the low pelvis and the large distance between pelvis and rib cage that characterize not only humans but also gibbons and lower primates (see *Figure 14*, p. 31). Either this specialized pelvic structure was present in the common ancestor of the great apes and humans but later disappeared in the human line, or it evolved independently in at least two separate instances. However, human ontogeny shows no trace of an apelike pelvis; rather, it is the ape who starts with a low, humanlike pelvic structure. The problems inherent in the Darwinian view are well known by specialists, but they are systematically neglected when the latest evolutionary tree is presented to the layman. In my view, this systematic disregard for legitimate objections amounts to mass indoctrination.

The alternative view of evolution presented in this book stems from careful observation of morphological change as it occurs in individual development and in the process of evolution. A synthesis of this detailed research leads to the following conclusions:

- The phenomena of life and consciousness cannot be explained in terms of physical-chemical laws. Life and consciousness are subject to their own dynamic principles that prevail within the boundary conditions set by physical laws.

- Life manifests through unique forms. The ontogenetic development of an organism is a formative process subject to organizational principles such as growth, retardation, hypermorphosis, and specialization. These principles work together in the gradual formation of the adult organism, complementing one another so as to bring about a meaningful, yet highly unique form.

- The same formative principles are at work in animal evolution as a whole, yet here they are guided by a more comprehensive principle that leads evolution onward, manifesting its potential increasingly in evolution's successive stages and finally coming to its full expression in the emergence of the human gestalt.

- This movement toward the human form is present in animal evolution from the outset. As evolution progresses, the human prototype manifests more fully in the embryonic stages of organisms. However, the adult forms of these organisms diverge from their humanlike beginnings as they adapt to specific environmental conditions.

- This process can be seen most clearly in the primates where the human form is almost achieved in the embryonic stage but is later lost as the apes mature. Only in the case of the human being does the human potential finally persist into adulthood. In this sense, the emergence of humanity can be seen as the fulfillment of evolution's longstanding promise.

NOTES

1. I have explored this question in greater detail elsewhere (Verhulst, 1994a). To a large extent, the passages that follow are taken from an earlier paper of mine (Verhulst, 1994b).

2. The quantum description of an atom or a molecule is made in Hilbert space, i.e., a space with infinite dimensions and with the use of complex numbers. It is mathematically impossible to project this description unequivocally into three-dimensional Euclidean space. This means that no classical spatial interpretation can be deduced from the quantum description, although the latter permits the prediction of all observables connected with the atom or molecule. This state of affairs is not a shortcoming of quantum mechanics; rather, it reflects the fact that an atom or a molecule—within the context of an experiment where it enters the wave function—cannot be observed as a classical entity with a well-defined form and spatial position.

3. Quantum chemists must introduce the stereometric form of the molecule into their calculations as a set of *a priori* boundary conditions (for instance, by adopting the Born-Oppenheimer approximation, whereby the positions of the nuclei are fixed in advance). The inability of quantum mechanics to describe isomers is only one aspect of a more general principle that forms the cornerstone of the standard (Copenhagen) interpretation of quantum mechanics: as such, the spatial specificity of the experimental set-up cannot be incorporated into the wave function associated with the phenomenon under consideration. But in fact the opposite is true; the geometry of the apparatus employed, for example, in a single-photon double-slit interference experiment influences the wave function describing that phenomenon (Bohr, 1949). The apparatus is taken as a "classical object" whose spatial forms unambiguously fix the boundary conditions of the experiment. Hence, these forms can never be eliminated, because the very formalism of quantum mechanics presupposes their existence.

4. Reproducibility depends on controlling the spatial parameters defining the start configuration of the phenomenon, and the possibility of control is not guaranteed in all cases because amplification mechanisms may be part of the start configuration. Such a boundary condition implies an uncontrollable quantum interaction with the start configuration of the phenomenon under consideration. In the absence of amplification mechanisms, this interaction can be disregarded and the phenomenon is considered reproducible (for instance, reading the initial position of the pointer of a measuring device does not cause a relevant change in the device). In irreproducible phenomena, the uncontrollable interaction implied by any attempt to observe a supposedly well-defined start configuration is amplified to the point of changing a relevant domain of that configuration. Here, however, we touch on the crucial point: because the inducing quantum interaction is

uncontrollable, its precise impact on this domain remains unknowable in principle. Under these circumstances, no physical content can be ascribed to the concept of a "well-defined start configuration." Consequently, no precise boundary conditions can be determined, and quantum mechanics cannot be applied to the phenomenon in question. Paradoxically, the uncontrollable nature of quantum interaction limits the classes of phenomena that can be described by quantum mechanics.

Bibliography

Abbie, A. 1947. Headform and human evolution. *J. Anat. (London)* 81: 233-258.

— 1948. "No! No! A thousand times no!" *Aust. J. Sc.* 11: 39-42.

— 1958. Timing and human evolution. *Proc. Linn. Soc. New South Wales* 83: 197-213.

Abitbol, M. M. 1987a. Evolution of the lumbosacral angle. *Am. J. Phys. Anthrop.* 72: 61-372.

— 1987b. Obstetrics and posture in pelvic anatomy. *J. Hum. Evol.* 16: 243-255.

— 1988. Effect of posture and locomotion on energy expenditure. *Am. J. Phys. Anthrop.* 77: 191-200.

— 1990. The multiple obstacles to encephalization. *Behav. Brain Sc.* 13: 344 ff.

Aeby, C. 1880. *Der Bronchialbaum der Säugethiere und des Menschen.* Leipzig: W. Engelmann.

Aiello, L. 1981. The allometry of primate body proportions. In M. H. Day, ed., *Vertebrate locomotion.* London: Academic Press.

Aiello, L., and C. Dean. 1990. *An introduction to human evolutionary anatomy.* London; San Diego: Academic Press.

Albrecht, H. 1895. Beitrag zur vergleichenden Anatomie des Säugethier-Kehlkopfs. *Sitzungsber. K. Akad. Wiss. Wien* 104: 227-322.

Alexander, R. M., A. S. Jayes, G. M. O. Maloiy, and E. M. Wahuta. 1979. Allometry of limb bones of mammals from shrews to elephants. *J. Zool. (London)* 189: 305-314.

Altman, P. L., and D.S. Dittmer. 1974. *Biology Data Book, vol. 3.* Bethesda, MD: Federation of American Societies for Experimental Biology.

Amundsen, D. W., and C. L. Diers. 1969. The age of menarche in classical Greece and Rome. *Hum. Biol.* 41: 125-132.

— 1970. The age of menopause in medieval Europe. *Hum. Biol.* 42: 79-86.

— 1973. The age of menarche in medieval Europa. *Hum. Biol.* 45: 363-369.

— 1973. The age of menopause in classical Greece and Rome. *Hum. Biol.* 45: 605-612.

Anderson, M., M. Blais, and W. T. Green. 1956. Growth of the normal foot during childhood and adolescence. *Am. J. Phys. Anthrop.* 14: 287-308.

Antinucci, F., and E. Visalberghi. 1986. Tool use in *Cebus apella*: A case study. *Int. J. Primatol.* 7: 351-363.

Arey, L. B. 1965. *Developmental Anatomy*. Philadelphia: Saunders.

Ariëns-Kappers, C.U. 1931. *Inleiding op de derde druk van L. Bolk "Hersenen en Cultuur."* Amsterdam: Scheltema & Holkema.

— 1942. Orthogenesis and progressive appearance of early-ontogenetic form relations in the adult stages during human evolution with a possible explanation for them. *Acta biotheor.* 6: 165-183.

Ayala, F. J., and T. Dobzhansky. 1974. *Studies in the philosophy of biology*. Berkeley: University of California Press.

Balinsky, B. I. 1970. *An introduction to embryology*. Philadelphia: Saunders.

Barriel, V., and P. Darlu. 1990. Approche moléculaire de la phylogénie des Hominoidea. L'exemple de la pseudo êta-globine. *Bull. Mém. Soc. Anthrop. Paris* 2: 3-24.

Baughan, B. A. Demirjian, G. Y. Levesque, and L. Chaput-Lapalme. 1979. The pattern of facial growth before and during puberty as shown by French-Canadian girls. *Ann. Hum. Biol.* 6: 59-76.

Beecher, R. M. 1977. Function and fusion of the mandibular symphysis. *Am. J. Phys. Anthrop.* 47: 325-336.

— 1979. Functional significance of the mandibular symphysis. *J. Morphol.* 159: 117-130.

Behe, M. J. 1998. *Darwin's black box: the biochemical challenge to evolution*. New York: Simon & Schuster.

Békésy, G. Von. 1932. Zur Theorie des Hörens bei der Schallaufnahme durch Knochenleitung. *Ann. Physik* 13: 111-136.

— 1941. Über das Hören der eigenen Stimme. *Arch. ges. Phonetik* Abt. 2.5: 117-140.

Bennett, C. H. 1990. Undecidable dynamics. *Nature* 356: 606-607.

Benton, M. J. 1990. *The reign of the reptiles*. New York: Crescent Books.

— 1991. *The rise of mammals*. New York: Crescent Books.

Bernhardt, D. B. 1988. Prenatal and postnatal growth and development of the foot and ankle. *Phys. Ther.* 68: 1831-1839.

Bernstein, H. 1923. Über das Stimmorgan der Primaten. *Abhandl. Senckenberg. Naturf. Ges. Frankfurt* 38: 105-128.

Betticher, D. C., and J. Geiser. 1989. Resistance of mammalian red blood cells of different size to hypertonic milieu. *Comp. Biochem. Physiol.* 93A: 429-432.

Bishop, M. J., and A. E. Friday. 1987. Tetrapod relationships: The molecular evidence. In C. Patterson, ed., *Molecules and morphology in evolution: Conflict or compromise?* Cambridge; New York: Cambridge University Press.

Blackburn, D. G. 1984. From whale toes to snake eyes: Comments on the reversibility of evolution. *Syst. Zool.* 33: 241-245.

Blechschmidt, E. 1961. *Die vorgeburtlichen Entwicklungsstadien des Menschen.* Basel: Karger.

Boas, J. E. V. 1884. Ein Beitrag zur Morphologie der Nägel, Krallen, Hufe und Klauen der Säugetiere. *Morph. Jb.* 9: 389-400 + Fig.

— 1894. Zur Morphologie der Wirbeltierkralle. *Morph. Jb.* 21: 281-310 + Fig.

Bogin, B. 1988. *Patterns of human growth.* Cambridge; New York: Cambridge University Press.

Bohr, N. 1933. Light and life. *Nature* 128: 421-423 & 457-459.

— 1937. Kausalität und Komplementarität. *Erkenntnis* 6: 293-303.

— 1949. Discussions with Einstein on epistemological problems in atomic physics. In Schlipp, P.A., ed. *Albert Einstein: Philosopher-scientist.* LaSalle, IL: Open Court.

— 1952. Medical research and natural philosophy. *Acta Medica Scandinavica* 142: 67-972.

Bok, S. T. 1924. Een functie-uitstellend beginsel in de embryonale ontwikkeling. *Nederl. Tijdschr. Geneesk.* 68.1: 210-225.

— 1926. Assimilatie tegenover additie van vormeigenschappen. *Nederl. Tijdschr. Geneesk.* 70.1: 850-858 and 1045-1052.

Böker, H. 1935. *Vergleichende biologische Anatomie der Wirbeltiere.* Jena: G. Fischer.

Bolk, L. 1902. Über die Persistenz fötaler Formerscheinungen bei einem erwachsenen Manne. *Morph. Jb.* 29: 79-84.

— 1912. Über die Obliteration der Nähte am Affenschädel, zugleich ein Beitrag zur Kenntnis der Nahtanomalien. *Nederl. Tijdschr. Geneesk.*15: 1-206.

— 1915a. Über Lagerung, Verschiebung und Neigung des Foramen Magnum am Schädel der Primaten. *Z. Morph. Anthrop.* 17: 611-692.

— 1915b. Over het voortijdige sluiten der pijlnaad in verband met scaphocephalie. *Geneeskundige Bladen* 18: 27-55.

— 1917a. Über das kaudale Rumpfende eines Fetus vom Schimpansen. *Anat. Anz.* 50: 354-358.

— 1917b. On metopism. *Am. J. Anat.* 22: 27-47.

— 1918. *Hersenen en Cultuur*. Amsterdam: Scheltema & Holkema.

— 1920. Over de grondvorm van de menselijke maag en over megacolon (ziekte van Hirschsprung). *Nederl. Tijdschr. Geneesk.* 64,II: 1073-1080.

— 1921. The part played by the endocrine glands in the evolution of man. *Lancet*, 10 Sept.: 588-592.

— 1922. Aangeboren afwijkingen beschouwd in het licht der foetalisatie-theorie. *Nederl. Tijdschr. Geneesk*. 66, II: 1536-1543.

— 1923a. The problem of orthognatism. *Proc. Kon. Nederl. Acad. Wet. A'dam* 25: 71-380.

— 1923b. On the significance of the supra-orbital ridges in the primates. *Proc. Kon. Nederl. Acad. Wet. A'dam* 25: 16-20.

— 1924. The chin problem. *Proc. Kon. Nederl. Acad. Wet. A'dam* 27: 329-344.

— 1926a. *Das Problem der Menschwerdung.* Jena: G. Fischer.

— 1926b. Vergleichende Untersuchungen an einem Fetus eines Gorillas und eines Schimpansen. *Zeitschrift für Anatomie und Entwicklungsgeschichte* 81: 1-89.

— 1926c. Over de oorzaak en de betekenis van het niet sluiten der schedel-naden bij de mens. *Nederl. Tijdschr. Geneesk*. 70, II: 2328-2339.

— 1926d. Le problème de l'anthropogenèse. *Bull. Assoc. Anat. Comp. Rend.* 21: 80-92.

— 1927. De biologische grondslag der menswording. *Nederl. Tijdschr. Geneesk*. 71, II: 2216-2225.

— 1929. Origin of racial characteristics in man. *Am. J. Phys. Anthrop.* 13: 1-28.

Bolk, L., E. Göppert, E. Kallius, and W. Lubosh. 1967. *Handbuch der vergleichenden Anatomie der Wirbeltiere*. Amsterdam: Asher. Reprint of the 1931-39 editions (Berlin; Vienna).

Born, M. 1955. Statistical interpretation of quantum mechanics. *Science* 122: 675-679.

— 1969. *Physics in my generation* (also includes Born, 1955). Berlin: Springer.

Bounos, G., P. A. L. Kongshavn, A. Taveroff, and P. Gold. 1988. Evolutionary traits in human milk proteins. *Med. Hypoth*. 27: 133-140.

Bramble, D. M., and D. R. Carrier. 1983. Running and breathing in mammals. *Science* 219: 251-256.

Braune, W., and O. Fischer. 1890. Über den Schwerpunkt des menschlichen Körpers mit Rücksicht auf die Ausrüstung des deutschen Infanteristen. *Abhandl. math.-phys. Kl. königl. sächs. Ges. Wiss*. 15: 561-672 + Fig.

Braus, H. 1929. *Anatomie des Menschen. Band 1*. Berlin: Springer.

Britten, R. J. 1986. Rates of DNA sequence evolution differ between taxonomic groups. *Science* 231: 1393-1398.

Bromage, T. 1987. The biological and chronological maturation of early hominids. *J. Hum. Evol.* 16: 257-272.

— 1992. Faces from the past. *New Scientist* 12: 38-41.

Bromage, T., and M. C. Dean. 1985. Reevaluation of the age at death of immature fossil hominids. *Nature* 317: 525-527.

Brown, F. M. 1954. Persistent activity rhythms in the oyster. *Am. J. Phys.* 178: 510-514.

— 1988. Common 30-day multiple in gestation time of terrestrial placentals. *Chronobiol. Int.* 5: 195-210.

Brown, F. M., and Y. H. Park. 1975. A persistent monthly variation in responses of planarians to light, and its annual modulation. *Int. J. Chronobiol.* 3: 57-62.

Brown, W. M., E. M. Prager, A. Wang, and A. C. Wilson. 1982. Mitochondrial DNA sequences of primates: Tempo and mode of evolution. *Journal of Molecular Evolution* 18: 225-239.

Bruce, J. A. 1941. Time and order of appearance of ossification centers and their development in the skull of the rabbit. *Am. J. Anat.* 68: 41-67.

Bruhns, F. 1910. Der Nagel der Halbaffen und Affen. *Gegenbaurs Morph. Jb.* 40: 501-609.

Buizer, D. A. G. 1983. *De Nederlandse zakpijpen (manteldieren) en mantelvisjes*. Hoogwoud: Natuurhistorische Koninklijke Nederlandse Vereniging.

Burdi, A. R 1968. Morphogenesis of mandibular dental arch shape in human embryos. *J. Dent. Res.* 47: 50-58.

Burry, P. H. 1982. Postnatal growth and maturation of the lung. *Thorax* 37: 561-563.

Butler, G. E., M. Mckie, S. G. Ratcliffe. 1990. The cyclical nature of prepubertal growth. *Ann. Hum. Biol.* 17: 177-198.

Campana, S. E. 1984. Lunar cycles of otolith growth in the juvenile starry flounder. *Mar. Biol.* 80: 239-246.

Campbell, B. G. 1974. *Human evolution.* Chicago: Aldine.

Carey, S. W. 1996. *Earth, universe, cosmos.* Hobart: University of Tasmania Press.

— 1988. *Theories of the Earth and Universe: A History of Dogma in the Earth Sciences.* Stanford CA: Stanford University Press.

Carlsöö, S. 1972. *How man moves.* London: Heinemann.

Caro, T. M 1987. Human breasts: Unsupported hypotheses reviewed. *Hum. Evol.* 2: 271-282.

Carrier, D. R. 1984. The energetic paradox of human running and hominid evolution. *Curr. Anthrop.* 25: 483-495.

Carter, J. 1989. Grouper sex in Belize. *Nat. Hist.* 1989: 61-68.

Chaline, J., D. Marchand, and C. Berge. 1986. L'evolution de l'homme: Un modèle gradualiste ou ponctualiste? *Bull. Soc. Roy. Belge. Anthrop. Préhist.* 97: 77-97.

Charig, A. 1985. *A new look at the dinosaurs.* London: British Museum.

Christiansen, K., And E.-M. Winkler. 1990. Sexualhormonspiegel und Körperform bei den Kung San: Ein Beitrag zur Erklärung der Pädomorphie sanider Populationen. *Anthrop. Anz.* 48: 267-277.

Ciochon, R. L., and J.G. Fleagle, eds. 1985. *Primate evolution and human origins.* Menlo Park, CA: Benjamin/Cummings.

Clark, W. E. (Le Gros), 1936. The problem of the claw in primates. *Proc. Zool. Soc. London* 1936: 1-24.

— 1965. *History of the primates.* Chicago: University of Chicago Press.

— 1967. *Man-apes or ape-men? The story of discoveries in Africa.* New York: Holt, Rinehart & Winston.

Clauser, G. 1971. *Die vorgeburtliche Entstehung der Sprache als anthropologisches Problem.* Stuttgart: F. Enke.

Cleland, J. 1887. Terminal forms of life. *J. Anat. Physiol.* 18: 355-362.

Coates, M. I., and J. A. Clack. 1990. Polydactyly in the earliest known tetrapod limbs. *Nature* 347: 66-69.

Compton, A. H. 1935. *The freedom of man.* New York: Greenwood.

Congdon, E. D. 1902. Transformation of the aortic-arch system during the development of the human embryo. *Contrib. Embryol.* 14: 49-109.

Conroy, G. C., G. W. Weber, H. Seidler, P. V. Tobias, A. Kane, and B. Brunsden. 1998. Endocranial capacity in an early hominid cranium from Sterkfontein, South Africa. *Science* 280: 1730-1731.

Corruccini, R., and R. L. Ciochon. 1976. Morphometric affinities of the human shoulder. *Am. J. Phys. Anthrop.* 45: 19-38.

Cutler, W. B. 1980. Lunar and menstrual phase locking. *Am. J. Obstet. Gynec.* 137: 834-839.

Cutler, W. B., W. M. Schleidt, E. Friedmann, G. preti, and R. Stine. 1987. Lunar influences on the reproductive cycle in woman. *Hum. Biol.* 59: 959-972.

Dabelow, A. 1928. Über Art und Ursachen der Entstehung des Kiefergelenkes der Säugetiere. *Gegenbaurs Morph. Jb.* 59: 493-560.

Dahl, J. F. 1985. The external genitalia of female pygmy chimpanzees. *Anat. Rec.* 211: 24-28.

Dahlberg, A. A. 1962. On the teeth of early sapiens. In G. Kurth, ed. *Evolution and Hominization.* Stuttgart: G. Fischer.

Dambricourt-Malassé, A. 1988. Hominisation et foetalisation (Bolk, 1926). *CR Acad. Sc. Paris* 307.2: 199-204.

Danforth, C. H. 1921. Distribution of hair on the digits in man. *Am. J. Phys. Anthrop.* 4: 189-204.

Daniels, D. 1983. The evolution of concealed ovulation and self-deception. *Eth. Sociob.* (now called: *Evolution and Human Behavior*) 4: 69-87.

Dao, A. H. and M.G. Netsky. 1984. Human tails and pseudotails. *Hum. Pathol.* 15: 449-453.

Darwin, C. 1859. *On the origin of species by means of natural selection.* London: John Murray.

— 1862. *On the various contrivances by which British and foreign orchids are fertilised by insects.* London: John Murray.

— 1871. *The descent of man*. London: John Murray.

Davenport, C. B. 1934. The thoracic index. *Hum. Biol.* 6: 1-23.

— 1935. Variation in proportion among mammals, with special reference to man. *J. Mammal.* 16: 291-296.

Davies, P. 1990. Chaos frees the universe. *New Scientist* 10: 48-51.

Delmas, A., and H. Pineau. 1984. Le poids des vertèbres présacrées; Essai sur la signification des poids absolus et relatifs chez les mammifères. *Cahiers d'Anthropologie et Biométrie Humaine* 2.3: 1-122.

Dennett, D. C. 1991. *Consciousness explained*. New York: Little, Brown.

Dembski, W. 1998. *Mere creation: Faith, science, and intelligent design*. Downers Grove, IL: Intervarsity Press.

Devillers, C., J. Chaline, and B. Laurin. 1990. Plaidoyer pour une embryologie évolutive. *La Recherche* 21: 802-807.

Dollo, L. 1893. Les lois de l'évolution. *Bull. Soc. Belg. Géol.* 7: 164-166.

— 1905. Les dinosauriens adaptés à la vie quadrupède secondaire. *Bull. Soc. Belg. Géol.* 19: 441-448.

Dörig, G. K. 1969. The incidence of anovular cycles in women. *J. Reprod. Fert. Suppl.* 6: 77-81.

Driessens, F. C. M., R. M. H. Verbeeck, and J. W. E. Van Dijk. 1989. Plasma calcium difference between man and vertebrates. *Comp. Biochem. Physiol.* 93: 651-654.

Dubrul, E. L. 1958. *Evolution of the speech apparatus*. Springfield, IL: Thomas.

— 1976. Biomechanics of speech sounds. *Ann. NY Acad. Sc.* 280: 631-642.

— 1977. Origin of the speech apparatus and its reconstruction in fossils. *Brain Lang.* 4: 365-381.

Duchesne-Guillemin, J. 1986. Origines de la nomenclature astrale. *Ciel et Terre* 102: 139-148.

Dullemeijer, P. 1975. Bolk's Foetalization theory. *Acta Morphol. Neerl.- Scand.* 13: 77-86.

Dunnill, M. S. 1982. The problem of lung growth. *Thorax* 37: 561-563.

Eaton, C. F. 1910. Osteology of the pteranodon. *Memoirs of the Connecticut Academy of Arts and Sciences* 2: 1-38.

Eccles, J. C. 1976. Brain and free will. In G. Globus, G. Maxwell, and I. Savodnik, eds. *Consciousness and the brain: A scientific and philosophical inquiry*. New York: Plenum.

Eggeling, H. Von. 1920. Inwieweit ist der Wurmfortsatz am menschlichen Blinddarm ein rudimentäres Gebilde? *Anat. Anz.* 53: 401-428.

Elftman, H., and J. Manter. 1935. Chimpanzee and human feet in bipedal walking. *Am. J. Phys. Anthrop.* 20: 69-79.

Ellison, P. T. 1981. Morbidity, mortality, and menarche. *Hum. Biol.* 53: 635-643.

Engel, S. 1959. The lobulation of the mammalian lung. *Anat. Anz.* 106: 86-89.

England, M. A. 1983, 1990. *A colour atlas of life before birth: Normal fetal development.* London: Wolfe Med. Pub.

Evans, F. G., and V.E. Krahl. 1945. The torsion of the humerus: A phylogenetic study from fish to man. *Am. J. Anat.* 76: 303-337.

Falk, D. 1998. Hominid brain evolution: Looks can be deceiving. *Science* 280: 1714.

Falkner, F., and J. M. Tanner. 1986. *Human Growth: A comprehensive treatise.* New York: Plenum.

Farbridge, K. J., and J. F. Leatherland. 1987. Lunar cycles of coho salmon. *J. Exp. Biol.* 129: 165-189.

Ferro-luzzi, G. (Eichinger). 1991. On lactose tolerance and the selective paradigm. *Curr. Anthrop*. 32: 447-448.

Finlay, B. L., and R. B. Darlington. 1995. Linked regularities in the development and evolution of mammalian brains. *Science* 268: 1578-1584.

Fisher-Homberger, E. 1977. Hebamen und Hymen. *Sudhoffs Archiv* 61: 75-94.

Fleagle, J. G. 1988. *Primate adaptation and evolution.* San Diego: Academic Press.

Flügel, C., and J. W. Rohen. 1991. The craniofacial proportions and laryngeal position in monkeys and man of different ages: A morphometric study based on CT-scans and radiographs. *Mech. Ageing Dev.* 61: 65-83.

Flügel, C., K. Schram, and J. W. Rohen. 1983. Postnatal development of skull base, neuro- and viscerocranium in man and monkey. *Acta Anat.* 146: 71-80.

Ford, J. 1983. How random is a coin toss? *Phys. Today* 4: 40-47.

Frechkop, S. 1948. De la formation sporadique d'un menton de type humain chez le siamang. *Bull. Inst. Roy. Sc. Nat. Belge* 24: 1-8.

— 1949. Le crâne de l'homme en tant que crâne de mammifère. *Bull. Inst. Roy. Sc. Nat. Belge* 25: 1-12.

— 1954. Le port de la tête et la forme du crâne chez les singes. *Bull. Inst. Roy. Sc. Nat. Belge* 30: 1-11.

Frick, H. 1956. Morphologie des Herzens. In J. G. Helmcke, H. Von Lengerken, and D. Starck, eds. *Handbuch der Zoologie VIII, 5. Teil.* Berlin: De Gruyter.

— 1960. Das Herz der Primaten. In H. Hofer, A. H. S. Schultz, and D. Starck, eds. *Primatologia* (III/2). Basel: Karger.

Friedmann, E. 1981. Menstrual and lunar cycles. *Am. J. Obstet. Gynec.* 140: 350.

Fuiman, L.A. 1983. Growth gradients in fish larvae. *J. Fish Biol.* 23: 117-123.

Fuss, F. K. 1990. Kinematik und Dynamik des Kiefergelenks und der Bewegungen der Mandibula. *Gegenbaurs Morph. Jb.* 136: 37-68.

Gallagher, K.T. 1989. Natural selection: A tautology? *Int. Philos. Quart.* 29: 17-31.

Gans, A. 1923. Alte Hinweise auf die Menschenähnlichkeit junger Affen. *Anat. Anz.* 62: 553-555.

Gardner, M. 1981. *Science: Good, bad and bogus.* Buffalo, N. Y.: Prometheus.

Garstang, W. 1928. The morphology of the tunicata, and its bearing on the phylogeny of the chordata. *Quart. J. Microsc. Sc.* 72: 51-187.

Gasser, T., A. Kneip, A. Binding, A. Prader, and L. Molinari. 1991. The dynamics of linear growth in distance, velocity and acceleration. *Ann. Hum. Biol.* 18: 187-205.

Gasser, T., A. Kneip, P. Ziegler, R. Largo, L. Molinari, and A. Prader. 1991. The dynamics of growth of width in distance, velocity and acceleration. *Ann. Hum. Biol.* 18: 449-461.

Gautherie M., And C. Gros. 1977. Circadian rhythm alternation of skin temperature in breast cancer. *Chronobiologia* 4: 1.

Gee, H. 1989. Four legs to stand on. *Nature* 342: 738-739.

— 1996. *Before the backbone: Views on the origin of the vertebrates.* London; New York: Chapman & Hall.

Gehlen, A. 1988. *Man: His nature and place in the world.* New York: Columbia University Press.

Genet-Varcin, E. 1974. Platyrrhine contributions to the phylogeny of the primates. *J. Hum. Evol.* 3: 259-263.

Gilson, E. 1971. *D'Aristote à Darwin et retour.* Paris: Vrin.

Goerttler, K. 1950. Die Anordnung, Histologie und Histogenese der quergestreiften Muskulatur im menschlichen Stimmband. *Zeitschrift für Anatomie und Entwicklungsgeschichte* 115: 352-401.

— 1954. Die Entwicklung der menschlichen Glottis als deszendenztheoretisches Problem. *Homo* 5: 104-108.

Goethe, J. W. Von. 1786. Dem Menschen wie den Tieren ist ein Zwischenknochen der obern Kinnlade zuzuschreiben. In D. Kuhn, ed. 1966. *Goethes Werke, Bd. 13.* Hamburg: Wegner Verlag.

— 1795. *Erster Entwurf einer allgemeinen Einleitung in die vergleichende Anatomie, ausgehend von der Osteologie.* In H. J. Becker, H. Müller, and J. Neubauer, eds. 1989. *J. W. Goethe: Sämtliche Werke nach Epochen seines Schaffens, Bd. 12.* Munich: Hanser.

— 1796. *Vorträge über die drei ersten Kapitel des Entwurfs einer allgemeinen Einleitung in die vergleichende Anatomie, ausgehend von der Osteologie.* In H. J. Becker, H. Müller, and J. Neubauer, eds. 1989. *J. W. Goethe: Sämtliche Werke nach Epochen seines Schaffens, Bd. 12.* Munich: Hanser.

Golani, I. 1992. A mobility gradient in the organization of vertebrate movement: The perception of movement through symbolic language. *Behav. Brain Sc.* 15: 249-308.

Goldberger, A. L., B. J. West, T. Dresselhaus, and V. Bhargava. 1985. Bronchial asymmetry and Fibonacci scaling. *Experientia* 41: 1537-1538.

Golub, D. M. 1931. Die Entwicklung des Aortenbogens und der von ihm ausgehenden Äste bei Säugetieren: 1. Entwicklung der Kiemenbogen bei Kalb und Menschen. *Zeitschrift für Anatomie und Entwicklungsgeschichte* 95: 447-472.

Goodman, J. R., R. H. Wolf, And J. A. Roberts. 1977. The unique kidney of the spider monkey (*Ateles geoffroyi*). *J. Med. Primatol.* 6: 232-236.

Goodman, M. 1982. Biomolecular evidence on human origins from the standpoint of Darwinian theory. *Hum. Biol.* 54: 247-264.

— 1985. Rates of molecular evolution: The humanoid slowdown. *Biol. Essays* 3: 9-14.

— 1996. Epilogue: A personal account of the origins of a new paradigm. *Mol. Phylogen. Evol.* 5: 269-285.

Goodman, M., G. Braunitzer, A. Stangl, and B. Schrank. 1983. Evidence on human origins from haemoglobins of African apes. *Nature* 303: 546-548.

Goodman, M., D. A. Tagle, D. H. Fitch, W. Bailey, J. Czelusniak, B.F. Koop, P. Benson, and J. L. Sliztom. 1990. Primate evolution at the DNA level and a classification of hominoids. *J. Mol. Evol.* 30: 260-266.

Goodwin, B. C., N. Holder and C. C. Wylie. 1983. *Development and evolution.* Cambridge; New York: Cambridge University Press.

Gould, S. J. 1970. Dollo on Dollo's law: Irreversibility and the status of evolutionary laws. *J. Hist. Biol.* 3: 189-212.

— 1977a. *Ontogeny and phylogeny.* Cambridge, MA: Harvard University Press.

— 1977b. *Ever since Darwin.* New York: Norton.

— 1980. *The panda's thumb.* New York: Norton.

— 1983. *Hen's teeth and horse's toes.* New York: Norton.

— 1985. *The flamingo's smile.* New York: Norton.

— 1987. *An urchin in the storm.* New York: Norton.

— 1989. Full of hot air. *Natural History* 10: 28-35.

— 1990. Eight (or fewer) little piggies. *Natural History* 1: 22-29.

Grant, P. G., and C. J. Hoff. 1975. The skin of primates: Numerical taxonomy of primate skin. *Am. J. Phys. Anthrop.* 42: 151-166.

Grau, E. G., W. N. Dickhoff, R. S. Nishioka, H. A. Bern, and L. C. Folmar. 1981. Lunar phasing of the thyroxine surge preparatory to seaward migration of salmonid fish. *Science* 211: 607-609.

Greenfield, L. O. 1992. Origin of the human canine: A new solution to an old enigma. *Yearbook of Physical Anthropology* 35: 153-185.

Grehan, J. R., and R. Ainsworth. 1985. Orthogenesis and evolution. *Systematic Zoology* 34: 174-192.

Grigson, C. 1990. Missing links in the Piltdown fraud. *New Scientist* 125: 55-58.

Groves, C. P. 1986. Systematics of the great apes. In: D. R. Swindler and J. Erwin, eds. *Comparative Primate Biology, Volume 1.* New York: A. R. Liss.

Grzimek, B., ed. 1972-75. *Grzimek's animal life encyclopedia.* New York: Van Nostrand Reinhold.

Gutt, R. W. 1975. Menschlicher Scharfsinn, anatomische Forschung und Sittengeschichte. *Med. Klin.* 70: 1150-1152.

Haeckel, E. 1866. *Generelle Morphologie der Organismen.* Berlin: G. Reimer.

Haegerman, S. F., and R. A. Schenck. 1989. Evolution of lactase persistence. *Lancet* 3: 493.

Hagen, A. 1901. *Die sexuelle Osphresiologie.* Berlin: H. Barsdorf.

Hajnis, K., J. Bruzek, and V. Blazek. 1985. Wachstum des Rumpfes und seine Proportionsänderungen während des Kindesalters. *Anthropologie (Brünn)* 23: 25-47.

Hall, B. K. 1984. Developmental mechanisms underlying the formation of atavisms. *Biol. Rev.* 59: 89-124.

Harris, E. F., K. Aksharanugraha, and R. G. Behrents. 1992. Metacarpophalangeal length changes in humans during adulthood: A longitudinal study. *Am. J. Phys. Anthrop.* 87: 263-275.

Harrison, G. G. 1975. Primary adult lactase defficiency: a problem in anthropological genetics. American Anthropologist 77: 812-835.

Hasegawa, M., H. Kishino, and T. Yano. 1985. Dating of the human-ape splitting by a molecular clock of mitochondrial DNA. *J. Mol. Evol.* 22: 160-174.

— 1987. Man's place in Hominoidea as inferred from molecular clocks of DNA. *J. Mol. Evol.* 26: 132-147.

— 1989. Estimation of branching dates among primates by molecular clocks of nuclear DNA which slowed down in Hominoidea. *J. Hum. Evol.* 18: 461-476.

Hatschek, R. 1908. Beitrag zur Frage der Menschenähnlichkeit des Ateles-Gehirns. *Anat. Anz.* 32: 389-394.

Hawkes, O. A. M. 1914. On the relative length of the first and second toes of the human foot, from the point of view of occurrence, anatomy and heredity. *J. Genet.* 3: 249-274.

Hayek, H. Von. 1939. Die Läppchen und Septa interlobularia der menschlichen Lunge. *Zeitschrift für Anatomie und Entwicklungsgeschichte* 110: 405-411.

— 1956. Die Lunge. *Kükentals Hdb. Zool.* 7.2 (August 1956): 1-24.

— 1960. Die Lunge und Pleura der Primaten. In: H. Hofer, A. H. Schultz, and D. Starck, eds. *Primatologia* (III/2). Basel: Karger.

— 1970. *Die menschliche Lunge.* Berlin: Springer.

Heads, M. 1985. On the nature of ancestors. *Systematic Zoology* 34: 205-215.

Heberer, G. 1959. *Evolution der Organismen.* Stuttgart: G. Fischer.

Heine, H. 1979. Zur Evolution der Beziehungen zwischen Gehirnentfaltung, sekundärem Kiefergelenk und Gesichtsentwicklung bei Säugetieren. *Gegenbaurs Morph. Jb.* 125: 49-53.

— 1983. Gehirnentfaltung, sekundäres Kiefergelenk und Gesichtsentwicklung der Säugetiere: Ein synergetisches und synchronistisches Ereignis. *Gegenbaurs Morph. Jb.* 129: 699-706.

Heisenberg, W. 1927. Über den anschaulichen Inhalt der quantentheoretischen Kinematik und Mechanik. *Zeitschrift für Physik* 43: 172-198.

Helmuth, H. 1985. Biomechanics, evolution and upright stature. *Anthrop. Anz.* 43: 1-9.

Heuvelmans, B., and B. Porchnev. 1974. *L'homme de Néanderthal est toujours vivant.* Paris: Plon.

Hildebrandt, G. 1977. Hygiogenese: Grundlinien einer therapeutischen Physiologie. *Therapiewoche* 27: 5384-5397.

Hill, W. C. (Osman). 1946. Note on the male external genitalia of the chimpanzee. *Proc. Zool. Soc. London* 116: 129-132.

— 1949. Some points in the enteric anatomy of the great apes. *Proc. Zool. Soc. London* 119: 19-32.

— 1958. External genitalia. In H. Hofer, A. H. Schultz, and D. Starck, eds. *Primatologia* (III/1). Basel: Karger.

Hill, W. C. (Osman), and L. Harrison-Matthews. 1949. The male genitalia of the gorilla. *Proc. Zool. Soc. London* 119: 363-377.

— 1950. Supplementary note on the male external genitalia of gorilla. *Proc. Zool. Soc. London* 120: 311-316.

— 1954-55. Ontogenetic changes in the mesogastric viscera of the Colobidae. *Proc. Zool. Soc. London* 124: 163-183.

Hillson, S. W. 1986. *Teeth.* Cambridge; New York: Cambridge University Press.

Hinrichsen, K. 1959. Morphologische Untersuchungen zur Topogenese der mandibularen Nagezähne der Maus. *Anat. Anz.* 107: 59.

— 1990. *Humanembryologie: Lehrbuch und Atlas der vorgeburtlichen Entwicklung des Menschen.* Berlin: Springer.

Hinton, R. J., and D. S. Carlson. 1983. Histological changes in the articular eminence and mandibular fossa during growth of the rhesus monkey. *Am. J. Anat.* 166: 99-116.

Hislop, A., and L. Reid. 1974. Growth and development of the respiratory system: Anatomical development. In J. A. Davis and J. Dobbing, eds. *Scientific Foundations of Pediatrics*. Philadelphia: Saunders.

Hockett, C. F. 1967. The foundations of language in man, the small mouthed animal. *Scient. Am.* 11: 141-144.

Hodgson, D. 1991. *The mind matters: Consciousness and choice in a quantum world.* Oxford: Clarendon Press.

Hofman, M. A. 1985. Size and shape of the cerebral cortex in mammals: The cortical surface. *Brain, Beh. Evol.* 27: 28-40.

Holder, N. 1983. The vertebrate limb: Patterns and constraints in development and evolution. In Goodwin et al., eds. *Development and evolution*. Cambridge; New York: Cambridge University Press.

Holmes, E. B. 1975. A reconsideration of the phylogeny of the tetrapod heart. *J. Morphol.* 147: 209-228.

Holmes, S. J. 1944. Recapitulation and its supposed causes. *Quart. Rev. Biol.* 18: 319-331.

Hsiao, S.-M., and A.H. Meier. 1989. Comparison of semilunar cycles of spawning activity in *Fundulus grandis* and *F. heteroclitus* held under constant laboratory conditions. *Quart. Rev. Biol.* 252: 213-218.

Huizinga, J., and A.D. De Vetten. 1967. Preliminary study of the foot of the Dogon. *Proc. Kon. Nederl. Acad. Wet. Series C* 70: 97-109.

Hulanicka, B., and K. Kotlarz. 1983. The final phase of growth in height. *Ann. Hum. Biol.* 10: 429-434.

Ihle, J. E. W., P. N. Van Kampen, H. F. Nierstrasz, and J. Versluys. 1924. *Leerboek der vergelijkende ontleedkunde van de Vertebraten.* Utrecht (Netherlands): Oosthoek.

Inman, V. T., J. B. Saunders, and L. C. Abbot. 1944. Observations on the function of the shoulder joint. *J. Bone Joint Surg.* 26: 1-30.

Inman, V. T., H. J. Ralston, and F. Todd. 1981. *Human walking.* London: Williams & Wilkins.

Izor, R. J., S. L. Walchuk, and L. Wilkins. 1981. Anatomy and systematic significance of the penis of the pygmy chimpanzee, Pan paniscus. *Folia Primatol.* 35: 218-224.

J. Jaekel. 1927. Die Atemorgane der Wirbeltiere in phylogenetischer Beleuchtung. *Zool. Anz.* 70: 273-303.

James, T. N. 1968. Anatomy of the sinus node, AV node and os cordis of the beef heart. *Anat. Rec.* 153: 361-372.

Jarvik, E. 1980. *Basic structure and evolution of vertebrates.* London, San Diego: Academic Press.

Jefferies, R. P. S. 1986. *Ancestry of the vertebrates.* Cambridge; New York: Cambridge University Press.

Jelley, J. V. 1986. The tides, their origins and behaviour. *Endeavour.* 10: 184-190.

Jenny, C., M. L. D. Kuhns, and F. Azakawa. 1987. Hymens in newborn female infants. *Pediatrics* 80: 399-400.

Jerison, H. J. 1979. Brain, body and encephalization in early primates. *J. Hum. Evol.* 8: 615-635.

Jolicoeur, P., G. Baron, and T. Cabana. 1988. Cross-sectional growth and decline of human stature and brain weight in nineteenth-century Germany. *Growth, Development & Aging* 52: 201-206.

Jones, H. G. 1947. The primary dentition in homo sapiens and the search for primitive features. *Am. J. Phys. Anthrop.* 5: 251-274.

Jones, F. W. 1929. *Man's place among the mammals.* London: Edward Arnold.

Josephson, B. D. 1988. Limits to the universality of quantum mechanics. *Foundations of Physics* 18: 1195-1204.

Jouffroy, F. K., M. Godinot, and Y. Nakano. 1993. Biometrical characteristics of primate hands. In H. Preuschoft and D.J. Chivers, eds. *Hands of Primates.* Berlin: Springer.

Kahle, W., H. Leonhardt, and W. Platzer. 1975. *Atlas der Anatomie.* Stuttgart: Thieme.

Keith, A. 1895. The modes of origin of the carotid and subclavian arteries from the arch of the aorta in some of the higher primates. *J. Anat. Physiol.* 29: 453-458.

— 1923. Man's posture: Its evolution and disorders. *Brit. Med. J.* 3: 451-454, 499-502, 545-548, 624-626, 669-672.

— 1949. *A new theory of human evolution.* London: Watts.

Keith, D. A. 1982. Development of the human temporomandibular joint. *Brit. J. Oral Surg.* 20: 217-224.

Kingsbury, B. F. 1924. The significance of the so-called law of cephalocaudal differential growth. *Anat. Rec.* 27: 305-321.

Kinzey, W. G. 1970. Basic rectangle of the mandible. *Nature* 228: 289-290.

— 1971. Evolution of the human canine tooth. *Am. Anthrop.* 73: 680-694.

Kipp, F. A. 1955. Die Entstehung der menschlichen Lautbildungsfähigkeit als Evolutionsproblem. *Experientia* 11: 89-128.

— 1980. *Die Evolution des Menschen im Hinblick auf seine lange Jugendzeit.* Stuttgart: Verlag Freies Geistesleben.

Kjaer, I., T.W. Kjaer, and N. Graem. 1993. Ossification sequence of occipital bone and vertebrae in human fetuses. *J. Craniofac. Gen. Dev. Biol.* 13: 83-88.

Klaatsch, H. 1920. *Der Werdegang der Menschheit und die Entstehung der Kultur.* Berlin: Bong.

Kleindienst, M. R. 1975. On new perspectives on ape and human evolution. *Curr. Anthrop.* 16: 644-646.

Knussmann, R. 1967. *Humerus, Ulna und Radius der Simiae.* Basel: Karger.

Koop, B. F., M. Goodman, P. Xu, K. Chan, And J.L. Slighton. 1986. Primate èta-globin DNA sequences and man's place among the great apes. *Nature* 319: 234ff.

Koop, B. F., D. A. Tagle, M. Goodman, and J.L. Slighton. 1989. A molecular view of primate phylogeny and important systematic and evolutionary questions. *Mol. Biol. Evol.* 6: 580-612.

Kortlandt, A. 1975. Reply (to Kleindienst). *Curr. Anthrop.* 16: 647-649.

Kortlandt, A. and M. Kooij. 1963. Protohominid behaviour in primates. *Symp. Zool. Soc. London* 10: 61-88.

Kranich, E. M. 1999. *Thinking beyond Darwin: The idea of the type as a key to vertebrate evolution.* Hudson, N. Y.: Lindisfarne Books.

— 1995. *Wesensbilder der Tiere: Einführung in die goetheanistische Zoologie.* Stuttgart: Freies Geistesleben.

Krantz, G. S. 1963. The functional significance of the mastoid process in man. *Am. J. Phys. Anthrop.* 21: 591-593.

— 1965. Evolution of the human hand and the great hand-axe tradition. *Kroeber Anthrop. Soc. Papers* 23: 114-129.

— 1988. Laryngeal descent in 40,000 year old fossils. In M. E. Landsberg, ed. *The genesis of language: A different judgement of evidence.* Berlin; New York: Mouton de Gruyter.

Krauss, L. M. 1986. Dark matter in the universe. *Scient. Am.* 12: 50-60.

Laitman, J. T. 1986. L'origine du language articulé. *La Recherche* 17: 1164-1173.

Landsmeer, J. M. F. 1986. A comparison of fingers and hand in Varanus, opossum and primates. *Acta Morphol. Neerl.-Scand.* 24: 193-221.

Lang, J., and M. Öder. 1984. Über die Biomorphose der Mandibula. *Gegenbaurs Morph. Jb.* 130: 185-234.

Langman, J. 1966. *Medical Embryology.* London: Williams & Wilkins.

Latour, J. M. 1958. *Het paard.* Brussels: Koninkl. Belg. Instituut voor Natuurwetenschappen.

Laurikainen, K. V. 1988. *Beyond the atom: The philosophical thought of Wolfgang Pauli.* Berlin: Springer.

Ledley, F. D. 1982. Evolution and the human tail. *New Engl. J. Med.* 306: 1212-1215 and 307: 1089-1090.

Lenneberg, E. H. 1967. *Biological foundations of language.* New York: Wiley.

Lessertisseur, J., and F. K. Jouffroy. 1975. Comparative osteometry of the foot of man and facultatively bipedal primates. In R.H. Tuttle, ed. *Primate functional morphology and evolution.* The Hague: Mouton.

Levi, F., and F. Halberg. 1982. Circaseptan bioperiodicity – spontaneous and reactive – and the search for pacemakers. *La Ricerca Clin. Lab.* 12: 323-370.

Lewis, F. T. 1923. A note on symmetry as a factor in the evolution of plants and animals. *Am. Natur.* 57: 5-41.

Li, W. H., and M. Tanimura. 1987. The molecular clock runs more slowly in man than in apes and monkeys. *Nature* 326: 93-96.

Lieberman, P. 1984. *The biology and evolution of language.* Cambridge, MA: Harvard University Press.

Lighthill, J. 1986. The recently recognized failure of predictability in Newtonian dynamics. *Proc. Roy. Soc. London Series A.* 407: 35-50.

Long, J. A. 1990. Heterochrony and the origin of tetrapods. *Lethaia* 23: 157-166.

Lovejoy, C. (Owen). 1988. The evolution of human walking. *Scient. Am.* 11: 82-89.

Løvtrup, S. 1978. On von Baerian and Haeckelian recapitulation. *Syst. Zool.* 27: 348-352.

Luboga, S. A., and B. A. Wood. 1990. Position and orientation of the foramen magnum in higher primates. *Am. J. Phys. Anthrop.* 81: 67-76.

Marcus, H. 1928. Vergleichende Untersuchungen über die respiratorische Oberfläche und ihr Verhältnis zum Körpergewicht. *Gegenbaurs Morph. Jb.* 59: 561-566.

Marsh, O. C. 1892. Recent polydactyl horses. *Am. J. Sc.* 43: 339-355.

Martin, R. D. 1990. *Primate origins and evolution: A phylogenetic reconstruction.* Princeton, N J: Princeton University Press.

Martin, R., and K. Saller. 1959. *Lehrbuch der Anthropologie, Band 2.* Stuttgart: G. Fischer.

Masters, A. V., D. Falk, and T. B. Gage. 1991. Effects of age and gender on the location and orientation of the foramen magnum in rhesus macaques (*Macaca mulatta*). *Am. J. Phys. Anthrop.* 86: 75-80.

May, E., and M. Martins. 1985. Zur Differenzierung menschlicher und nichtmenschlicher Primatenrippen. *Anat. Anz.* 160: 179-202.

Mayr, E. 1982. *The growth of biological thought.* Cambridge MA: Harvard University Press.

McCracken, R. D. 1971. Lactase deficiency: An example of dietary evolution. *Curr. Anthrop.* 12: 479-517.

McFarland, W. N., F. H. Pough, T. J. Cade, and J. B. Heiser. 1985. *Vertebrate Life.* New York: Macmillan.

McKinney, M. L., and K. J. McNamara. 1991. *Heterochrony: The evolution of ontogeny.* New York: Plenum Press.

McNamara, K. J. 1997. *Shapes of Time.* Baltimore, MD: John Hopkins University Press.

Meinertz, T. 1975. Weitere Untersuchungen über den Sinus coronarius cordis, die V. cordis media und den Arcus aortae sowie den Ductus (Lig.) Botalli bei einer Anzahl von Säugetierherzen. *Gegenbaurs Morph. Jb.* 121: 139-154.

Menaker, W., and A. Menaker. 1959. Lunar periodicity in human reproduction: A likely unit of biological time. *Am. J. Obst. Gynec.* 77: 905-913.

Michejda, M., and D. Lamey. 1971. Flexion and metric age changes of the cranial base in the Macaca mulatta: Infants and juveniles. *Folia Primatol.* 34: 133-141.

Mijsberg, W. A. 1922. Het ontstaan van de opgerichte gang in de voorgeschiedenis van de mens. *Nederl. Tijds. Geneesk.* 66: 34-41.

Miles, L. E. M., D. M. Raynal, and M. A. Wilson. 1977. Blind man living in normal society has circadian rhythms of 24.9 hours. *Science* 198: 421-423.

Milgrom, M. 1988. La modification des lois de Newton. *La Recherche* 19: 182-190.

Mills, J. R. E. 1978. The relationship between tooth patterns and jaw movements in the hominoidea. In: P. M. Butler and K. A. Joysey, eds. *Development, function and evolution of teeth.* London, New York: Academic Press.

Mivart, St. G. 1874. *Man and apes.* New York: D. Appleton.

Moerman, M. L. 1982. Growth of the birth canal in adolescent girls. *Am. J. Obstet. Gynec.* 143: 528-532.

Mohssine, E. H., M. Bounias, and J. M. Cornuet. 1990. Lunar phase influence on the glycemia of worker honeybees. *Chronobiologia* 17: 201-207.

Montagna, W. 1965. The skin. *Scient. Am.* 2: 56-66.

Montagna, W., H. Machida, and E. Perkins. 1966. The skin of primates: The stump-tail macaque. *Am. J. Phys. Anthrop.* 24: 71-86.

Montagu, A. 1931. On the primate thumb. *Am. J. Phys. Anthrop.* 15: 291-314.

— 1935. The premaxilla in the primates. *Quart. Rev. Biol.* 10: 32-59 and 181-208.

— 1937. The medio-frontal suture and the problem of metopism in primates. *J. Roy. Anthrop. Inst.* 67: 157-201.

— 1955. Time, morphology and neoteny in the evolution of man. *Am. Anthrop.* 57: 13-27.

— 1981. *Growing Young.* New York: McGraw-Hill.

Moore, K. L. 1982. *The developing human: Clinically oriented embryology.* Philadelphia: Saunders.

Morgan, E. 1982. *The aquatic ape.* New York: Stein & Day.

Mottershedd, S. 1988. Sesamoid bones and cartilages: An enquiry into their function. *Clin. Anat.* 1: 59-62.

Naef, A. 1926a. Über die Urformen der Anthropomorphen und die Stammesgeschichte des Menschenschädels. *Naturwiss.* 14: 445-452.

— 1926b. Zur Morphologie und Stammesgeschichte des Affenschädels. *Naturwiss.* 14: 89-97.

Napier, J. 1962. The evolution of the hand. *Scient. Am.* 12: 56-62.

— 1967. The antiquity of human walking. *Scient. Am.* 4: 56-66.

— 1993. *Hands* (revised by R.H. Tuttle). Princeton, NJ: Princeton University Press.

Nath, S., and M. Chacko. 1988. Maturational sequence in the arm and the leg segments among the Danig females of Udaipur, Rajasthan. *Anthropologie (Brünn)* 26: 7-11.

Negus, V. E. 1949. *The comparative anatomy and physiology of the larynx.* London: Heinemann.

Nelson, T. R., B. J. West, and A. L. Goldberger. 1990. The fractal lung: Universal and species-related scaling patterns. *Experientia* 46: 251-254.

Neuville, H. 1927. De certains caractères de la forme humaine et de leurs causes. *L'anthropologie* 37: 305-328 and 491-515.

O'Rahilly, R., F. Muller, and D. B. Meyer. 1980. The human vertebral column at the end of the embryonic period proper: 1. The column as a whole. *J. Anat.* 131: 565-575.

Ortmann, R. 1984. Über die Beziehung der Ellipse zur horizontalen Schädelkontur beim Menschen und einigen Primaten. *Gegenbaurs Morph. Jb.* 130: 111-123.

Owen, H. G. 1983. *Atlas of continental displacement: 200 million years to the present.* Cambridge: Cambridge University Press.

Oxnard, C. E. 1969. Evolution of the human shoulder: Some possible pathways. *Am. J. Phys. Anthrop.* 30: 319-332.

Pannella, G. 1971. Fish otoliths: Daily growth layers and periodical patterns. *Science* 173: 1124-1127.

Panzer, W. 1932. Beiträge zur biologischen Anatomie des Baumkletterns der Säugetiere: 1. Das Nagel-Kralle-Problem. *Zeitschrift für Anatomie und Entwicklungsgeschichte* 98: 147-198.

Parsons, F. G. 1902. The blood-vessels of mammals in relation to those of man. *Lancet* 3: 651-653.

Passingham, R. E. 1982. *The human primate.* Oxford; San Francisco: Freeman.

— 1985. Rates of brain development in mammals including man. *Brain and Behavioral Science* 26: 167-175.

Patten, B. M. 1968. *Human Embryology.* New York: McGraw-Hill.

Pavelka, M. S. M., and L. M. Fedigan. 1991. Menopause: A comparative life history perspective. *Yearbook of Physical Anthropology* 34: 13-38.

Payne, S. R., D. J. Deardon, G. F. Abercrombie, and G. L. Carlson. 1989. Urinary retention and the lunisolar cycle: Is it a lunatic phenomenon? *Brit. Med. J.* 299: 1560-1562.

Petersen, H. 1924. Über die Bedeutung der aufrechten Körperhaltung für die Eigenart des menschlichen Umweltbildes. *Naturwiss.* 10: 186-191.

Petrovits, L. 1930. Die Übereinstimmung des Kiefergelenkes des neugeborenen Kindes mit dem Kiefergelenk der Anthropoiden. *Anat. Anz.* 69: 136-144.

Phalen, R. P., and M. J. Oldham. 1983. Tracheobronchial airway structure as revealed by casting techniques. *Am. Rev. Resp. Dis.* 128: S1-S4.

Pinkus, H. 1958. Embryology of hair. In W. Montagna and R.A. Ellis, eds. *Biology of hair growth.* New York: Academic Press.

Pilbeam, D., and S. J. Gould. 1974. Size and scaling in human biology. *Science.* 186: 892-901.

Piperno, E. 1978. Morphogenesis and histogenesis of the cartilaginous tracheal rings in the domestic fowl (*Gallus domesticus*). *Anat. Anz.* 143: 167-175.

Poissonnet, C. M., A. R. Burdi, and S. M. Garn. 1984. The chronology of adipose tissue appearance and distribution in the human fetus. *Early Hum. Dev.* 10: 1-11.

Pöllmann, L. 1984. Wound healing: A study on circaseptan reactive periodicity. *Chronobiology Intern.* 1: 151-157.

Pond, C. M. 1987. Fat and figures. *New Scientist* 6: 62-66.

— 1991. Adipose tissue in human evolution. In: M. Roede, J. Wind, J. Patrick, and V. Reynolds, eds. *The aquatic ape: Fact or fiction?* London: Souvenir Press.

Poppelbaum, H. 1960. *Man and animal: their essential difference.* London: Anthroposophical Publishing Co.

Popper, K. 1976. *Unended Quest.* Glasgow: Fontana/Colins.

— 1983. Is determinism self-refuting? *Mind* 92: 103-104.

Popper, K., and J. C. Eccles. 1977. *The self and its brain.* New York: Springer International.

Portmann, A. 1967. *Animal forms and patterns: A study of the appearance of animals.* New York: Schocken.

— 1976. *Einführung in die vergleichende Morphologie der Wirbeltiere.* Basel: Schwabe.

Preuss, W.H. 1882. *Geist und Stoff.* Ed. R. Riemeck and W. Schad (1980). Stuttgart: Freies Geistesleben.

Preuschoft, H., S. Hayama, and M. M. Günther. 1988. Curvature of the lumbar spine as a consequence of mechanical necessities in Japanese macaques trained for bipedalism. *Folia Primatol.* 50: 42-58.

Rabkin, S. W., and F. A. L. Mathewson. 1980. Chronobiology of cardiac sudden death in men. *J.A.M.A.* 244: 1357-1358.

Rak, Y., and C. Howell. 1978. Cranium of a juvenile *Australopithecus boisei* from the lower Oma basin, Ethiopia. *Am. J. Phys. Anthrop.* 48: 345-366.

Reiche, C., and P. Schwartze. 1989. Makroskopisch-anatomische Veränderungen des Schädels, der Wirbelsäule, des Rückenmarks, der Extremitäten und des Schwanzes von postnatal wachsenden Albinoratten. *Gegenbaurs Morph. Jb.* 135: 479-490.

Reiter, A. 1944. Die Frühentwicklung der menschlichen Wirbelsäule: III. Die Entwicklung der Lumbal-, Sacral- und Coccygealwirbelsäule. *Zeitschrift für Anatomie und Entwicklungsgeschichte* 113: 204-227.

Richard, A. F. 1985. *Primates in nature.* New York: W.H. Freeman.

Richter, C. P. 1968. Periodic phenomena in man and animals: Their relation to neuroendocrine mechanisms. In R. P. Michael, ed. *Endocrinology and human behaviour.* London; New York: Oxford University Press.

Richter, J., and G. Kern. 1987. Secular changes in the development of children born in Görlitz, German Democratic Republic, 1956 to 1967. *Hum. Biol.* 59: 345-355.

Riesenfeld, A. 1969. The adaptive mandible: An experimental study. *Acta Anat.* 72: 246-262.

Robinson, J. T., L. Freedman, and B. A. Sigmon. 1972. Some aspects of pongid and hominid bipedality. *J. Hum. Evol.* 1: 361-369.

Rockel, A. J., R. W. Hiorns, and T. P. S. Powell. 1980. The basic uniformity in the structure of the neocortex. *Brain* 103: 221-224.

Romball, C. G., and W. C. Weigle. 1973. A cyclical appearance of antibody-producing cells after a single injection of serum protein antigen. *J. Exp. Med.* 138: 1426-1441.

Romer, A. S. 1962. *The vertebrate body.* Philadelphia: Saunders.

Romero-Herrera, A. E., H. Lehmann, O. Castillo, K. A. Josysey, and A. E. Friday. 1976. Myoglobin of the orangutan as a phylogenetic enigma. *Nature* 261: 162-164.

Roscher, W. E. 1906. Die Hebdomadenlehren der griechischen Philosophen und Ärzte. *Abhandl. phil.-hist. Kl. königl. sächs. Ges. Wiss. Leipzig.* 24: 1-239.

Rosenberg, K. R. 1992. The evolution of modern childbirth. *Yearbook of Physical Anthropology* 35: 89-124.

Roth, M., and J. Krkoska. 1978. The adaptive mandible: A product of the relative osteo-neural growth. *Gegenbaurs Morph. Jb.* 124: 765-783.

Ruge, G. 1893. Die Grenzlinien der Pleurasäcke bei Primaten. *Morph. Jb.* 19: 149-250.

Sanchez De La Pena, S., F. Halberg, H.-G. Schweiger, J. Eaton, and J. Sheppard. 1984. Circadian temperature rhythm and circadian-circaseptan aspects of murine death from malaria. *Proc. Soc. Exp. Biol. Med.* 175: 196-204.

Schad, W. 1992. *Der Heterochronie-Modus in der Evolution der Wirbeltierklassen und Hominiden.* Dissertation, Universität Witten/Herdecke.

— 1993. Heterochronical patterns of evolution in the transitional stages of vertebrate classes. *Acta Biotheor.* 41: 383-389.

Schillling, R. 1952. Die Umwandlung unserer stimmphysiologischen Vorstellungen aufgrund der Goerttlerschen anatomischen Untersuchungen am menschlichen Stimmband. *Zeitschrift für Phonetik und allgemeine Sprachwissenschaft* 6: 94-105.

Schindewolf, O. H. 1928. Das Problem der Menschwerdung: Ein paläontologischer Lösungsversuch. *Jahrbuch der Preußischen Geologischen Landesanstalt* 49: 716-766.

— 1972. Phylogenie und Anthropologie in paläontologischer Sicht. In: H. G. Gadamer and P. Vogler, eds. *Neue Anthropologie, Bd. 1.: Biologische Anthropologie, Teil 1.* Munich: Deutscher Taschenbuch-Verlag.

Schultz, A. 1922. Zygodactyly and its inheritance. *J. Hered.* 13: 113-117.

— 1924. Growth studies on primates bearing upon man's evolution. *Am. J. Phys. Anthrop.* 7: 149-164.

— 1925. Embryological evidence of the evolution of man. *J. Wash. Acad. Sc.* 15: 247-263.

— 1926a. Fetal growth of man and other primates. *Quart. Rev. Biol.* 1: 465-521.

— 1926b. Variations in man and their evolutionary significance. *Am. Natur.* 60: 297-323.

— 1926c. Studies on the variability of Platyrrhine monkeys. *J. Mammal* 7: 286-305.

— 1927. Studies on the growth of the gorilla and of other higher primates. *Memoirs of the Carnegie Museum* 11: 1-88.

— 1930. The skeleton of the trunk and limbs of higher primates. *Hum. Biol.* 2: 303-409.

— 1931. Man as a primate. *Scientific Monthly* 33: 385-412.

— 1933. Die Körperproportionen der erwachsenen catarrhinen Primaten, mit spezieller Berücksichtigung der Menschenaffen. *Anthrop. Anz.* 10: 154-185.

— 1934. Some distinguishing characters of the mountain gorilla. *J. Mammal.* 15: 51-61.

— 1936. Characters common to higher primates and characters specific for man. *Quart. Rev. Biol.* 11: 259-283 and 425-455.

— 1937. Fetal growth and development of the rhesus monkey. *Contr. Embryol.* 26: 73-97.

— 1938a. Genital swelling in the female orang-utan. *J. Mammal.* 19: 363-366.

— 1938b. The relative length of the regions of the spinal column in Old World primates. *Am. J. Phys. Anthrop.* 24: 1-22.

— 1940. Growth and development of the chimpanzee. *Contr. Embryol.* 28: 1-63.

— 1941. Growth and development of the orang-utan. *Contr. Embryol.* 29: 57-110.

— 1948. The relation in size between premaxilla, diastema and canine. *Am. J. Phys. Anthrop.* 6: 163-180.

— 1949a. Ontogenetic specializations in man. *Archiv Julius Klaus-Stiftung* 24: 197-216.

— 1949b. Sex differences in the pelvis of primates. *Am. J. Phys. Anthrop.* 7: 401-424.

— 1950a. The physical distinctions of man. *Proc. Am. Philos. Soc.* 94: 428-449.

— 1950b. Morphological observations on gorillas. In W. K. Gregory, ed. *The anatomy of the gorilla.* New York: Columbia University Press.

— 1951. The specializations of man and his place among the catarrhine primates. *Cold Spring Harbor Symposia on Quantitative Biology* 15: 37-53.

— 1952. Über das Wachstum der Warzenfortsätze beim Menschen und den Menschenaffen. *Homo* 3: 105-109.

— 1953. The relative thickness of the long bones and the vertebrae in primates. *Am. J. Phys. Anthrop.* 11: 277-311.

— 1955. The position of the occipital condyles and of the face relative to the skull base in primates. *Am. J. Phys. Anthrop.* 13: 97-120.

— 1956. Postembryonic changes. In H. Hofer, A. H. S. Schultz, and D. Starck, eds. *Primatologia* (I). Basel: Karger.

— 1961. Vertebral column and thorax. In H. Hofer, A. H. S. Schultz, and D. Starck, eds. *Primatologia* (IV/5). Basel: Karger.

— 1962. The relative weight of the skeletal parts in adult primates. *Am. J. Phys. Anthrop.* 20: 1-10.

— 1963. Relations between the length of the main parts of the foot skeleton in primates. *Folia Primatol.* 1: 150-171.

— 1968. The recent hominoid primates. In S. L. Washburn and P. C. Jay, eds. *Perspectives on human evolution 1.* New York: Holt, Rinehart & Winston.

— 1969a. *The life of primates.* London: Weidenfeld & Nicolson.

— 1969b. The skeleton of the chimpanzee. *The chimpanzee* 1: 50-103.

Schultz, J. 1986. *Movements and rhythms of the stars: A guide to naked-eye observation of sun, moon, and planets.* Spring Valley, NY: Anthroposophic Press.

Schwabe, C. 1994. Theoretical limitations of molecular phylogenetics and the evolution of relaxins. *Comparative Biochemistry and Physiology B.* 107: 167-177.

Schwartz, J. H. 1987. *The red ape: Orang-utans and human origins.* Boston: Houghton Mifflin.

Schweiger, H.-G., S. Berger, H. Kretschmer, H. Mörler, E. Halberg, R. B. Sothern, and F. Halberg. 1986. Evidence for a circaseptan and a circasemiseptan growth response to light-dark cycle shifts in nucleated and enucleated Acetabularia cells, respectively. *Proc. Nat. Acad. Sci. USA* 83: 8619-8623.

Scott, J. H. 1957. The shape of the dental arches. *J. Dent. Res.* 36: 996-1003.

— 1967. Dento-facial development and growth. *Pergamon Series on Dentistry* 6.

Seidl, W., and G. Steding. 1981. Contribution to the development of the heart, Part. III: The aortic arch complex. Normal development and morphogenesis of congenital malformation. *Thorac. Cardiovasc. Surgeon* 29: 359-368.

Seiler, R. 1974. Die Muskeln des außeren Ohres und ihre Funktion bei Menschen, Schimpansen und Makaken. *Gegenbaurs Morph. Jb.* 120: 78-122.

Shaner, R. F. 1962. Comparative development of the bulbus and ventricles of the vertebrate heart with special reference to Spitzer's theory of heart malformations. *Anat. Rec.* 142: 519-529.

Shea, B. T. 1986. Scapula form and locomotation in chimpanzee evolution. *Am. J. Phys. Anthrop.* 70: 475-488.

— 1989. Heterochrony in human evolution: The case for neoteny reconsidered. *Yearbook of Physical Anthropology* 32: 69-101.

Short, R. V. 1976. The evolution of human reproduction. *Proc. Roy. Soc. London Series B* 195: 3-24.

Shubin, N. H., and P. Alberch. 1986. A morphogenetic approach to the origin and basic organization of the tetrapod limb. *Evol. Biol.* 20: 319-387.

Simon, G., L. Reid, J. M. Tanner, H. Goldstein, and B. Benjamin. 1972. Growth of radiologically determined heart diameter, lung width and lung length from 5 - 19 years, with standards for clinical use. *Arch. Dis. Childh.* 47: 373-381.

Sinclair, D. 1973. *Human growth after birth.* London, New York: Oxford University Press.

Sissons, H. A. 1949. Intermittent periosteal activity. *Nature.* 163: 1001.

Slijper, E. J. 1942. Biologic-anatomical investigations on the bipedal gait and upright posture in mammals, with special reference to a little goat, born without forelegs. *Proc. Kon. Nederl. Acad. Wet.* 45: 288-295 and 407-415.

Smith, B. H. 1987. Reply to Mann, Lampl and Monge (letter). *Nature* 328: 674-675.

Snell, K. 1863. *Die Schöpfung des Menschen.* Ed. F. A. Kipp (1981). Stuttgart: Freies Geistesleben. [Snell '63 and Snell '87 are contained in the same volume.]

— 1887. *Vorlesungen über die Abstammung des Menschen.* In *Die Schöpfung des Menschen*. Ed. F. A. Kipp (1981). Stuttgart: Freies Geistesleben.

Snipes, R. L., and A. Kriete. 1991. Quantitative investigation of the area and volume in different compartments of the intestine of 18 mammalian species. *Z. Säugetierkunde* 56: 225-244.

Sokolov, V. E. 1982. *Mammal skin*. Berkeley: University of California Press.

Stahl, W. R. 1962. Similarity and dimensional methods in biology. *Science* 137: 205-212.

— 1965. Organ weights in primates and other mammals. *Science* 150: 1039-1042.

Stanyon, R., B. Chiarelli, K. Gottlieb, and W. H. Patton. 1986. The phylogenetic and taxonomic status of *Pan paniscus*: A chromosomal perspective. *Am. J. Phys. Anthrop.* 69: 489-498.

Starck, D. 1975. *Embryologie*. Stuttgart: Thieme.

Starck, D., and B. Kummer. 1962. Zur Ontogenese des Schimpansenschädels. *Anthrop. Anz.* 25: 204-215.

Steiner, B. 1938. Über das biogenetische Grundgesetz (idealistische Morphogenese). *Acta Biotheor.* 4: 65-72.

Steiner, R. 1992. *Das Ewige in der Menschenseele* (ten public lectures given in Berlin, 1918). Dornach: Rudolf Steiner Verlag.

— 1995. Intuitive thinking as a spiritual path: A philosophy of freedom. Hudson, NY: Anthroposophic Press.

Stoddart, D. M. 1990. *The scented ape: The biology and culture of human odour.* Cambridge; New York: Cambridge University Press.

Stratz, C. H. 1909. Wachstum und Proportionen des Menschen vor und nach der Geburt. *Arch. f. Anthrop.* 8: 287-297.

Straus, W. L. 1940. The posture of the great ape hand in locomotion, and its phylogenetic implications. *Am. J. Phys. Anthrop.* 27: 199-207.

— 1942. Rudimentary digits in primates. *Quart. Rev. Biol.* 17: 228-243.

— 1949. The riddle of man's ancestry. *Quart. Rev. Biol.* 24: 200-229.

Straus, W. L., and J. A. Arcadi. 1958. Urinary system. In H. Hofer, A. H. Schultz, and D. Starck, eds. *Primatologia* (III/1). Basel: Karger.

Sullivan, P. G. 1986. Skull, jaw and teeth growth patterns. In F. Falkner and J. M. Tanner, eds. *Human growth: A comprehensive treatise.* New York: Plenum.

Tanner, J. M. 1962. *Growth at adolescence.* Oxford: Blackwell Scientific Publications.

Tax, H. R. 1980. *Podopediatrics.* Baltimore: Williams & Wilkins.

Thom, R. 1992. Sensorimotor reference frames and physiological attractors. In I. Golani. 1992. A mobility gradient in the organization of vertebrate movement: the perception of movement through symbolic language. *Behav. Brain Sc.* 15: 249-308.

Thompson, D. W. 1942. *On growth and form.* Cambridge (Eng.): The University Press.

Thomson, K. S. 1988. Ontogeny and phylogeny recapitulated. *Am. Scientist* 76: 273-275.

Tredgold, A. F. 1897. Variations of ribs in the primates with special reference to the number of sternal ribs in man. *J. Anat. Physiol.* 31: 288-302.

Treolar, A. E. et al. 1967. Human menstrual cycle variation. *Int. J. Fert.* 12: 77-125.

Tuttle, R. H. 1969. Knuckle-walking and the evolution of hominoid hands. *Am. J. Phys. Anthrop*. 26: 171-206.

— 1974. Darwin's apes, dental apes, and the descent of man: Normal science in evolutionary anthropology. *Curr. Anthrop.* 15: 389-398.

— ed., 1975. *Primate functional morphology and evolution.* The Hague: Mouton.

Uno, H., A. Cappas, and C. Schlagel. 1985. Cyclic dynamics of hair follicles and the effect of minoxidil on the bald scalp of stumptailed macaques. *Am. J. Dermatopath.* 7: 283-297.

Vandebroek, G. 1961. *Beginselen van de vergelijkende anatomie der chordata.* (This is an edited university course published in Leuven, Belgium.)

Van Den Broek, A. J. P. 1908. Über einige anatomische Merkmale von *Ateles*, in Zusammenhang mit der Anatomie der Platyrrhinen. *Anat. Anz.* 33: 111-124.

Verhaegen, M. J. B. 1985. The aquatic ape theory: Evidence and a possible scenario. *Med. Hypoth.* 16: 17-32.

— 1991. Human regulation of body temperature and water balance. In: M. Roede, J. Wind, J. Patrick, and V. Reynolds, eds. *The aquatic ape: Fact or fiction?* London: Souvenir Press.

Verhulst, J. 1993a. Louis Bolk revisited: I. Is the human lung a retarded organ? *Med. Hypoth*. 40: 311-320.

— 1993b. Louis Bolk revisited: II. Retardation, hypermorphosis and body proportions in humans. *Med. Hypoth.* 41: 100-114.

— 1994a. *Der Glanz von Kopenhagen.* Stuttgart: Freies Geistesleben.

— 1994b. Speech and the retardation of the human mandible: A Bolkian view. *Journal of Social and Evolutionary Systems* 17: 307-337.

— 1996. Atavisms in *Homo sapiens*: A Bolkian heterodoxy revisited. *Acta Biotheor.* 44: 59-73.

— 1999a. Nonuniform distribution of the ecliptical longitudes of sun and moon at the birthdays of top scientists. *Psych. Rep.* 85: 35-40.

— 1999b. Bolkian and Bokian retardation in *Homo sapiens*. *Acta Biotheor*. 47: 7-28.

Verhulst, J., and N. Jaspers. 1997. Does human retardation occur at the molecular level? In J. Wirz and E. Lammerts Van Bueren, eds. *The future of DNA.* Boston: Kluwer Academic Publishers.

Verhulst, J., and P. Onghena. 1996. Periodic birth-year pattern of the founders of quantum physics. *Psych. Rep.* 78: 19-25.

— 1997. Cranial suture closing in *Homo sapiens*: Evidence for circaseptennian periodicity. *Ann. Hum. Biol.* 24: 141-156.

— 1998a. Circaseptennian (about 7 year) periodicity in the birth year distribution of Nobel laureates for physics. *Psych. Rep.* 82: 127-130.

— 1998b. Periodic birth year pattern of the founders of statistical sciences. *Psych. Rep.* 83: 235-242.

Vigener, J. 1896. Ein Beitrag zur Morphologie des Nagels. *Morphologische Arbeiten* (now *Z. Morph. Anthrop.)* 6: 555-604.

Vogel, L. 1979. *Der dreigliedrige Mensch.* Dornach: Philosophisch-Anthroposopischer Verlag.

Voit, M. 1921. Der Mensch als primitive Tierform. *Naturwiss.* 8: 140-144.

Vollmer. 1828. *Natur und Sittengemälde der Tropenländer.* Munich: F. W. Michaelis.

Vollrath, L., A. Kantarjian, and C. Howe. 1975. Mammalian pineal gland: 7-day rhythmic activity? *Experientia* 31: 458-460.

Walker, A., and M. Teaford. 1986. The hunt of proconsul. *Scient. Am.* 1: 58-64.

Walmsley, R. and W. S. Monkhouse. 1988. The heart of the newborn child: An anatomical study based upon transverse serial sections. *J. Anat.* 159: 93-111.

Washburn, S. L. 1963. *Classification and human evolution.* Chicago: Aldine.

— 1978. The evolution of man. *Scient. Am.* 9: 146-154.

Webster, J. H. D. 1951. The periodicity of the "sevens" in mind, man and nature: A neo-hippocratic study. *Brit. J. Med. Psychol.* 24: 277-282.

Weidenreich, F. 1904. Die Bildung des Kinnes und seine angebliche Beziehung zur Sprache. *Anat. Anz.* 24: 545-555.

Weiner, J. S. 1955. *The Piltdown forgery.* London; New York: Oxford University Press.

Wells, G. A. 1967. Goethe and evolution. *J. Hist. Ideas* 28: 537-550.

Went, F. W. 1972. Parallel evolution. *Taxon* 20: 197-226.

West, B. J., and A. L. Goldberger. 1987. Physiology in fractal dimensions. *Am. Scientist* 75: 354-365.

Wigner, E. P. 1983. Remarks on the mind-body question. In J.A. Wheeler and W.H. Zurek, eds. *Quantum theory and measurement.* Princeton, NJ: Princeton University Press.

Williamson, D. I. 1992. *Larvae and Evolution: Toward a New Zoology.* New York: Chapman & Hall.

Wind, J. 1970. *On the phylogeny and the ontogeny of the human larynx.* Groningen: Wolters-Noordhoff.

Wood, N. K., L. E. Wragg, and O. H. Stuteville. 1968. The premaxilla: Embryological evidence that it does not exist in man. *Anat. Rec.* 158: 485-490.

Woodhull, A. M., K. Maltrud, and B. L. Mello. 1985. Alignment of the human body in standing. *Eur. J. Appl. Physiol.* 54: 109-115.

Woolley, R.G. 1978. Must a molecule have a shape? *J. Amer. Chem. Soc.* 100: 1073-1078.

— 1985. The molecular structure conundrum. *J. Chem. Educ.* 62: 1082-1084.

— 1991. Quantum chemistry beyond the Born-Oppenheimer approximation. *J. Mol. Struct.* 230: 17-46.

Wu, J., G. Cornelissen, B. Tarquini, G. Mainardi, M. Cagnoni, J. R. Gernandez, R. C. Hermida, K. Tamura, J. Kato, K. Kato, and F. Halberg. 1990. Circaseptan and circannual modulation of circadian rhythms in neonatal blood pressure and heart rate. In D. K. Hayes, J. E. Pauly, and R. J. Reiter, eds. *Chronobiology: Its role in clinical medicine, general biology and agriculture.* New York: Wiley-Liss.

Zajonc, R. B., P. K. Adelmann, S. T. Murphy and P. M. Niedenthal. 1987. Convergence in the physical appearance of spouses. *Motiv. Emot.* 11: 335-346.

Zeltner, T. B., and P.H. Burri. 1987. The postnatal development and growth of the human lung: II. Morphology. *Resp. Physiol.* 67: 269-282.

Zilles, K., E. Armstrong, K.H. Moser, A. Schleicher, and H. Stephan. 1989. Gyrification in the cerebral cortex of primates. *Brain, Beh. Evol.* 34: 143-150.

Index

Page numbers in bold refer to figures

About the Author

Born in Londerzeel, Belgium in 1949, Jos Verhulst studied economics, chemistry, and philosophy and earned his doctorate in the field of theoretical chemistry. His writings include a number of papers on theoretical chemistry, evolutionary theory, and comparative anatomy and a major work on quantum physics: *Der Glanz von Kopenhagen: Geistige Perspektiven der modernen Physik* (The brilliance of Kopenhagen: spiritual perspectives in modern physics). Also a social activist, he has written a book about direct democracy. He teaches science at the Waldorf High School in Antwerp.